Introduction to UAV Systems

Aerospace Series List

UAS Integration into Civil Airspace: Policy, Regulations and Strategy	Douglas M. Marshall	2022
Introduction to UAV Systems, Fifth Edition	Paul G. Fahlstrom, Thomas J. Gleason, Mohammad H. Sadraey	March 2022
Sustainable Aviation Technology and Operations: Research and Innovation Perspectives	Roberto Sabatini	
Foundations of Space Dynamics	Ashish Tewari	2020
Essentials of Supersonic Commercial Aircraft Conceptual Design	Egbert Torenbeek	2020
Design of Unmanned Aerial Systems	Mohammad H. Sadraey	
Future Propulsion Systems and Energy Sources in Sustainable Aviation	Saeed Farokhi	2018
Introduction to Flight Testing	James W. Gregory, Tianshu Liu	2017
Flight Dynamics and Control of Aero and Space Vehicles	Rama K. Yedavalli	2017
Theory of Lift: Introductory Computational Aerodynamics with MATLAB®/Octave	McBain	August 2012
Sense and Avoid in UAS: Research and Applications	Angelov	April 2012
Morphing Aerospace Vehicles and Structures	Valasek	April 2012
Gas Turbine Propulsion Systems	MacIsaac and Langton	July 2011
Basic Helicopter Aerodynamics, Third Edition	Seddon and Newman	July 2011
Advanced Control of Aircraft, Spacecraft, and Rockets	Tewari	July 2011
Cooperative Path Planning of Unmanned Aerial Vehicles	Tsourdos et al.	November 2010
Principles of Flight for Pilots	Swatton	October 2010
Air Travel and Health: A Systems Perspective	Seabridge et al.	September 2010
Design and Analysis of Composite Structures: With Applications to Aerospace Structures	Kassapoglou	September 2010
Unmanned Aircraft Systems: UAVS Design, Development, and Deployment	Austin	April 2010
Introduction to Antenna Placement and Installations	Macnamara	April 2010
Principles of Flight Simulation	Allerton	October 2009
Aircraft Fuel Systems	Langton et al.	May 2009
The Global Airline Industry	Belobaba	April 2009
Computational Modelling and Simulation of Aircraft and the Environment: Volume 1 – Platform Kinematics and Synthetic Environment	Diston	April 2009
Handbook of Space Technology	Ley, Wittmann, and Hallmann	April 2009
Aircraft Performance Theory and Practice for Pilots	Swatton	August 2008
Surrogate Modelling in Engineering Design: A Practical Guide	Forrester, Sobester, and Keane	August 2008
Aircraft Systems, Third Edition	Moir and Seabridge	March 2008
Introduction to Aircraft Aeroelasticity and Loads	Wright and Cooper	December 2007
Stability and Control of Aircraft Systems	Langton	September 2006
Military Avionics Systems	Moir and Seabridge	February 2006
Design and Development of Aircraft Systems	Moir and Seabridge	June 2004
Aircraft Loading and Structural Layout	Howe	May 2004
Aircraft Display Systems	Jukes	December 2003
Civil Avionics Systems	Moir and Seabridge	December 2002

Introduction to UAV Systems

Paul Gerin Fahlstrom
UAV Manager, US Army Material Command

Thomas James Gleason
Gleason Research Associates, Inc.

Mohammad H. Sadraey
Southern New Hampshire University

Fifth Edition

This fifth edition first published 2022
© 2022 John Wiley & Sons, Inc

Edition History
John Wiley & Sons, Ltd. (4e, 2012)

All rights reserved. No part of this publication may be reproduced, stored in a retrieval system, or transmitted, in any form or by any means, electronic, mechanical, photocopying, recording or otherwise, except as permitted by law. Advice on how to obtain permission to reuse material from this title is available at http://www.wiley.com/go/permissions.

The right of Paul Gerin Fahlstrom, Thomas James Gleason, and Mohammad H. Sadraey to be identified as the authors of this work has been asserted in accordance with law.

Registered Office
John Wiley & Sons, Inc., 111 River Street, Hoboken, NJ 07030, USA

Editorial Office
111 River Street, Hoboken, NJ 07030, USA

For details of our global editorial offices, customer services, and more information about Wiley products visit us at www.wiley.com.

Wiley also publishes its books in a variety of electronic formats and by print-on-demand. Some content that appears in standard print versions of this book may not be available in other formats.

Limit of Liability/Disclaimer of Warranty
The contents of this work are intended to further general scientific research, understanding, and discussion only and are not intended and should not be relied upon as recommending or promoting scientific method, diagnosis, or treatment by physicians for any particular patient. In view of ongoing research, equipment modifications, changes in governmental regulations, and the constant flow of information relating to the use of medicines, equipment, and devices, the reader is urged to review and evaluate the information provided in the package insert or instructions for each medicine, equipment, or device for, among other things, any changes in the instructions or indication of usage and for added warnings and precautions. While the publisher and authors have used their best efforts in preparing this work, they make no representations or warranties with respect to the accuracy or completeness of the contents of this work and specifically disclaim all warranties, including without limitation any implied warranties of merchantability or fitness for a particular purpose. No warranty may be created or extended by sales representatives, written sales materials or promotional statements for this work. The fact that an organization, website, or product is referred to in this work as a citation and/or potential source of further information does not mean that the publisher and authors endorse the information or services the organization, website, or product may provide or recommendations it may make. This work is sold with the understanding that the publisher is not engaged in rendering professional services. The advice and strategies contained herein may not be suitable for your situation. You should consult with a specialist where appropriate. Further, readers should be aware that websites listed in this work may have changed or disappeared between when this work was written and when it is read. Neither the publisher nor authors shall be liable for any loss of profit or any other commercial damages, including but not limited to special, incidental, consequential, or other damages.

Library of Congress Cataloging-in-Publication Data

Names: Fahlstrom, Paul Gerin, author. | Gleason, Thomas J., author. |
    Sadraey, Mohammad H., author.
Title: Introduction to UAV systems / Paul Gerin Fahlstrom, Thomas James
    Gleason, Mohammad H. Sadraey.
Description: Fifth edition. | Hoboken, NJ : Wiley, 2022. | Series:
    Aerospace series | Includes bibliographical references and index.
Identifiers: LCCN 2022001629 (print) | LCCN 2022001630 (ebook) | ISBN
    9781119802617 (cloth) | ISBN 9781119802631 (adobe pdf) | ISBN
    9781119802624 (epub)
Subjects: LCSH: Drone aircraft. | Cruise missiles.
Classification: LCC UG1242.D7 F34 2022 (print) | LCC UG1242.D7 (ebook) |
    DDC 623.74/69–dc23/eng/20220202
LC record available at https://lccn.loc.gov/2022001629
LC ebook record available at https://lccn.loc.gov/2022001630

Cover Design: Wiley
Cover Image: © MQ-9 Reaper unmanned aerial vehicle, public domain, Wikimedia Commons

Set in 9.5/12.5pt STIXTwoText by Straive, Pondicherry, India

SKY10033879_032222

This book is dedicated to our wives, Beverly Ann Evans Fahlstrom, Archodessia Glyphis Gleason, and Seyedeh Zafarani, who have provided support and encouragement throughout the process of its preparation.

vii

Contents

Preface *xvii*
Series Preface *xxi*
Acknowledgments *xxiii*
List of Acronyms *xxv*
About the Companion Website *xxix*

Part I   Introduction *1*

1   History and Overview *3*
1.1   Overview *3*
1.2   History *4*
    1.2.1   Early History *4*
    1.2.2   The Vietnam War *4*
    1.2.3   Resurgence *5*
    1.2.4   Joint Operations *6*
    1.2.5   Desert Storm *6*
    1.2.6   Bosnia *6*
    1.2.7   Afghanistan and Iraq *7*
    1.2.8   Long-Range Long-Endurance Operations *7*
1.3   Overview of UAV Systems *8*
    1.3.1   Air Vehicle *9*
    1.3.2   Mission Planning and Control Station *9*
    1.3.3   Launch and Recovery Equipment *10*
    1.3.4   Payloads *10*
    1.3.5   Data Links *11*
    1.3.6   Ground Support Equipment *11*
1.4   The Aquila *12*
    1.4.1   Aquila Mission and Requirements *12*
    1.4.2   Air Vehicle *13*
    1.4.3   Ground Control Station *13*
    1.4.4   Launch and Recovery *14*
    1.4.5   Payload *14*
    1.4.6   Other Equipment *14*
    1.4.7   Summary *14*

viii   *Contents*

1.5   Global Hawk *16*
       1.5.1   Mission Requirements and Development *16*
       1.5.2   Air Vehicle *16*
       1.5.3   Payloads *17*
       1.5.4   Communications System *17*
       1.5.5   Development Setbacks *18*
1.6   Predator Family *19*
       1.6.1   Predator Development *19*
       1.6.2   Reaper *19*
       1.6.3   Features *20*
1.7   Top UAV Manufacturers *21*
1.8   Ethical Concerns of UAVs *21*
       Questions *22*

2   **Classes and Missions of UAVs** *27*
2.1   Overview *27*
2.2   Classes of UAV Systems *27*
       2.2.1   Classification Criteria *27*
       2.2.2   Classification by Range and Endurance *28*
       2.2.3   Classification by Missions *29*
       2.2.4   The Tier System *32*
2.3   Examples of UAVs by Size Group *33*
       2.3.1   Micro-UAVs *34*
       2.3.2   Mini-UAVs *34*
       2.3.3   Very Small UAVs *35*
       2.3.4   Small UAVs *36*
       2.3.5   Medium UAVs *37*
       2.3.6   Large UAVs *40*
2.4   Expendable UAVs *42*
       Questions *43*

       Part II   The Air Vehicle *47*

3   **Aerodynamics** *49*
3.1   Overview *49*
3.2   Aerodynamic Forces *49*
3.3   Mach Number *51*
3.4   Airfoil *51*
3.5   Pressure Distribution *54*
3.6   Drag Polar *57*
3.7   The Real Wing and Airplane *58*
3.8   Induced Drag *58*
3.9   Boundary Layer *60*
3.10   Friction Drag *63*
3.11   Total Air-Vehicle Drag *63*

Contents | ix

3.12	Flapping Wings *64*	
3.13	Aerodynamic Efficiency *66*	
	Questions *67*	

4 Performance *71*

4.1	Overview *71*	
4.2	Cruising Flight *72*	
4.3	Range *73*	
	4.3.1	Range for a Non-Electric-Engine Propeller-Driven Aircraft *74*
	4.3.2	Range for a Jet-Propelled Aircraft *76*
4.4	Endurance *78*	
	4.4.1	Endurance for a Non-Electric-Engine Propeller-Driven Aircraft *78*
	4.4.2	Endurance for a Jet-Propelled Aircraft *79*
4.5	Climbing Flight *80*	
4.6	Gliding Flight *82*	
4.7	Launch *83*	
4.8	Recovery *84*	
	Questions *85*	

5 Flight Stability and Control *89*

5.1	Overview *89*	
5.2	Trim *90*	
	5.2.1	Longitudinal Trim *91*
	5.2.2	Directional Trim *93*
	5.2.3	Lateral Trim *93*
	5.2.4	Summary *94*
5.3	Stability *94*	
	5.3.1	Longitudinal Static Stability *95*
	5.3.2	Directional Static Stability *97*
	5.3.3	Lateral Static Stability *99*
	5.3.4	Dynamic Stability *100*
5.4	Control *101*	
	5.4.1	Aerodynamic Control *101*
	5.4.2	Pitch Control *102*
	5.4.3	Directional Control *104*
	5.4.4	Lateral Control *105*
	Questions *106*	

6 Propulsion *109*

6.1	Overview *109*	
6.2	Propulsion Systems Classification *109*	
6.3	Thrust Generation *110*	
6.4	Powered Lift *111*	
6.5	Sources of Power *114*	
	6.5.1	Four-Cycle Engine *115*
	6.5.2	Two-Cycle Engine *116*

x Contents

6.5.3 Rotary Engine *118*
6.5.4 Gas Turbine Engines *119*
6.5.5 Electric Motors *120*
6.6 Sources of Electric Energy *122*
6.6.1 Batteries *122*
6.6.2 Solar Cells *124*
6.6.3 Fuel Cells *126*
6.7 Power and Thrust *127*
6.7.1 Relation Between Power and Thrust *128*
6.7.2 Propeller *128*
6.7.3 Variations of Power and Thrust with Altitude *130*
Questions *131*

7 Air Vehicle Structures *135*
7.1 Overview *135*
7.2 Structural Members *135*
7.2.1 Skin *136*
7.2.2 Fuselage Structural Members *136*
7.2.3 Wing and Tail Structural Members *137*
7.2.4 Other Structural Members *138*
7.3 Basic Flight Loads *139*
7.4 Dynamic Loads *143*
7.5 Structural Materials *145*
7.5.1 Overview *145*
7.5.2 Aluminum *145*
7.6 Composite Materials *146*
7.6.1 Sandwich Construction *146*
7.6.2 Skin or Reinforcing Materials *147*
7.6.3 Resin Materials *147*
7.6.4 Core Materials *148*
7.7 Construction Techniques *148*
7.8 Basic Structural Calculations *149*
7.8.1 Normal and Shear Stress *150*
7.8.2 Deflection *152*
7.8.3 Buckling Load *153*
7.8.4 Factor of Safety *154*
7.8.5 Structural Fatigue *155*
Questions *155*

Part III Mission Planning and Control *159*

8 Mission Planning and Control Station *161*
8.1 Introduction *161*
8.2 MPCS Subsystems *161*
8.3 MPCS Physical Configuration *165*

Contents xi

8.4   MPCS Interfaces   *169*
8.5   MPCS Architecture   *170*
      8.5.1   Fundamentals   *170*
      8.5.2   Local Area Networks   *172*
      8.5.3   Levels of Communication   *172*
      8.5.4   Bridges and Gateways   *173*
8.6   Elements of a LAN   *174*
      8.6.1   Layout and Logical Structure (Topology)   *174*
      8.6.2   Communications Medium   *175*
      8.6.3   Network Transmission and Access   *175*
8.7   OSI Standard   *175*
      8.7.1   Physical Layer   *175*
      8.7.2   Data-Link Layer   *175*
      8.7.3   Network Layer   *176*
      8.7.4   Transport Layer   *176*
      8.7.5   Session Layer   *176*
      8.7.6   Presentation Layer   *176*
      8.7.7   Application Layer   *176*
8.8   Mission Planning   *176*
8.9   Pilot-In-Command   *179*
        Questions   *180*

9   **Control of Air Vehicle and Payload**   *183*
9.1   Overview   *183*
9.2   Levels of Control   *184*
9.3   Remote Piloting the Air Vehicle   *185*
      9.3.1   Remote Manual Piloting   *186*
      9.3.2   Autopilot-Assisted Control   *188*
      9.3.3   Complete Automation   *188*
      9.3.4   Summary   *189*
9.4   Autopilot   *190*
      9.4.1   Fundamental   *190*
      9.4.2   Autopilot Categories   *191*
      9.4.3   Inner and Outer Loops   *191*
      9.4.4   Overall Modes of Operation   *192*
      9.4.5   Control Process   *193*
      9.4.6   Control Axes   *193*
      9.4.7   Controller   *194*
      9.4.8   Actuator   *195*
      9.4.9   Open-Source Commercial Autopilots   *195*
9.5   Sensors Supporting the Autopilot   *196*
      9.5.1   Altimeter   *196*
      9.5.2   Airspeed Sensor   *196*
      9.5.3   Attitude Sensors   *197*
      9.5.4   GPS   *198*
      9.5.5   Accelerometers   *199*

xii

9.6   Navigation and Target Location *199*
9.7   Controlling Payloads *201*
    9.7.1   Signal Relay Payloads *202*
    9.7.2   Atmospheric, Radiological, and Environmental Monitoring *202*
    9.7.3   Imaging and Pseudo-Imaging Payloads *203*
9.8   Controlling the Mission *204*
9.9   Autonomy *206*
    Questions *208*

Part IV   Payloads *213*

10   Reconnaissance/Surveillance Payloads *215*
10.1   Overview *215*
10.2   Imaging Sensors *216*
10.3   Target Detection, Recognition, and Identification *217*
    10.3.1   Sensor Resolution *218*
    10.3.2   Target Contrast *221*
    10.3.3   Transmission Through the Atmosphere *222*
    10.3.4   Target Signature *225*
    10.3.5   Display Characteristics *225*
    10.3.6   Range Prediction Procedure *226*
    10.3.7   A Few Considerations *228*
    10.3.8   Pitfalls *230*
10.4   The Search Process *231*
    10.4.1   Types of Search *231*
    10.4.2   Field of View *232*
    10.4.3   Search Pattern *234*
    10.4.4   Search Time *235*
10.5   Other Considerations *237*
    10.5.1   Location and Installation *237*
    10.5.2   Stabilization of the Line of Sight *238*
    10.5.3   Gyroscope and Gimbal *238*
    10.5.4   Gimbal-Gyro Configuration *240*
    10.5.5   Thermal Design *241*
    10.5.6   Environmental Conditions Affecting Stabilization *241*
    10.5.7   Boresight *242*
    10.5.8   Stabilization Design *242*
    Questions *243*

11   Weapon Payloads *247*
11.1   Overview *247*
11.2   History of Lethal Unmanned Aircraft *249*
11.3   Mission Requirements for Armed Utility UAVs *251*
11.4   Design Issues Related to Carriage and Delivery of Weapons *252*
    11.4.1   Payload Capacity *253*
    11.4.2   Structural Issues *253*

Contents   xiii

11.4.3 Electrical Interfaces *255*
11.4.4 Electromagnetic Interference *256*
11.4.5 Launch Constraints for Legacy Weapons *257*
11.4.6 Safe Separation *257*
11.4.7 Data Links *258*
11.4.8 Payload Location *258*
11.5 Signature Reduction *258*
11.5.1 Acoustical Signatures *259*
11.5.2 Visual Signatures *263*
11.5.3 Infrared Signatures *264*
11.5.4 Radar Signatures *265*
11.5.5 Emitted Signals *269*
11.5.6 Active Susceptibility Reduction Measures *269*
11.6 Autonomy for Weapon Payloads *270*
11.6.1 Fundamental Concept *270*
11.6.2 Rules of Engagement *272*
Questions *273*

12 Other Payloads *277*
12.1 Overview *277*
12.2 Radar *277*
12.2.1 General Radar Considerations *277*
12.2.2 Synthetic Aperture Radar *280*
12.3 Electronic Warfare *280*
12.4 Chemical Detection *281*
12.5 Nuclear Radiation Sensors *282*
12.6 Meteorological and Environmental Sensors *282*
12.7 Pseudo-Satellites *283*
12.8 Robotic Arm *286*
12.9 Package and Cargo *287*
12.10 Urban Air Mobility *288*
Questions *288*

Part V   Data Links *291*

13 Data Link Functions and Attributes *293*
13.1 Overview *293*
13.2 Background *293*
13.3 Data-Link Functions *295*
13.4 Desirable Data-Link Attributes *296*
13.4.1 Worldwide Availability *297*
13.4.2 Resistance to Unintentional Interference *298*
13.4.3 Low Probability of Intercept (LPI) *298*
13.4.4 Security *298*
13.4.5 Resistance to Deception *299*
13.4.6 Anti-ARM *299*

xiv *Contents*

      13.4.7   Anti-Jam *300*
      13.4.8   Digital Data Links *301*
      13.4.9   Signal Strength *301*
13.5   System Interface Issues *302*
      13.5.1   Mechanical and Electrical *302*
      13.5.2   Data-Rate Restrictions *302*
      13.5.3   Control-Loop Delays *303*
      13.5.4   Interoperability, Interchangeability, and Commonality *305*
13.6   Antennas *307*
      13.6.1   Omnidirectional Antenna *307*
      13.6.2   Parabolic Reflectors *307*
      13.6.3   Array/Directional Antennas *308*
      13.6.4   Lens Antennas *308*
13.7   Data Link Frequency *309*
    Questions *311*

14   **Data-Link Margin** *315*
14.1   Overview *315*
14.2   Sources of Data-Link Margin *316*
      14.2.1   Transmitter Power *316*
      14.2.2   Antenna Gain *316*
      14.2.3   Processing Gain *322*
14.3   Anti-Jam Margin *327*
      14.3.1   Definition of Anti-Jam Margin *327*
      14.3.2   Jammer Geometry *328*
      14.3.3   System Implications of AJ Capability *331*
      14.3.4   Anti-Jam Uplinks *334*
14.4   Propagation *334*
      14.4.1   Obstruction of the Propagation Path *334*
      14.4.2   Atmospheric Absorption *336*
      14.4.3   Precipitation Losses *336*
14.5   Data-Link Signal-to-Noise Budget *336*
    Questions *339*

15   **Data-Rate Reduction** *343*
15.1   Overview *343*
15.2   Compression Versus Truncation *343*
15.3   Video Data *344*
      15.3.1   Gray Scale *344*
      15.3.2   Encoding of Gray Scale *345*
      15.3.3   Effects of Bandwidth Compression on Operator Performance *346*
      15.3.4   Frame Rate *348*
      15.3.5   Control Loop Mode *348*
      15.3.6   Forms of Truncation *350*
      15.3.7   Summary *351*
15.4   Non-Video Data *352*
15.5   Location of the Data-Rate Reduction Function *353*
    Questions *354*

Contents xv

16   Data-Link Tradeoffs _357_
16.1   Overview _357_
16.2   Basic Tradeoffs _357_
16.3   Pitfalls of "Putting Off" Data-Link Issues _359_
16.4   Future Technology _360_
        Questions _360_

Part VI   Launch and Recovery _363_

17   Launch Systems _365_
17.1   Overview _365_
17.2   Conventional Takeoff _365_
17.3   Basic Considerations _366_
17.4   Launch Methods for Fixed-Wing Air Vehicles _370_
        17.4.1   Overview _370_
        17.4.2   Rail Launchers _372_
        17.4.3   Pneumatic Launchers _373_
        17.4.4   Hydraulic-Pneumatic Launchers _374_
        17.4.5   Zero Length RATO Launch of UAVs _375_
        17.4.6   Tube Launch _375_
17.5   Rocket-Assisted Takeoff _376_
        17.5.1   RATO Configuration _376_
        17.5.2   Ignition Systems _376_
        17.5.3   Expended RATO Separation _376_
        17.5.4   Other Launch Equipment _377_
        17.5.5   Energy (Impulse) Required _377_
        17.5.6   Propellant Weight Required _378_
        17.5.7   Thrust, Burning Time, and Acceleration _379_
17.6   Vertical Takeoff _379_
        Questions _379_

18   Recovery Systems _383_
18.1   Overview _383_
18.2   Conventional Landing _383_
18.3   Vertical Net Systems _384_
18.4   Parachute Recovery _385_
18.5   VTOL UAVs _386_
18.6   Mid-Air Retrieval _388_
18.7   Shipboard Recovery _389_
18.8   Break-Apart Landing _392_
18.9   Skid and Belly Landing _393_
18.10   Suspended Cables _394_
        Questions _395_

19   Launch and Recovery Tradeoffs _397_
19.1   UAV Launch Method Tradeoffs _397_
19.2   Recovery Method Tradeoffs _400_

19.3 Overall Conclusions *402*
Questions *402*

20 Rotary-Wing UAVs and Quadcopters *405*
20.1 Overview *405*
20.2 Rotary-Wing Configurations *406*
    20.2.1 Single Rotor *406*
    20.2.2 Twin Co-axial Rotors *407*
    20.2.3 Twin Tandem Rotors *408*
    20.2.4 Multicopters *408*
20.3 Hybrid UAVs *409*
    20.3.1 Tilt Rotor *409*
    20.3.2 Tilt Wing *409*
    20.3.3 Thrust Vectoring *410*
    20.3.4 Fixed-Wing Quadcopter Combination *410*
20.4 Quadcopters *411*
    20.4.1 Overview *411*
    20.4.2 Aerodynamics *413*
    20.4.3 Control *414*
    Questions *416*

References *419*

Index *421*

Preface

Introduction to UAV Systems, Fifth Edition, has been written to meet the needs of both newcomers to the world of unmanned aerial vehicle (UAV) systems and experienced members of the UAV community who desire an overview and who, though they may find the treatment of their particular discipline elementary, will gain valuable insights into the other disciplines that contribute to a UAV system. The material has been presented such that it is readily understandable to college freshman and to both technical and nontechnical persons working in the UAV field, and is based on standard engineering texts as well as material developed by the authors while working in the field. Most equations have been given without proof and the reader is encouraged to refer to standard texts of each discipline when engaging in actual design or analysis as no attempt is made to make this book a complete design handbook.

This book is also not intended to be the primary text for an introductory course in aerodynamics or imaging sensors or data links. Rather, it is intended to provide enough information in each of those areas, and others, to illustrate how they all play together to support the design of complete UAV systems and to allow the reader to understand how the technology in all of these areas affect the system-level tradeoffs that shape the overall system design. As such, it might be used as a supplementary text for a course in any of the specialty areas to provide a system-level context for the specialized material.

For a beginning student, we hope that it will whet the appetite for knowing more about at least one of the technology areas and demonstrate the power of even the simplest mathematical treatment of these subjects in allowing understanding of the tradeoffs that must occur during the system design process.

For a UAV user or operator, we hope that it will provide understanding of how the system technology affects the manner in which the UAV accomplishes its objectives and the techniques that the operator must use to make that happen.

For a "subject matter expert" in any of the disciplines involved in the design of a UAV system, we hope that it will allow better understanding of the context in which his or her specialty must operate to produce success for the system as a whole and why other specialists may seem preoccupied with things that seem unimportant to him or her.

Finally, for a technology manager, we hope that this book can help him or her understand how everything fits together, how important it is to consider the system-integration issues early in the design process so that the integration issues are considered during the basic selection of subsystem designs, and help him or her understand what the specialists are talking about and, perhaps, ask the right questions at critical times in the development process.

Part One contains a brief history and overview of UAVs in Chapter 1 and a discussion of classes and missions of UAVs in Chapter 2.

Part Two is devoted to the design of the air vehicle including basic aerodynamics, performance, stability and control, propulsion and loads, structures and materials in Chapters 3 through 7.

Part Three discusses the mission planning, control function, and autopilot in Chapter 8 and operational control in Chapter 9.

Part Four has three chapters addressing payloads. Chapter 10 discusses the most universal types of payloads, reconnaissance, and surveillance sensors. Chapter 11 discusses weapons payloads, a class of payloads that has become prominent since its introduction about 10 years ago. Chapter 12 discusses a few of the many other types of payloads that may be used on UAVs.

Part Five covers data links, the communication subsystems used to connect the air vehicle to the ground controllers, and delivers the data gathered by the air-vehicle payloads. Chapter 13 describes and discusses basic data-link functions and attributes. Chapter 14 covers the factors that affect the performance of a data link, including the effects or intentional and unintentional interference. Chapter 15 addresses the impact on the operator and system performance of various approaches to reducing the data-rate requirements of the data link to accommodate limitations on available bandwidth. Chapter 16 summarizes data-link tradeoffs, which are one of the key elements in the overall system tradeoffs.

Part Six describes approaches for the launch and recovery of UAVs, including ordinary takeoff and landing, but extending to many approaches not used for manned aircraft. Chapter 17 describes launch systems and Chapter 18 recovery systems. Chapter 19 summarizes the tradeoffs between the many different launch and recovery approaches. Chapter 20 – a new chapter in this edition – is dedicated to fundamentals, control, and characteristics of rotary-wing UAVs and quadcopters.

Introduction to UAV Systems was first published in 1992. Much has happened in the UAV world in the 30 years since the first edition was written. In the Preface to the second edition (1998), we commented that there had been further problems in the development process for tactical UAVs but that there had been some positive signs in the use of UAVs in support of the Bosnian peace-keeping missions and that there even was some talk of the possible use of "uninhabited" combat vehicles within the US Air Force that was beginning for the first time to show some interest in UAVs. At that time, we concluded that "despite some interest, and real progress in some areas, however, we believe that the entire field continues to struggle for acceptance, and UAVs have not come of age and taken their place as proven and established tools."

In the 30 years since we made that statement, the situation has changed dramatically. UAVs have been widely adopted in the military world, unmanned combat vehicles have been deployed and used in highly visible ways, often featured on the evening news, and unmanned systems now appear to be serious contenders for the next generation of fighters and bombers.

While civilian applications still lag, impeded by the very-real issues related to mixing manned and unmanned aircraft in the general airspace, the success of military applications has encouraged attempts to resolve these issues and establish unmanned aircraft in nonmilitary roles.

The Fifth Edition has been extensively revised and restructured. The revisions have, we hope, made some of the material clearer and easier to understand and have added a number of new subjects in areas that have become more prominent in the UAV world during the last two decades, such as quadcopter, automatic flight control systems, new payloads, and the various levels of autonomy that may be given to an air vehicle. It also revises a number of details that have clearly been overtaken by events, and all chapters have been brought up to date to introduce some of the new terminology, concepts, and specific UAV systems that have appeared over the last 12 years. However, the basic subsystems that make up a UAV "system of systems" have not greatly changed, and at the level that this text addresses them, the basic issues and design principles have not changed since the first edition was published.

The first two authors – sadly, now deceased - met while participating in a "red team" that was attempting to diagnose and solve serious problems in an early UAV program. The eventual diagnosis was that there had been far too little "systems engineering" during the design process and that various subsystems did not work together as required for system-level success. This book grew out of a desire to write down at least some of the "lessons learned" during that experience and make them available to those who design UAV systems in the future.

Supplementary materials, including answers to end-of-chapter questions, are available for registered instructors who adopt this book as a course text. Please visit www.wiley.com/go/fahlstrom/uavsystems5e for information and to register for access to these resources.

We believe that most of those lessons learned are universal enough that they are just as applicable today as they were when they were learned years ago, and hope that this book can help future UAV system designers apply them and avoid having to learn them again the "hard way."

Mohammad Hashem Sadraey
September 2021

The Kettering Bug (Photograph courtesy of Norman C. "Dutch" Heilman)

Series Preface

This book is a welcome addition to the Aerospace Series, continuing the tradition of the Series in providing clear and practical advice to practitioners in the field of aerospace. This book will appeal to a wide readership and is an especially good introduction to the subject by extending the range of titles on the topic of unmanned air vehicles, and more importantly presenting a systems viewpoint of unmanned air systems. This is important as the range of vehicles currently available provides a diverse range of capabilities with differing structural designs, propulsions systems, payloads, ground systems, and launch/recovery mechanisms. It is difficult to see any rationalization or standardization of vehicles or support environment in the range of available solutions.

The book covers the history of unmanned flight and describes the range of solutions available world-wide. It then addresses the key aspects of the subsystems such as structure, propulsion, navigation, sensor payloads, launch and recovery, and associated ground systems in a readable and precise manner, pulling them together as elements of a total integrated system. In this way it is complementary to other systems books in the Series.

It is important for engineers and designers to visualize the totality of a system in order to gain an understanding of all that is involved in designing new vehicles or in writing new requirements to arrive at a coherent design of vehicle and infrastructure. Even more important if the new vehicle needs to interact and inter-operate with other vehicles or to operate from different facilities.

If unmanned air systems are going to become accepted in civilian airspace and in commercial applications then it is vital that a set of standards and design guidelines is in place to ensure consistency, to aid the certification process, and to provide a global infrastructure similar to that existing for today's manned fleets. Without that understanding, certification of unmanned air vehicles to operate in civilian controlled airspace is going to be a long and arduous task.

This book sets the standard for a definitive work on the subject of unmanned air systems by providing a measure of consistency and a clear understanding of the topic.

Acknowledgments

We would like to thank Engineering Arresting System Corporation (ESCO) (Aston, PA), Division of Zodiac Aerospace and General Atomics Aeronautical Systems, Inc., for providing pictures and diagrams and/or other information relating to their air vehicles and equipment.

The Joint UAV Program Office (Patuxent River Naval Air Station, MD) and the US Army Aviation and Missile Command (Huntsville, AL) both provided general information during the preparation of the first edition.

We especially thank Mr. Robert Veazey, who provided the original drafts of the material on launch and recovery while an employee of ESCO, and Mr. Tom Murley, formerly of Lear Astronics, and Mr. Bob Sherman for their critical reading of the draft and constructive suggestions.

Great appreciation goes to Mrs. Lauren Poplawski, Acquisitions Editor for Mechanical and Aerospace Engineering at Wiley, who suggested the third author to prepare the Fifth Edition, and provided him with guidance through the preparation of the manuscript. We are also thankful for Kimberly Monroe-Hill, Managing Editor and Gabby Robles, Associate Managing Editor, who provided some helpful information when the manuscript for this new edition was sculpted. The authors are also very grateful to Copy Editor Mrs. Patricia Bateson for her careful reading of the manuscript for the Fifth Edition and for many helpful suggestions related to style and grammar. People at Wiley were very patient with us throughout the process of working out the details of how that might be accomplished.

List of Acronyms

AC	alternating current, aerodynamic center
ADC	analog-to-digital converter
ADT	air data terminal
AFCS	automatic flight control systems
AGL	above ground level
Ah	ampere-hours
AI	artificial intelligence
AJ	anti jam
AOA	angle of attack
AR	aspect ratio
ARM	antiradiation munition
AV	air vehicle
BD	bi-directional
BIT	built-in test
BVLOS	beyond visual line of sight
BLOS	beyond line of sight
C2	command and control
CARS	common automatic recovery system
CBR	chemical, biological, radiological
CCD	charge-coupled device
CFD	computational fluid dynamics
CG	center of gravity
CLRS	central launch and recovery section
CP	center of pressure
CPU	central processing unit
COMINT	communication intelligence
CW	continuous wave
DAA	detect and avoid
DAC	digital-to-analog converter
dB	decibel
dBA	dBs relative to the lowest pressure difference that is audible to a person
dBmv	dBs relative to 1 mV
dBsm	dB relative to 1 m^2
DC	direct current
DF	direction finding

DOF	degrees of freedom
ECCM	electronic counter-countermeasures
ECM	electronic countermeasure
ELINT	electronic intelligence
EMI	electromagnetic interference
EO	electro optic
ERP	effective radiated power
ESM	electronic support measure
eVTOL	electric vertical take-off and landing
EW	electronic warfare
FAA	Federal Aviation Administration
FAR	Federal Aviation Regulations
FCS	forward control section
FEM	finite element method
FLIR	forward-looking infrared
FLOT	Forward Line of Own Troops
FOV	field of view
fps	frames per second
FPV	first-person view
FSED	Full Scale Engineering Development
GCS	ground control station
GDT	ground data terminal
GHz	giga hertz
GNC	Guidance-Navigation-Control
GPS	global positioning system
GSE	ground support equipment
Gyro	gyroscope
HALE	high-altitude, long endurance
HELLFIRE	helicopter launched fire and forget missile
HERO	Hazards of Electromagnetic Radiation to Ordnance
HMMWV	High Mobility Multipurpose Wheeled Vehicle
HP	hydraulic-pneumatic
HTOL	horizontal takeoff and landing
IAI	Israeli Aircraft Industries
IC	integrated circuit
IFF	identification friend or foe
IMC	Image Motion Compensation
iOS	iPhone operating system
IR	infrared
ISR	Intelligence, Surveillance, Reconnaissance
ISO	International Organization for Standardization
JATO	Jet Assisted Take-Off
JII	Joint Integration Interface
JPEG	Joint Photographic Experts Group
JPO	joint project office
JSTARS	Joint Surveillance Target Attack Radar System
KE	kinetic energy

LAN	local area network
Li-ion	lithium ion
Li-poly	lithium polymer
LOS	line of sight
LPI	low-probability of intercept
LQR	linear quadratic regulator
mAh	milli ampere hour
MALE	medium-altitude, long endurance
MARS	mid-air recovery system
MART	Mini Avion de Reconnaissance Telepilot
Mbps	mega bits per second
MDO	multidisciplinary design optimization
MEMS	Micro-Eectro-Mechanical System
MET	meteorological
MICNS	Modular Integrated Communication and Navigation System
MPCS	mission planning and control station
MRC	minimum resolvable contrast
MRDT	minimum resolvable delta in temperature
MRT	minimum resolvable temperature
MSL	mean sea level
MTF	modulation transfer function
MTI	Moving Target Indicator
NACA	National Advisory Committee for Aeronautics
NAS	National Airspace System
NASA	National Aeronautics and Space Administration
NDI	non-developmental item
NiCd	nickel cadmium
NiMH	nickel metal hydride
nm	nautical mile
NOAA	National Oceanic and Atmospheric Administration
NP	neutral point
OBC	optical bar camera
OSD	on-screen display
OSI	Open System Interconnection
OT	operational test
PGM	precision guided munition
PIC	Pilot In Command
PID	Proportional, Integral, Derivative
PIN	positive intrinsic negative
PLSS	Precision Location and Strike System
QFT	quantitative feedback theory
RAM	radar-absorbing material
RAP	radar-absorbing paint
RATO	rocket-assisted takeoff
RC Plane	radio-controlled airplane, remotely controlled airplane
R&D	Research and Development
RCS	radar cross-section

RF	radio frequency
RGT	remote ground terminal
RMS	root mean square
ROC	rate of climb
RPG	rocket propelled grenade
RPM	revolutions per minute
RPA	remotely piloted aircraft
RPV	remotely piloted vehicle
SAR	synthetic aperture radar
SEAD	Suppression of Enemy Air Defense
SF	safety factor
shp	shaft horsepower
SIGINT	signal intelligence
SLAR	side-looking airborne radar
SOTAS	Stand-Off Target Acquisition System
SPARS	Ship Pioneer Arresting System
sUAS	small unmanned aircraft systems
TADARS	Target Acquisition/Designation and Aerial Reconnaissance System
TUAV	tactical UAV
UAM	Urban Air Mobility
UAS	unmanned aerial system
UAV	unmanned aerial vehicle
UCAV	unmanned combat aerial vehicle
UD	unidirectional
VLOS	visual line-of-sight
VTOL	vertical takeoff and landing

About the Companion Website

This book is accompanied by a companion website:

https://www.wiley.com/go/fahlstrom/uavsystems5e

The website includes:

- Solutions manual

Part I

Introduction

Part One provides a general background for an introduction to the technology of unmanned aerial vehicle systems, called "UAV systems" or "unmanned aerial systems" (UAS). This part is comprised of two chapters, 1 and 2.

Chapter 1 presents a brief history of UAVs. It then identifies and describes the functions of the major elements (subsystems) that may be present in a generic UAS. Finally, it provides a short history of a major UAV development program that failed to produce a fielded UAS, despite significant success in many of the individual subsystems, and teaches useful lessons about the importance of understanding the inter-relationship and interactions of the subsystems of the UAS and the implications of system performance requirements at a total-systems level. This story is told here to emphasize the importance of the word "system" in the terms "UAV System" and "UAS."

Chapter 2 contains a survey of UAS that have been or presently are in use and discusses various schemes that are used to classify UAV systems according to their size, endurance, and/or mission. The information in this chapter is subject to becoming dated because the technology of many of the subsystems of a UAS is evolving rapidly as they become more and more part of the mainstream after many years of being on the fringes of the aeronautical engineering world. Nonetheless, some feeling for the wide variety of UAS concepts and types is needed to put the later discussion of design and system integration issues into context. Currently about 100 countries are employing military drones.

Introduction to UAV Systems, Fifth Edition. Paul Gerin Fahlstrom, Thomas James Gleason, and Mohammad H. Sadraey.
© 2022 John Wiley & Sons, Inc. Published 2022 by John Wiley & Sons, Inc.
Companion website: www.wiley.com/go/fahlstrom/uavsystems5e

1

History and Overview

1.1 Overview

The first portion of the chapter reviews the history of UAV systems from the earliest and crudest "flying objects" through the events of the last decade, which has been a momentous period for UAV systems.

The second portion of the chapter describes the subsystems that comprise a complete UAV system configuration to provide a framework for the subsequent treatment of the various individual technologies that contribute to a complete UAS. The air vehicle itself is a complicated system including structures, aerodynamic elements (wings and control surfaces), propulsion systems, and control systems. The complete system includes, in addition, sensors and other payloads, communication packages, and launch and recovery subsystems.

Finally, a cautionary tale is presented to illustrate why it is important to consider the UAV *system* as a whole rather than to concentrate only on individual components and subsystems. This is the story of a UAS that was developed between about 1975 and 1985 and that may be the most ambitious attempt at completeness, from a system standpoint, that has so far been undertaken in the UAS community.

It included every key UAS element in a totally self-contained form, all designed from scratch to work together as a portable system that required no local infrastructure beyond a relatively small open field in which a catapult launcher and a net recovery system could be located. This system, called the Aquila remotely piloted vehicle (RPV) system, was developed and tested over a period of about a decade at a cost that approached a billion dollars. It eventually could meet most of its operational requirements.

The Aquila UAS turned out to be very expensive and required a large convoy of 5-ton trucks for transportation. Most importantly, it did not fully meet some unrealistic expectations that had been built up over the decade during which it was being developed. It was never put in production or fielded. Nonetheless, it remains the only UAS of which the authors are aware that attempted to be complete unto itself and it is worth understanding what that ambition implied and how it drove costs and complexity in a way that eventually led the system to be abandoned in favor of less complete, self-sufficient, and capable UAV systems that cost less and required less ground support equipment.

Introduction to UAV Systems, Fifth Edition. Paul Gerin Fahlstrom, Thomas James Gleason, and Mohammad H. Sadraey.
© 2022 John Wiley & Sons, Inc. Published 2022 by John Wiley & Sons, Inc.
Companion website: www.wiley.com/go/fahlstrom/uavsystems5e

1.2 History

1.2.1 Early History

Throughout their history, UAV systems have tended to be driven by military applications, as is true of many areas of technology, with civilian applications tending to follow once the development and testing had been accomplished in the military arena.

One could say that the first UAV was a stone thrown by a caveman in prehistoric times or perhaps a Chinese rocket launched in the thirteenth century. These "vehicles" had little or no control and essentially followed a ballistic trajectory. If we restrict ourselves to vehicles that generate aerodynamic lift and/or have a modicum of control, the kite would probably fit the definition of the first UAV.

In 1883, an Englishman named Douglas Archibald attached an anemometer to the line of a kite and measured wind velocity at altitudes up to 1,200 ft. Mr. Archibald attached cameras to kites in 1887, providing one of the world's first reconnaissance UAVs. William Eddy took hundreds of photographs from kites during the Spanish–American war, which may have been one of the first uses of UAVs in combat.

It was not until World War I, however, that UAVs became recognized systems. Charles Kettering (of General Motors fame) developed a biplane UAV for the Army Signal Corps. It took about 3 years to develop and was called the Kettering Aerial Torpedo, but is better known as the "Kettering Bug" or just plain "Bug." The Bug could fly nearly 40 mi at 55 mi/h and carry 180 lb of high explosives. The air vehicle was guided to the target by pre-set controls and had detachable wings that were released when over the target, allowing the fuselage to plunge to the ground as a bomb. Also, in 1917, Lawrence Sperry developed a UAV, similar to Kettering's, for the Navy, called the Sperry-Curtis Aerial Torpedo. It made several successful flights out of Sperry's Long Island airfield, but was not used in the war.

We often hear of the UAV pioneers who developed the early aircraft, but other pioneers were instrumental in inventing or developing important parts of the system. One was Archibald Montgomery Low, who developed data links. Professor Low, born in England in 1888, was known as the "Father of Radio Guidance Systems." He developed the first data link and solved interference problems caused by the UAV engine. His first UAVs crashed, but on September 3, 1924, he made the world's first successful radio-controlled flight. He was a prolific writer and inventor and died in 1956.

In 1933, the British flew three refurbished Fairey Queen biplanes by remote control from a ship. Two crashed, but the third flew successfully, making Great Britain the first country to fully appreciate the value of UAVs, especially after they decided to use one as a target and couldn't shoot it down.

In 1937 another Englishman, Reginald Leigh Denny, and two Americans, Walter Righter and Kenneth Case, developed a series of UAVs called RP-1, RP-2, RP-3, and RP-4. They formed a company in 1939 called the Radioplane Company, which later became part of the Northrop-Ventura Division. Radioplane built thousands of target drones during World War II. (One of their early assemblers was Norma Jean Daugherty, later known as Marilyn Monroe.) Of course, the Germans used lethal UAVs (V-1's and V-2's) during the later years of the war, but it was not until the Vietnam War era that UAVs were successfully used for reconnaissance.

1.2.2 The Vietnam War

The first real use of UAVs by the United States in a combat reconnaissance role began during the Vietnam War. UAVs, such as the AQM-34 Firebee developed by Teledyne Ryan, were used for a wide range of missions, such as intelligence gathering, decoys, and leaflet dropping.

1.2 History 5

During the Vietnam War era, UAVs were used extensively in combat, but for reconnaissance missions only. The air vehicles were usually air launched from C-130's and recovered by parachute. The air vehicles were what might be called deep penetrators and were developed from existing target drones.

The impetus to operations in Southeast Asia came from activities during the Cuban Missile Crisis when UAVs were developed for reconnaissance but not used because the crisis ended before they became available. One of the first contracts was between Ryan and the Air Force, known as 147A, for vehicles based on the Ryan Firebee target drone (stretched versions). This was in 1962 and they were called Fireflys. Although the Fireflys were not operational during the Cuban crisis, they set the stage for Vietnam. Northrop also improved their early designs, which were essentially model airplanes, to jet-propelled deep penetrators, but stuck mostly to target drones. The Ryan Firefly was the primary air vehicle used in Southeast Asia.

A total of 3,435 sorties were flown, and most of these (2,873, or nearly 84%) were recovered. One air vehicle, the TOMCAT, successfully completed 68 missions before it was lost. Another vehicle completed 97.3% of its missions of low-altitude, real-time photography. By the end of the Vietnam War in 1972, air vehicles were experiencing 90% success rates [1].

1.2.3 Resurgence

At the end of the Vietnam War, general interest in UAVs dwindled until the Israelis neutralized the Syrian air defense system in the Bekaa Valley in 1982 using UAVs for reconnaissance, jamming, and decoys. The Israeli Air Force pioneered several UAVs in the early 1980s. In 1982, United States observers noted Israel's use of UAVs in Lebanon and persuaded the Navy to acquire a UAV capability. One of the early UAVs acquired by the Navy was the RQ-2 Pioneer. It was developed jointly by AAI Corporation and Israeli Aircraft Industries and became a very useful air vehicle during Desert Storm for collecting tactical intelligence.

Actually, the Israeli UAVs were not as technically successful as many people believe, with much of their operational success being achieved through the element of surprise rather than technical sophistication. The air vehicle was basically unreliable and couldn't fly at night, and the data-link transmissions interfered with the manned fighter communications. However, they proved that UAVs could perform valuable, real-time combat service in an operational environment.

The United States began to work again on UAVs in August 1971 when the Defense Science Board recommended mini-RPVs for artillery target spotting and laser designation. In February 1974, the Army's Material Command established an RPV weapons system management office and by the end of that year (December) a "Systems Technology Demonstration" contract was awarded to Lockheed Aircraft Company, with the air vehicle subcontracted to Developmental Sciences Incorporated (later DSC, Lear Astronics, Ontario, CA). The launcher was manufactured by All American Engineering (later ESCO-Datron), and the recovery net system by Dornier of the then still-partitioned West Germany. Ten bidders competed for the program. The demonstration was highly successful, proving the concept to be feasible. The system was flown by Army personnel and accumulated more than 300 flight hours.

In September 1978, the so-called Target Acquisition/Designation and Aerial Reconnaissance System (TADARS) required operational capability (ROC) was approved, and approximately 1 year later, in August 1979, a 43-month Full Scale Engineering Development (FSED) contract was awarded to Lockheed as the sole source. The system was given the name "Aquila" and is discussed in more detail at the end of this chapter. For a number of reasons that provide important lessons to

6   *1 History and Overview*

UAV system developers, Aquila development stretched out for many years and the system was never fielded.

In 1984, partly as a result of an urgent need and partly because the Army desired some competition for Aquila, the Army started a program called Gray Wolf, which demonstrated, for the first time for a UAV, hundreds of hours of night operations in what could be called "combat conditions." This program, still partly classified, was discontinued because of inadequate funding.

1.2.4 Joint Operations

The US Navy and Marine Corps entered the UAV arena in 1985 by purchasing the Mazlat/ Israeli Aircraft Industries (IAI) and AAI Pioneer system, which suffered considerable growing pains but still remains in service. However, Congress by this time became restless and demanded that a joint project office (JPO) be formed so that commonality and interoperability among the services would be maximized. The JPO was put under the administrative control of the Department of the Navy. This office has developed a master plan that not only defines the missions but also describes the desirable features for each kind of system needed by the services. Some elements of this plan will be discussed in Chapter 2 in the section called "Classes of UAV Systems."

The US Air Force was initially reluctant to embrace UAVs, notwithstanding their wealth of experience with target-drone unmanned aircraft. However, this attitude changed significantly during the 1990s and the Air Force not only has been very active in developing and using UAVs for a variety of purposes but also has been the most active of the four US services in attempting to take control of all UAV programs and assets within the US military.

1.2.5 Desert Storm

The invasion of Iraq to Kuwait in 1990–1991 allowed military planners an opportunity to use UAVs in combat conditions. They found them to be a highly desirable asset even though the performance of the systems then available was less than satisfactory in many ways. Five UAV systems were used in the operation: (1) the Pioneer by US forces, (2) the Ex-Drone by US forces, (3) the Pointer by US forces, (4) the "Mini Avion de Reconnaissance Telepilot" (MART) by French forces, and (5) the CL 89, a helicopter UAV, by British forces.

Although numerous anecdotal stories and descriptions of great accomplishments have been cited, the facts are that the UAVs did not play a decisive or a pivotal role in the war. For example, the Marines did not fire upon a single UAV-acquired target during the ground offensive according to a Naval Proceedings article published in November 1991 [2]. What was accomplished, however, was the awakening in the mind of the military community of a realization of "what could have been." What was learned in Desert Storm was that UAVs were potentially a key weapon system, which assured their continuing development.

1.2.6 Bosnia

The NATO UAV operation in Bosnia was one of surveillance and reconnaissance. Bomb-damage assessment was successfully accomplished after NATO's 1995 air attacks on Bosnian-Serb military facilities. Clearly shown in aerial photographs are Serbian tanks and bomb-damaged buildings. Night reconnaissance was particularly important as it was under the cover of darkness that most

1.2 History

clandestine operations took place. The Predator was the primary UAV used in Bosnia, flying from an airbase in Hungary.

1.2.7 Afghanistan and Iraq

The war in Iraq (which lasted from 2003 to 2011) has transformed the status of UAVs from a potential key weapons system searching for proponents and missions to their rightful place as key weapon systems performing many roles that are central to the operations of all four services. At the beginning of the war, UAVs were still under development and somewhat "iffy," but many developmental UAVs were committed to Operation Iraqi Freedom.

The Global Hawk was effectively used during the first year despite being in the early stages of developmental. The Pioneer, the Shadow, the Hunter, and the Pointer were used extensively.

The Marines flew hundreds of missions using Pioneers during the battle for Fallujah in 2004 to locate and mark targets and keep track of insurgent forces. They were especially effective at night and could be considered one of the decisive weapons in that battle.

The armed version of the Predator, mini-UAVs such as the Dragon Eye, and a wide range of other UAV systems have been used on the battlefields of Afghanistan and Iraq and have proven the military value of UAVs.

Predator UAVs have been operational in Bosnia since 1995 and as part of Operation Enduring Freedom in Afghanistan and Operation Iraqi Freedom, flying more than 500,000 flight hours on over 50,000 flights. It is interesting to know that in 2002, an Iraqi MiG-25 intercepted an air-to-air equipped Predator. Both fired missiles at each other, but MiG-25 evaded but Predator was shot down.

During Operation Iraqi Freedom, the Global Hawk flew 15 missions, collecting over 4,800 images from March 18 to April 23, 2003.

1.2.8 Long-Range Long-Endurance Operations

After the 9/11/2001 terrorist attack on the World Trade Center in New York City, and during wars in Iraq and Afghanistan, US military explored the high value of UAVs in wars and even peace-time operations. Afterward, there were multiple UAV applications around the world including the war with ISIS and Al-Qaeda (2013–2017), many of which were long-range long-endurance wars. Three top UAVs that were employed for long-range long-endurance operations around the world are RQ-4 Global Hawk, RQ-1 Predator, and MQ-9 Reaper. The Reaper with its missiles, sensors, and relatively long endurance is a UAV that transformed military combat.

During Operation Enduring Freedom (from November to September 2002 in Afghanistan), the Global Hawk provided more than 17,000 near-real-time, high-resolution intelligence, surveillance, and reconnaissance images, flying more than 60 combat missions and logging more than 1,200 combat hours. By 2015, the Global Hawk surpassed 140,000 flight hours and 100,000 combat flight hours and exceeded operational performance targets. The following summarizes three recent projects for HALE UAVs with a pseudo-satellite mission.

In 2020, in partnership with NASA, the Swift Engineering [3] has developed the Swift HALE UAV to demonstrate how a successful high-altitude, long-endurance flight can expand science research in a cost-efficient and timely manner. This solar-powered UAV has a wingspan of 72 ft and weighs less than 180 lb, flies 10- to 15-lb payloads at a time, and is designed to operate at an altitude of 70,000 ft for 30 days or more.

Furthermore, BAE Systems is developing a solar-powered, stratosphere-flying UAV that can act as a backup option to disabled communications satellites. The Phasa-35 is designed to operate at altitudes of up to 70,000 ft. This HALE UAV with a wing span of 115 ft can remain in the air for up to a year without returning to Earth.

Moreover, the Airbus Zephyr unmanned aerial system has been designed and tested as an ultralight high-altitude pseudo-satellite. However, in 2020, the UAV broke up in flight after encountering unstable atmospheric conditions, which resulted in a series of uncommanded rolls and an uncontrolled spiral descent.

1.3 Overview of UAV Systems

The primary element of an unmanned aerial system is the air vehicle (AV). A number of titles are employed in the literature for the air vehicle, including: (1) unmanned aerial vehicle, (2) remotely piloted vehicles (RPV), (3) radio-controlled (RC) plane, (4) model airplane, (5) remotely piloted aircraft (RPA), and (6) drones. The term RPV is an older name, and currently not much used. The term "RC plane" and "model airplane" are primarily used by hobbyists, aeromodellers, and Academy of Model Aeronautics (World's largest model aviation association)[1], and "drone" is mainly employed by the media.

All, of course, are unmanned so the name "unmanned aerial vehicle" or UAV can be thought of as the generic title. To the purist, the "remotely piloted vehicle" is piloted or steered (controlled) from a remotely located position, so an RPV is always a UAV, but a UAV, which may perform autonomous or preprogrammed missions, need not always be an RPV.

In the past, these aircraft were all called drones, that is, a "pilotless airplane controlled by radio signals," according to *Webster's Dictionary*. Today the UAV developer and user community do not use the term "drone" except for vehicles that have limited flexibility for accomplishing sophisticated missions and fly in a persistently dull, monotonous, and indifferent manner, such as a target drone. This has not prevented the press and the general public from adopting the word "drone" as a convenient, if technically incorrect, general term for UAVs. Thus, even the most sophisticated air vehicle with extensive semiautonomous functions is likely to be headlined as a "drone" in the morning paper or on the evening news.

Whether the UAV is controlled manually or via a preprogrammed navigation system, it should not necessarily be thought of as having to be "flown," that is, controlled by someone that has piloting skills. UAVs used by the military usually have autopilots and navigation systems that maintain attitude, altitude, and ground track automatically.

Manual control usually means controlling the position of the UAV by manually adjusting the heading, altitude, speed, etc., through switches, a joystick, or some kind of pointing device (mouse or trackball) located in the ground control station, but allowing the autopilot to stabilize the vehicle and assume control when the desired course is reached. Navigation systems of various types (global positioning system (GPS), radio, inertial) allow for preprogrammed missions, which may or may not be overridden manually.

As a minimum, a typical UAV system is composed of air vehicles, one or more ground control stations (GCS) and/or mission planning and control stations (MPCS), payload, and data link. In addition, many systems include launch and recovery subsystems, air-vehicle carriers, and other ground handling and maintenance equipment. A very simple generic UAV system is shown in Figure 1.1.

---

1 https://www.modelaircraft.org

1.3 Overview of UAV Systems 9

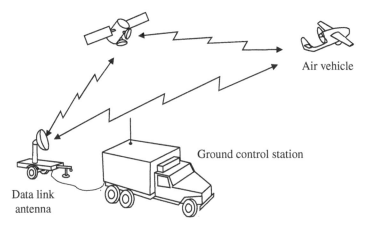

Figure 1.1 Generic UAV system

1.3.1 Air Vehicle

The air vehicle is the airborne part of the unmanned aerial system that includes the airframe, propulsion unit, flight controls, and electric power system. The air data terminal is mounted in the air vehicle, and is the airborne portion of the communications data link. The payload is also onboard the air vehicle, but it is recognized as an independent subsystem that often is easily interchanged with different air vehicles and uniquely designed to accomplish one or more of a variety of missions. The air vehicle can be a fixed-wing airplane, rotary wing (single or multiple), or a ducted fan. Lighter-than-air vehicles are also eligible to be termed UAVs.

1.3.2 Mission Planning and Control Station

The MPCS, also called the GCS, is the operational control center of the UAV system where video, command, and telemetry data from the air vehicle are processed and displayed. These data are usually relayed through a ground terminal, which is the ground portion of the data link. The MPCS shelter incorporates a mission planning facility, control and display consoles, video and telemetry instrumentation, a computer and signal processing group, the ground data terminal, communications equipment, and environmental control and survivability protection equipment.

The MPCS can also serve as the command post for the person who performs mission planning, receives mission assignments from supported headquarters, and reports acquired data and information to the appropriate unit, be it weapon fire direction, intelligence, or command and control, for example, the mission commander. The station usually has positions for both the air vehicle and mission payload operators to perform monitoring and mission execution functions.

In some small UAS, the ground control station is contained in a case that can be carried around in a back-pack and set up on the ground, and consists of little more than a remote control and some sort of display, probably augmented by embedded microprocessors or hosted on a ruggedized laptop computer.

At the other extreme, some ground stations are located in permanent structures thousands of miles away from where the air vehicle is flying, using satellite relays to maintain communications with the air vehicle. In this case, the operator's consoles might be located in an internal room of a large building, connected to satellite dishes on the roof. A cut-away view of a typical field MPCS for a long-range UAV is shown in Figure 1.2.

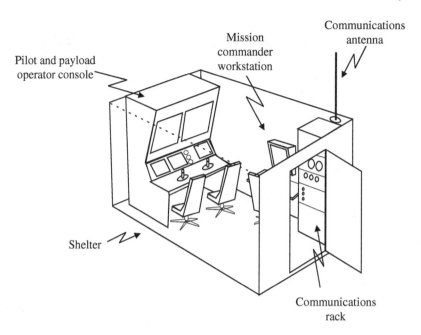

Figure 1.2 Mission planning and control station for a long-range UAV

1.3.3 Launch and Recovery Equipment

Launch and recovery can be accomplished by a number of techniques ranging from conventional takeoff and landing on prepared sites to vertical ascent/descent using rotary wing or fan systems. Catapults using either pyrotechnic (rocket) or a combination of pneumatic/hydraulic arrangements are also popular methods for launching air vehicles. Some small UAVs are launched by hand, essentially thrown into the air like a toy glider.

Nets and arresting gear are used to capture fixed-wing air vehicles in small spaces. Parachutes and parafoils are used for landing in small areas for point recoveries. One advantage of a rotary-wing or fan-powered vehicle is that elaborate launch and recovery equipment usually is not necessary. However, operations from the deck of a pitching ship, even with a rotary-wing vehicle, will require hold-down equipment unless the ship motion is minimal.

1.3.4 Payloads

Carrying a payload is the ultimate reason for having a UAV system, and the payload sometimes is the most expensive subsystem of the UAV. Payloads often include video cameras, either daylight or night (image-intensifiers or thermal infrared), for reconnaissance and surveillance missions. Film cameras were widely used with UAV systems in the past, but are largely replaced today with electronic image collection and storage, as has happened in all areas in which video images are used. Currently, video cameras are the most popular payloads in UAVs.

If target designation is required, a laser is added to the imaging device and the cost increases dramatically. Radar sensors, often using a Moving Target Indicator (MTI) and/or synthetic aperture radar (SAR) technology, are also important payloads for UAVs conducting reconnaissance missions. Another major category of payloads is electronic warfare (EW) systems. They include the full spectrum of signal intelligence (SIGINT) and jammer equipment. Other

1.3 Overview of UAV Systems

sensors such as meteorological and chemical sensing devices have been proposed as UAV payloads.

Armed UAVs carry weapons to be fired, dropped, or launched. "Lethal" UAVs carry explosive or other types of warheads and may be deliberately crashed into targets. As discussed elsewhere in this book, there is a significant overlap between UAVs, cruise missiles, and other types of missiles. The design issues for missiles, which are "one-shot" systems intended to destroy themselves at the end of one flight, are different from those of reusable UAVs and this book concentrates of the reusable systems, although much that is said about them applies as well to the expendable systems.

Another use of UAVs is as a platform for data and communications relays to extend the coverage and range of line-of-sight radio-frequency systems, including the data links used to control UAVs and to return data to the UAV users.

1.3.5 Data Links

The data link is a key subsystem for any UAV. The data link for a UAV system provides two-way communication (i.e., uplink and down link), either upon demand or on a continuous basis. An uplink with a data rate of a few kbps provides control of the air-vehicle flight path and commands to its payload. The downlink provides both a low data-rate channel to acknowledge commands and transmit status information about the air vehicle and a high data-rate channel (1–10 Mbps) for sensor data such as a video camera and radar.

The data link may also be called upon to measure the position of the air vehicle by determining its azimuth and range from the ground-station antenna. This information is used to assist in navigation and in accurately determining air-vehicle location (e.g., altitude). Other flight parameters, such as aircraft speed, climb rate, and direction, are often transmitted by a down link to MPCS.

Data links require some kind of anti-jam and anti-deception capability if they are to be sure of effectiveness in combat.

The ground data terminal is usually a microwave electronic system and antenna that provides line-of-sight communications, sometimes via satellite or other relays, between the MPCS and the air vehicle. It can be co-located with the MPCS shelter or remote from it. In the case of the remote location, it is usually connected to the MPCS by hard wire (often fiber-optic cables). The ground terminal transmits guidance and payload commands and receives flight status information (altitude, speed, direction, etc.) and mission payload sensor data (video imagery, target range, lines of bearing, etc.).

The air data terminal is the airborne part of the data link. It includes the transmitter and antenna for transmitting video and air-vehicle data and the receiver for receiving commands from the ground.

1.3.6 Ground Support Equipment

Ground support equipment (GSE) is becoming increasingly important because UAV systems are electronically sophisticated and mechanically complex systems. GSE for a long-range UAV may include: test and maintenance equipment, a supply of spare parts and other expendables, a fuel supply and any refueling equipment required by a particular air vehicle, handling equipment to move air vehicles around on the ground, if they are not man-portable or intended to roll around on landing gear, and generators to power all of the other support equipment.

12   *1 History and Overview*

If the UAS ground systems are to have mobility on the ground, rather than being a fixed ground station located in buildings, the GSE must include transportation for all of the things listed earlier, as well as transportation for spare air vehicles and for the personnel who make up the ground crew, including their living and working shelters and food, clothing, and other personal gear.

As can be seen, a completely self-contained, mobile UAS can require a lot of support equipment and trucks of various types. This can be true even for an air vehicle that is designed to be lifted and carried by three or four men.

1.4 The Aquila

The American UAS called the Aquila was a unique early development of a total integrated system. It was one of the first UAV systems to be planned and designed having unique components for launch, recovery, and tactical operation. The Aquila was an example of a system that contained all of the components of the generic system described previously. It also is a good example of why it is essential to consider how all the parts of a UAS fit together and work together and collectively drive the cost, complexity, and support costs of the system. Its story is briefly discussed here. Throughout this book, we will use lessons learned at great cost during the Aquila program to illustrate issues that still are important for those involved in setting requirements for UAS and in the design and integration of the systems intended to meet those requirements.

In 1971, more than a decade before the Israeli success in the Bekaa Valley, the US Army had successfully launched a demonstration UAV program, and had expanded it to include a high-technology sensor and data link. The sensor and data-link technology broke new ground in detection, communication, and control capability. The program moved to formal development in 1978 with a 43-month schedule to produce a production-ready system. The program was extended to 52 months because the super-sophisticated MICNS (Modular Integrated Communication and Navigation System) data link experienced troubles and was delayed. Then, for reasons unknown to industry, the Army shut the program down altogether. It was subsequently restarted by Congress (about 1982), but at the cost of extending it to a 70-month program. From then on everything went downhill.

In 1985, a Red Team formed to review the system came to the conclusion that not only had the system not demonstrated the necessary maturity to continue to production, but also that the systems engineering did not properly account for deficiencies in the integration of the data link, control system, and payload, and it probably would not work anyway. After two more years of intensive effort by the government and contractor, many of the problems were fixed, but nevertheless it failed to demonstrate all of the by then required capabilities during operational testing (OT) II and was never put into production.

The lessons learned in the Aquila program still are important for anyone involved in specifying operational requirement, designing, or integrating a UAS. This book refers to them in the chapters describing reconnaissance and surveillance payloads, and data links in particular, because the system-level problems of Aquila were largely in the area of understanding those subsystems and how they interacted with each other, with the outside world, and with basic underlying processes such as the control loop that connects the ground controller to the air vehicle and its subsystems.

1.4.1 Aquila Mission and Requirements

The Aquila system was designed to acquire targets and combat information in real time, beyond the line-of-sight of supported ground forces. During any single mission, the Aquila was capable of

1.4 The Aquila

performing airborne target acquisition and location, laser designation for precision-guided munitions (PGM), target damage assessment, and battlefield reconnaissance (day or night). This is quite an elaborate requirement.

To accomplish this, an Aquila battery needed 95 men, 25 five-ton trucks, 9 smaller trucks, and a number of trailers and other equipment, requiring several C-5 sorties for deployment by air. All of this allowed operation and control of 13 air vehicles. The operational concept utilized a central launch and recovery section (CLRS) where launch, recovery, and maintenance were conducted. The air vehicle was flown toward the Forward Line of Own Troops (FLOT), and handed off to a forward control section (FCS), consisting mainly of a ground control station, from which combat operations were conducted.

It was planned that eventually the ground control station with the FCS would be miniaturized and be transported by a High Mobility Multipurpose Wheeled Vehicle (HMMWV) to provide more mobility and to reduce target size when operating close to the FLOT. The Aquila battery belonged to an Army Corps. The CLRS was attached to Division Artillery because the battery supported a division. The FCS was attached to a maneuver brigade.

1.4.2 Air Vehicle

The Aquila air vehicle was a tail-less flying wing with a rear-mounted 26-horsepower, two-cycle piston engine, and a pusher propeller. Figure 1.3 shows the Aquila air vehicle. The fuselage was about 2 m long and the wingspan was 3.9 m. The airframe was constructed of kevlar-epoxy material but metalized to prevent radar waves from penetrating the skin and reflecting off the square electronic boxes inside. The gross takeoff weight was about 265 lb and it could fly between 90 and 180 km/h up to about 12,000 ft.

1.4.3 Ground Control Station

The Aquila ground control station contained three control and display consoles, video and telemetry instrumentation, a computer and signal processing group, internal/external communications equipment, ground data terminal control equipment, and survivability protection equipment.

The GCS was the command post for the mission commander and had the display and control consoles for the vehicle operator, payload operator, and mission commander. The GCS was powered by a 30-kW generator. A second 30-kW generator was provided as a backup. Attached to the GCS by 750 m of fiber-optic cable was the remote ground terminal (RGT). The RGT consisted of a tracking dish antenna, transmitter, receiver, and other electronics, all trailer-mounted

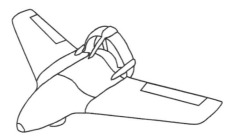

Figure 1.3 Aquila air vehicle

as a single unit. The RGT received downlink data from the air vehicle in the form of flight status information, payload sensor data, and video. The RGT transmitted both guidance commands and mission payload commands to the air vehicle. The RGT had to maintain line-of-sight contact with the air vehicle. It also had to measure the range and azimuth to the air vehicle for navigation purposes, and the overall accuracy of the system depended on the stability of its mounting.

1.4.4 Launch and Recovery

The Aquila launch system contained an initializer that was linked to the RGT and controlled the sequence of the launch procedure including initializing the inertial platform. The catapult was a pneumatic/hydraulic system that launched the air vehicle into the air with the appropriate airspeed.

The air vehicle was recovered in a net barrier mounted on a 5-ton truck. The net was supported by hydraulic-driven, foldout arms, which also contained the guidance equipment to automatically guide the air vehicle into the net.

1.4.5 Payload

The Aquila payload was a day video camera (Electro-Optic) with a bore-sighted laser rangefinder/designator for designating targets. Once locked on to a target, moving or stationary, it would seldom miss. The laser rangefinder/designator was optically aligned and automatically bore-sighted with the video camera. Scene and feature track modes provided line-of-sight stabilization and auto-tracking for accurate location and tracking of moving and stationary targets. An infrared (IR) night payload was also under development for use with Aquila.

1.4.6 Other Equipment

An air-vehicle handling truck was part of the battery ground support equipment and included a lifting crane. The lifting crane was necessary, not because the air vehicle was extremely heavy, but because the box in which it was transported contained lead to resist nuclear radiation. In addition, a maintenance shelter, also on a 5-ton truck, was used for unit-level maintenance and was a part of the battery.

1.4.7 Summary

The Aquila system had everything imaginable in what one could call "The Complete UAV System;" "zero-length" launcher, "zero-length" automatic recovery with a net, anti-jam data link, and day and night payload with designator. This came at a very high cost, however – not only in dollars but also in terms of manpower, trucks, and equipment. The complete system became large and unwieldy, which contributed to its downfall. All of this equipment was necessary to meet the elaborate operational and design requirements placed on the Aquila system by the Army, including a level of nuclear blast and radiation survivability (a significant contributor to the size and weight of shelters and the RGT mount). Eventually, it was determined that many of the components of the system could be made smaller and lighter and mounted on HMMWVs instead of 5-ton trucks, but by that time the whole system had gained a bad reputation for:

1.4 The Aquila

- Having been in development for over 10 years;
- Being very expensive;
- Requiring a great deal of manpower, a large convoy of heavy trucks for mobility, and extensive support;
- What was widely perceived to be a poor reliability record (driven by the complexity of the data link, air-vehicle subsystems, and the zero-length recovery system);
- Failure to meet some operational expectations that were unrealistic, but had been allowed to build up during the development program because the system developers did not understand the limitations of the system.

Foremost among the operational "disappointments" was that Aquila turned out to be unable to carry out large-area searches for small groups of infiltrating vehicles, let alone personnel on foot. This failing was due to limitations on the sensor fields of view and resolution and on shortcomings in the system-level implementation of the search capability. It also was partly driven by the failure to understand that searching for things using an imaging sensor on a UAV required personnel with special training in techniques for searching and interpretation of the images provided. The sources of these problems and some ways to reduce this problem by a better system-level implementation of area searches are addressed in the discussions of imaging sensors in Part Four and data links in Part Five.

The Aquila program was terminated as a failure, despite having succeeded in producing many subsystems and components that individually met all of their requirements. The US Army Red Team concluded that there had been a pervasive lack of systems engineering during the definition and design phases of the program. This failure set back US efforts to field a tactical UAS on an Army-wide basis, but opened the door for a series of small-scale "experiments" using less expensive, less-sophisticated air vehicles developed and offered by a growing "cottage industry" of UAV suppliers.

These air vehicles were generally conventionally configured oversized model airplanes or undersized light aircraft that tended to land and take off from runways if based on land, did not have any attempt at reduced radar signatures and little if any reduced infrared or acoustical signatures, and rarely had laser designators or any other way to actively participate in guidance of weapons.

They generally did not explicitly include a large support structure. Although they required most of the same support as an Aquila system, they often got that support from contractor personnel deployed with the systems in an ad hoc manner.

UAV requirements that have followed Aquila have acknowledged the cost of a "complete" stand-alone system by relaxing some of the requirements for self-sufficiency that helped drive the Aquila design to extremes. In particular, many land-based UAVs now are either small enough to be hand launched and recovered in a soft crash landing or designed to take off and land on runways. All or most use the global positioning system (GPS) for navigation. Many use data transmission via satellites to allow the ground station to be located at fixed installations far from the operational area and eliminate the data link as a subsystem that is counted as part of the UAS.

However, the issues of limited fields-of-view and resolution for imaging sensors, data-rate restrictions on downlinks, and latencies and delays in the ground-to-air control loop that were central to the Aquila problems are still present and can be exacerbated by use of satellite data transmission and control loops that circle the globe. Introducing UAV program managers, designers, system integrators, and users to the basics of these and other similarly universal issues in UAV system design and integration is one of the objectives of this textbook.

1.5 Global Hawk

1.5.1 Mission Requirements and Development

The requirement for a Global Hawk type of system grew out of Operation Desert Storm (in 1991). The Global Hawk was intended to compliment or replace the aging U-2 spy plane fleet. The Global Hawk is an advanced intelligence, surveillance, and reconnaissance air system (i.e., ISR mission). The strategy for this UAV program involved four phases, which were to be completed between 1994 and 1999.

It flew for the first time at Edwards Air Force Base, California, on Saturday, February 28, 1998. The first flight of the Global Hawk became the first UAV to cross the Pacific Ocean in April 2001 when it flew from the United States to Australia. The entire mission, including the takeoff and landing, was performed autonomously by the UAV as planned.

A total of 21 sorties of flight tests were conducted over 16 months using two air vehicles accumulating 158 total flight hours. It entered service in 2001 and reached the serial production stage in 2003. The Global Hawks, monitored by shifts of pilots in a ground control station in California, fly 24-hour missions, and they are cheaper to operate than the manned aircraft Lockheed U-2.

1.5.2 Air Vehicle

The Global Hawk unmanned aerial system consists of the aircraft, payloads, data links, ground stations, and logistics support package. Global Hawk is the largest active current UAV with successful flights, and with a high altitude and long endurance (HALE). A Global Hawk (Figure 1.4) is equipped with a single AE 3007H turbofan engine – mounted on top of the rear fuselage – supplied by Rolls-Royce. The engine is mounted on the top surface of the rear fuselage section with the engine exhaust between the V tail.

The wing and tail are made of graphite composite materials. The wing has structural hard points for external stores. The aluminum fuselage contains pressurized payload and avionics compartments. The V-configuration of the tail provides a reduced radar and infrared signature. Some mass and geometry features as well as the flight performance of Global Hawk are provided in Table 1.1.

Figure 1.4 Global Hawk (Source: Bobbi Zapka / Wikimedia Commons / Public Domain)

Table 1.1 RQ-4B Global Hawk data and performance

No.	Parameter	Value (unit)
1	Wingspan	39.9 m
2	Length	14.5 m
3	Maximum takeoff mass	14,628 kg
4	External payload weight	3,000 lb
5	Internal payload weight	750 lb
6	Turbofan engine thrust	34 kN
7	Maximum speed	340 knots
8	Range	22,779 km
9	Endurance	32+ hours
10	Service ceiling	60,000 ft

The prime navigation and control system consists of two systems, the Inertial Navigation System and the Global Positioning System (INS/GPS). The aircraft is flown by entering specific way points into the mission plan. The GCS consists of two elements, the Launch and Recovery Element (LRE) and the Mission Control Element (MCE). The LRE is located at the air vehicle base. It launches and recovers the air vehicle and verifies the health and status of the various onboard systems. The MCE is employed to conduct the entire flight from after takeoff until before landing.

Many changes have been applied in the design of Northrop Grumman RQ-4B Global Hawk as compared with RQ-4A. For instance, the RQ-4B Global Hawk has a 50% payload increase, larger wingspan (130.9 ft) and longer fuselage (47.6 ft), and a new generator to provide 150% more electrical output. Although RQ-4B carries more fuel than RQ-4A, it has a slightly shorter range and endurance, due to a heavier maximum takeoff weight.

1.5.3 Payloads

Originally RQ-4A had three sensors (as payload): an Electro-Optical/Infrared sensor and two Synthetic Aperture Radar Sensors – which are located under the fuselage belly in the integrated sensor suit – have been enhanced for RQ-4B. The main thrust of the air vehicle changes over time has involved the sensors. The enhancement improves the range of both the SAR and infrared system by approximately 50%.

1.5.4 Communications System

The Global Hawk has a wide-band satellite data link and a line-of-sight data link. Data is transferred by: (1) Ku-band satellite communications, (2) X-band line-of-sight links, and (3) both Satcom and line-of-sight links at the UHF-band. The synthetic aperture radar and ground moving target indicator operates at the X-band with a 600 MHz bandwidth.

The air traffic control (ATC) and command-and-control (C2) of the NASA Global Hawk from the Dryden Flight Research Center is applied in two distinct regions: (1) The line-of-sight (LOS) and (2) The beyond line-of-sight (BLOS). The communications link used for LOS are through UHF/VHF links. The primary communications links used for BLOS are two Iridium Satcom links. However, an Inmarsat Satcom link provides a backup communications capability.

The NASA Global Hawk payload communications architecture is independent of the communications links utilized to control the aircraft. Four dedicated Iridium SatCom communication links are used for continuous narrow band communications between the ground station and the UAV payloads. Moreover, two additional Iridium links are used to monitor power consumption by individual payloads, and to control features such as lasers and dropsonde. The use of the Iridium system provides a complete global coverage, including the Polar regions.

1.5.5 Development Setbacks

During Global Hawk flight tests programs and long operations, there were a number of setbacks [4], where a few resulted in the loss of the air vehicle and one caused damage to the sensor suite of another air vehicle.

The major setback during flight testing was the destruction of air vehicle 2 on March 29, 1999. The aircraft experienced an uneventful liftoff from the runway at Edwards Air Force Base (AFB). As it was climbing, the air vehicle unexpectedly flipped over on its back, shut down its engine, and locked the flight controls into a death spin. The aircraft executed the termination command and crashed. The crash was due to a lack of proper frequency coordination between the Nellis AFB and Edwards AFB flight test ranges.

In December 1999, a software problem caused another Global Hawk to accelerate to an excessive taxi speed after a successful, full stop landing on Edwards' main runway. An error in software code to coordinate between the mission planning system and the aircraft commanded the vehicle to taxi at 155 knots. The nose gear collapsed causing $5.3 million worth of damage to the electro-optical/ infra-red (EO/IR) sensors. The primary cause of this mishap was the execution of a commanded ground speed of 155 knots for a taxi on the contingency mission plan.

During the deployment phase, two of the prototype air vehicles were lost and sustained. The first loss occurred on December 30, 2001, when the Global Hawk was returning from a truncated operational mission in support of Operation Enduring Freedom. To help a descent at 54,000 ft, four spoilers were raised to the maximum deflection (45 degrees), which caused a turbulent air-induced flutter. The subsequent energy of the resultant flutter was absorbed by the right V-tail main spar. The right outboard ruddervator actuator control rod failed, allowing the ruddervator to travel unrestrained beyond its normal range. Then, the vehicle departed controlled flight, entered a right spin, and crashed. The loss was attributed to a structural failure of the right ruddervator assembly of the V-tail (massive delamination of the main spar).

The second loss occurred on July 10, 2002, when a Global Hawk was flying an operational mission in Operation Enduring Freedom. The mishap vehicle experienced a catastrophic engine failure and glided for about half an hour. The vehicle impacted the ground during the attempted emergency landing. The mishap was attributed to a fuel nozzle failure in the high flow position that eventually led to the engine internal failure.

Another loss was when on June 20, 2019, Iran shot down a Global Hawk with a surface-to-air missile over the Strait of Hormuz. Iran said that the UAV violated its airspace, while US officials responded that the air vehicle was flying in international airspace.

These real stories provided valuable lessons and presented expensive experiences for young UAV designers. As typical of any development program, the Global Hawk design changed as the result of flight tests.

1.6 Predator Family

1.6.1 Predator Development

RQ-1 Predator is a long-endurance, medium-altitude unmanned aircraft system for surveillance, reconnaissance, and attack missions, designed and manufactured by General Atomics Aeronautical Systems.

The Predator had an unconventional development cycle with origins going back to a project by Abraham E. Karem. He is a pioneer in innovative fixed and rotary-wing unmanned vehicles and is regarded as one of the founding fathers of UAV technology. Initially, by 1983, a small long-endurance tactical reconnaissance UAV prototype was developed called the Albatross for the DARPA. Then, by 1988, further development resulted in a more advanced design, the Amber, which was followed by the GNAT 750. Karem's company (Karem Aircraft, Inc.) and its UAV were soon acquired by General Atomics.

The CIA utilized the GNAT 750 in military operations over Bosnia in 1993 and 1994. The program suffered from a few weaknesses, but it held enough promise that the DOD expressed interest in a larger, more advanced version of the GNAT 750 for medium-altitude reconnaissance, then designated RQ-1 Predator. By 1995, it became operational over Bosnia. In parallel, the Air Force saw the Predator as a new tool in tactical reconnaissance with the added benefit of a live satellite data link.

In the late 1990s, Predator's capability was expanded to include a laser designator to illuminate targets and guide weapons dropped from other aircraft. In 1999, the UAV had its first significant test during Operation Allied Force in Kosovo. By 2000, due to concern over the rising threat of al Qaeda, the Predator was scheduled for arming with the Hellfire laser-guided missile.

After the September 11, 2001 attacks, the armed Predator become fully operational, and by January 2003, flew 164 missions over Afghanistan. The armed Predator – capable of both reconnaissance and attack missions – has continued to have a pivotal role in combat operations. In 2002, the Air Force adapted a Predator to carry Stinger missiles and attempted an air-to-air engagement with an Iraqi MiG-25, but resulted in the loss of the Predator.

Predator UAVs have been operational since 1995 in support of NATO, UN, and US operations, and as part of Operation Enduring Freedom in Afghanistan and Operation Iraqi Freedom, flying more than 500,000 flight hours. The US Air Force Predator production ended in 2011 with 268 air vehicles manufactured. Hundreds of Predators have been sold to a number of countries including Italy, Spain, France, UK, Australia, Netherlands, Canada, and Germany.

This military UAV has been used in the Balkans, Afghanistan, Iraq, and other global locations. By 2011, the US military had nearly 11,000 UAVs on their inventory, including hundreds of Predators. The Predator was retired in 2018. The Predator-series family encompasses MQ-1 Predator, MQ-1C Gray Eagle, MQ-9 Reaper (Predator B), MQ-9B SkyGuardian, and Predator C Avenger.

1.6.2 Reaper

After about 10 years of Predator operations, and when some weaknesses were identified, new challenges arose in employing Predator. DOD decided to have a new version of Predator with enhanced performance features and an advanced design. The operation requirements included such performance items as a faster cruising speed and higher flight altitude, and also heavier and more advanced payloads. The conceptual design and the air vehicle configuration were almost kept. The only major configuration change was to have a V-tail instead of an inverted V-tail.

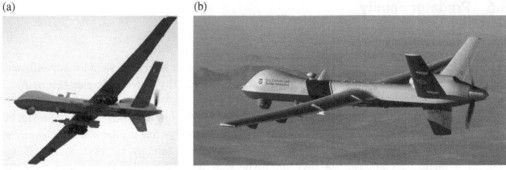

Left: A British MQ-9A Reaper operating over Afghanistan in 2009 (Source: Tam McDonald / Wikimedia Commons / OGL v1.0). Right: CBP's Reaper (Source: Gerald L. Nino / Wikimedia Commons / Public Domain)

Figure 1.5 General Atomics MQ-9 Reaper

The US Air Force first deployed the MQ-9 Reaper (developed by General Atomics, then called Predator B) to Afghanistan in October 2007 for precision airstrikes and it slowly began replacing the Predator. The General Atomics MQ-9 Reaper (Figure 1.5) flew its first operational mission in Iraq in July 2008. In the meantime, the Army began development of a refined derivative, the MQ-1C Gray Eagle, which began operations in 2012. The US Air Force retired the Predator in 2018, replacing it with the Reaper.

The first version of Predator (A) had a piston engine, but the upgraded Predator B, or MQ-9 Reaper, is equipped with a turboprop engine (with a greater power). Predator B is larger, much heavier, with an improved flight performance (e.g., faster cruise speed, longer range, and longer endurance) than the earlier MQ-1 Predator.

There are two groups of Payloads: (1) surveillance imagery sensors, which include a synthetic aperture radar, electro-optic video, and forward-looking infrared (FLIR) cameras, (2) weapon payloads, which include four anti-armor missiles AGM-114 Hellfire), two laser-guided bombs (GBU-12), and 500 lb joint direct attack munition. Other payload options include a laser designator and rangefinder, electronic support and countermeasures, a moving target indicator (MTI), and an airborne signals intelligence payload.

1.6.3 Features

Reaper UAV has a single turboprop engine on the rear fuselage, a fixed tricycle landing gear, a high aspect ratio wing, with a V-tail. It has an aileron for roll control and a ruddervator for longitudinal and directional control. The air vehicle is equipped with UHF and VHF radio relay links, a C-band line-of-sight data link, which has a range of 150 nm, and UHF and Ku-band satellite data links.

The ground control station (GCS) is built into a single 30 ft trailer, containing pilot and payload operator consoles, three data exploitation and mission planning consoles, and two synthetic aperture radar workstations together with satellite and line-of-sight ground data terminals. The GCS also includes a data distribution system, which is equipped with a 5.5 m dish antenna or Ku-band ground data terminal and a 2.4 m dish antenna for data dissemination. The flight can be controlled through line-of-site data links or through Ku-band satellite links to produce a continuous video. Some mass and geometry features as well as flight performance of Reaper are provided in Table 1.2.

Predator and Reaper still have two major weaknesses: (1) the inability to operate in contested airspace with effective enemy air defenses and (2) jamming. These highlight the advances required

1.8 Ethical Concerns of UAVs

Table 1.2 Reaper data and performance

No.	Parameter	Value (unit)
1	Wingspan	65 ft 7 in
2	Length	36 ft 1 in
3	Maximum takeoff weight	10,494 lb
4	External payload weight	3,000 lb
5	Internal payload weight	800 lb
6	Turboprop engine power	900 hp
7	Maximum speed	260 knots
8	Range	1,200 mi
9	Endurance – fully loaded	14 hours
10	Service ceiling	50,000 ft

for future Reaper versions to maintain its operational significance. Jamming can pose a significant threat to the Predator's data links and GPS navigation.

1.7 Top UAV Manufacturers

The global unmanned aerial vehicle market is witnessing a strong compounded annual growth, even in 2020 where the COVID-19 emerged as a global pandemic. By January 2019, at least 60 countries were using or developing over 1,300 various UAVs. Top unmanned aerial systems and air vehicles in the market are Northrop Grumman (US), General Atomics (US), AeroVironment (US), Lockheed Martin (US), Elbit Systems (Israel), Israel Aerospace Industries (Israel), BAE Systems (UK), Parrot (France), Microdrones (Germany), SZ DJI Technology (China), Ehang (China), Yuneec International (China), Textron (US), Saab (Sweden), and Raytheon (US). The overall market is expected to reach $21.8 billion by 2027.

A number of European countries (France, Italy, Greece, Spain, Switzerland, and Sweden) are collectively developing the next generation of UCAVs (most notably the nEUROn) and the MALE unmanned aircraft. It is interesting to note that, as of March 2020, DJI accounts for around 70% of the world's consumer UAV market. The dominant US UAV manufacturers include Boeing, Lockheed Martin, Aurora Flight Sciences, General Atomics, Northrop Grumman, and AeroVironment.

1.8 Ethical Concerns of UAVs

All engineering products share some ethical issues, but the ethical concerns in UAVs are new and not yet regulated. Like other engineering products, there are many ways that UAVs are utilized unlawfully or unethically (e.g., drug trafficking). There are basically two ethical issues in employing UAVs: (1) invasion of privacy and (2) killing innocent individuals (lethal use). For instance, between 2004 to 2010, the US drone program in Pakistan [3] has killed several hundred civilians accidentally.

There are many unanswered questions regarding these two major areas. There are much less ethical concerns regarding mechanical robots than UAVs. When a UAV is flying over houses and is collecting data, this is a case for ethical concern. Due to these concerns, when a user of an UAV does not feel accountability, he/she may trespass the privacy of other citizens and cause loss of life of other individuals.

According to the US government accountability office, there are still four areas of concern for UAVs in using airspace: (1) the inability to recognize and avoid other aircraft, (2) lack of operational standards, (3) vulnerability in command and control of UAV, and (4) lack of Government regulations necessary to safely facilitate the accelerated integration of UAVs into the national airspace system. Moreover, the utilization of unmanned aerial systems for military applications is currently a contested and debated issue.

Having a center in ethics-informed interdisciplinary research and the integration of ethical literacy throughout the UAV curriculum is a valuable step toward removing ethical concerns. It is promising that AIAA has developed a code of ethics, and recommends all aeronautical engineers to observe these codes in designing and developing air vehicles.

There are concerns about the risks of flying the military UAVs outside war zones. There are reports that US UAVs have repeatedly crashed at civilian airports overseas throughout the world. Among the problems cited in the reports "are pilot error, mechanical failure, software bugs, and poor coordination with civilian air-traffic controllers. Since an initial report of a crash in January of 2011 at a US base in Djibouti, there have been 'at least six more Predators and Reapers' crashes" in the vicinity of civilian airports overseas.

In 2021, the FAA announced [3] two final rules for unmanned aircraft, which will require Remote Identification (Remote ID) of drones and allow remote operators of small drones to fly over people and at night under certain conditions. This is a major step toward further integrating UAVs into the National Air Space.

A great many arguments challenge the ethical justifiability of remote weapons (UAVs) by US military. One of the questions is: Does it really matter if a human kills an enemy or if a machine (UAV) kills that enemy as long as the enemy is eliminated? The issues at stake in the "UAV ethics" discussions are complex. The authors are not ready to answer these types of questions, but we invite scholars to dive deep into this topic and help law makers to cast laws to resolve these concerns.

Questions

1) What do UAV, RPV, RPA, GPS, UAS, MTI, SAR, MPCS, GCS, PGM, CLRS, HMMWV, EO/IR, LOS, BLOS, FLOT, and SIGINT stand for?
2) Write the titles that are employed in the literature for the unmanned air vehicle.
3) Briefly write what Chapter 1 presents.
4) When was the first real use of UAVs by the United States in a combat reconnaissance role?
5) What UAVs were used in the operation during the Kuwait/Iraq war?
6) What UAVs were used in the Operation Iraqi Freedom?
7) How many bidders were competing in 1971 on the UAV program for a call by US DOD?
8) Name four companies that worked on a UAV in the early 1970s in response to a call by the Defense Science Board.
9) When did the US Navy and Marine Corps enter the UAV arena? What did they purchase in that year?

10) What UAV systems – and by which countries – were used in combat operation during the invasion of Iraq to Kuwait in 1990–1991?
11) What was the mission of NATO UAV operation in Bosnia in 1995?
12) Which UAV was the primary UAV used in Bosnia in 1995? Where was the UAV flying from?
13) What UAVs were used extensively during the war in Iraq (from 2003 to 2011)?
14) What UAV was used during the battle for Fallujah in Iraq in 2004 to locate and mark targets and keep track of insurgent forces?
15) How many missions did the Global Hawk fly during Operation Iraqi Freedom in 2003?
16) What term is adopted by the press and the general public as a general term for UAVs?
17) What does manual control of a UAV usually mean?
18) What is a typical UAV system composed of?
19) Show a very simple generic UAV system in a figure.
20) What does the air vehicle as the airborne part of the unmanned aerial system include?
21) What is the mission planning and control station also called?
22) Briefly describe the function of the mission planning and control station.
23) Write two extreme forms (with brief features) of an MPCS.
24) Write a few popular types of launch and recovery techniques/equipment.
25) Write one advantage of a rotary-wing or fan-powered vehicle over a fixed-wing UAV in launch and recovery.
26) What is the ultimate reason for having a UAV system?
27) Briefly compare the cameras (as the UAV payload) that were used in the past versus current ones.
28) What is the most popular payload in UAVs today?
29) Name a few types of UAV payloads.
30) What does a "Lethal" UAV carry as the payload?
31) Briefly compare a lethal UAV with a missile.
32) What is the function of a UAV when used as a platform for data and communications relays?
33) What does a downlink provide?
34) What is the typical data rate of: (a) uplink and (b) downlink?
35) What typical flight parameters are often transmitted by a downlink to MPCS?
36) What feature is required by a data link, if it is to be sure of effectiveness in combat?
37) Briefly describe the features of the ground data terminal of a data link.
38) What does the air data terminal include as the airborne part of the data link?
39) What does a ground support equipment for a long-range UAV typically include?
40) When must a GSE include transportation?
41) What American UAS was a unique early development of a total integrated system?
42) In what year did the Aquila program move to formal development?
43) What was the initial and final durations of the Aquila program?
44) What were the Aquila mission and requirements?
45) List the number of personnel, trucks, sorties, and other equipment used to allow operation and control of an Aquila battery (i.e., 13 air vehicles).
46) Briefly describe the configuration of the Aquila air vehicle.
47) Write the fuselage length and wingspan of the Aquila air vehicle.
48) Write engine type and engine power of the Aquila air vehicle.
49) What material was the Aquila air vehicle made of?
50) Write the speed range of the Aquila air vehicle.
51) Write the flight altitude of the Aquila air vehicle.

52) Briefly describe the equipment used in the Aquila ground control station.
53) How was the connection made between the ground control station and the remote ground terminal of the Aquila UAS?
54) How much was the power of the generator for the ground control station of the Aquila UAS?
55) Did the remote ground terminal have to maintain line-of-sight contact with the Aquila air vehicle?
56) Briefly describe the launch and recovery systems of the Aquila.
57) What was the Aquila payload?
58) Why was the box – in which Aquila was transported – heavy?
59) Did the data link of Aquila have an anti-jam capability?
60) Did the Aquila system include a level of nuclear blast and radiation survivability?
61) Provide reasons why the whole Aquila system gained a bad reputation.
62) What was the foremost operational "disappointment" of the Aquila? What was this failing due to?
63) Was the Aquila program terminated as a failure or as a success?
64) Some of the UAV requirements – that have followed Aquila – have been relaxed. Provide at least three items.
65) The Global Hawk was intended to compliment or replace a manned aircraft. What was that?
66) When and where was the first flight of the Global Hawk?
67) When did the Global Hawk enter service? When did it reach the serial production stage?
68) Which unmanned aerial vehicle became the first UAV to cross the Pacific Ocean?
69) Which unmanned aerial vehicle is the largest active current UAV?
70) Briefly describe the configuration of the Global Hawk.
71) What propulsion system generates thrust for the Global Hawk?
72) Does the Global Hawk have pressurized compartments?
73) What material is the wing and tail of the Global Hawk made of?
74) What material is the fuselage of the Global Hawk made of?
75) What systems are employed for the navigation of the Global Hawk?
76) Name two elements for the GCS of the Global Hawk.
77) Write: (a) maximum takeoff mass, (b) length, and (c) engine thrust of the RQ-4B Global Hawk.
78) Write: (a) maximum speed, (b) range, (c) endurance, and (d) ceiling of the RQ-4B Global Hawk.
79) Briefly compare the RQ-4A Global Hawk with the RQ-4B Global Hawk.
80) List sensors of the RQ-4A Global Hawk.
81) What frequency bands are employed by the Global Hawk?
82) Briefly describe: (a) line-of-sight (LOS) and (b) beyond line-of-sight (BLOS) communications for the Global Hawk.
83) Briefly describe the major setback during flight testing of the Global Hawk.
84) What was the reason behind the mishap for the Global Hawk in the December 1999 flight?
85) Why were two Global Hawk prototype air vehicles lost during the deployment phase in 2002 and 2003? Explain.
86) Why was a Global Hawk lost on June 20, 2019?
87) Briefly describe the unconventional development cycle of the Predator during the 1980s.
88) An advanced version of what aircraft was designated as RQ-1 Predator in the early 1990s?
89) What UAV attempted an air-to-air engagement with an Iraqi MiG-25 in 2002?
90) When was the Predator UAV retired?
91) Briefly compare the primary differences between the Predator and the Reaper.
92) Compare the main difference between configurations of the Predator and the Reaper.

Questions 25

93) When was the first operational mission of the MQ-9 Reaper?
94) What are the payloads of the MQ-9 Reaper?
95) Briefly describe the characteristics of GCS of the MQ-9 Reaper.
96) What is the type of propulsion system for the MQ-9 Reaper?
97) Write: (a) maximum takeoff mass, (b) wingspan, and (c) engine Power of the MQ-9 Reaper.
98) Write: (a) maximum speed, (b) range, (c) endurance, and (d) ceiling of the MQ-9 Reaper.
99) List the dominant US UAV manufacturers.
100) Briefly discuss the ethical concerns of UAVs.
101) What is the mission of the Swift HALE UAV?
102) Name three recent UAV projects with a pseudo-satellite mission.

2

Classes and Missions of UAVs

2.1 Overview

This chapter provides general classes and missions of Unmanned Aerial Vehicles (UAV). It also describes a representative sample of unmanned aerial systems (UAS), including some of the earlier designs that had a large impact on current systems. The range of UAS sizes and types now runs from air vehicles (AVs) small enough to land on the palm of your hand to large lighter-than-air vehicles. This chapter mainly concentrates on those in the range from model[1] airplanes up to medium-sized aircraft, as does the rest of this book, where "unmanned aerial systems" and "UAV Systems" are utilized interchangeably.

Much of the early development of UAS was driven by government and military requirements, and the bureaucracies that manage such programs have made repeated efforts to establish a standard terminology for describing various types of UAS in terms of the capabilities of the air vehicles. While the "standard" terminology constantly evolves and occasionally changes abruptly, some of it has come into general use in the UAV community and is briefly described.

Finally, the chapter also attempts to summarize the applications for which UAS have been or are being considered, which provides a context for the system requirements that drive the design tradeoffs that are the primary topic of this book.

2.2 Classes of UAV Systems

2.2.1 Classification Criteria

There are a number of criteria for classification of UAVs. It is convenient to have a generally agreed upon scheme for classifying UAVs rather like the classification of military aircraft in general into such classes as transport, observation, fighter, attack, cargo, and so on. Table 2.1 provides some criteria for classification of UAVs, and their related classes.

Parts 48 and 107 of FAR regulate the application of small UAVs. For instance, you may not operate an sUAS at night, which is defined in the US (except Alaska) as the time between the end of

---

1 Model airplane – generally known collectively as the sport and pastime of aeromodelling – here is not an aircraft model for the wind tunnel application. It is a small remotely controlled (RC) plane that is often designed by non-aeroengineer traditional airplane enthusiasts and is usually built in a garage.

Introduction to UAV Systems, Fifth Edition. Paul Gerin Fahlstrom, Thomas James Gleason, and Mohammad H. Sadraey.
© 2022 John Wiley & Sons, Inc. Published 2022 by John Wiley & Sons, Inc.
Companion website: www.wiley.com/go/fahlstrom/uavsystems5e

Table 2.1 Criteria for classification of UAVs

No.	Classification Criterion	Class
1	Manufacturing location	1. Home-made (Model), 2. Industrial
2	User	1. Civil, 2. Military
3	Mission	1. Filming, 2. Package delivery, 3. Intelligence, Surveillance, and Reconnaissance (ISR), 4. Precision strike, 5. Combat (UCAV), 6. Teaming, 7. Meteorological measurements, 8. High-altitude platform, 9. Search and observation
4	Size	1. Micro, 2. Mini, 3. Very small, 4. Small, 5. Medium, 6. Large
5	Wing configuration	1. Fixed-wing, 2. Rotary-wing (includes multi-copter), 3. Hybrid
6	FAA [5]	Small UAVs (under FAR Parts 48 and 107)
7	Altitude/Range/Endurance	1. Very low-cost close range, 2. Close range, 3. Short range, 4. Mid-range, 5. Long range, 6. Medium-altitude, long endurance (MALE), 7. High-altitude, long endurance (HALE)
8	Number of uses	1. Reusable, 2. Expendable

evening civil twilight and beginning of morning civil twilight. This is due to the fact that, at night, there is no sufficient visibility to the remote pilot. A small unmanned aircraft is defined as an unmanned aircraft weighing more than 0.55 lb and less than 55 pounds on takeoff, including everything that is on board or otherwise attached to the aircraft.

In the following sections, three specific classifications based on: (1) range and endurance, (2) mission, and (3) tier (for US Air Force, Marine Corps, and Army) are presented.

2.2.2 Classification by Range and Endurance

Shortly after being appointed the central manager of US military UAV programs, the Joint UAV Program Office (JPO) defined classes of UAVs as a step toward providing some measure of standardization to UAV terminology. They were:

- *Very Low-Cost, Close-Range*: Required by the Marine Corps and perhaps the Army to have a range of about 5 km (3 miles) and cost about $10,000 per air vehicle. This UAV system fits into what could be called the "model airplane" type of system and its feasibility with regard to both performance and cost had not been proven but since has been demonstrated by systems such as the Raven and Dragon Eye.
- *Close Range*: Required by all of the services but its concept of operation varied greatly depending on the service. The Air Force usage would be in the role of airfield damage assessment and would operate over its own airfields. The Army and Marine Corps would use it to look over the next hill, and desired a system that was easy to move and operate on the battlefield. The Navy wanted it to operate from small ships such as frigates. It was to have a range of 50 km (31 miles), with 30 km (19 miles) forward of the FLOT. The required endurance was from 1 to 6 h depending on the mission. All services agreed that the priority mission was reconnaissance and surveillance, day and night.
- *Short Range*: The Short-Range UAV was also required by all of the services and, like the Close-Range UAV, had the day/night, reconnaissance, and surveillance mission as a top priority. It had a required range of 150 km (93 miles) beyond the FLOT, but 300 km (186 miles) was desired. The endurance time was to be 8–12 h. The Navy required the system to be capable of launch and recovery from larger ships of the Amphibious Assault Ship and Battleship class.

- *Mid-Range*: The Mid-Range UAV was required by all the services except the Army. It required the capability of being ground or air launched and was not required to loiter. The latter requirement suggested that the air vehicle was a high-speed deep penetrator and, in fact, the velocity requirement was high subsonic. The radius of action was 650 km (404 miles) and it was to be used for day/night reconnaissance and surveillance. A secondary mission for the mid-Range was the gathering of meteorological data.
- *Endurance*: The Endurance UAV was required by all services and, as the name suggested, was to have a loiter capability of at least 36 h. The air vehicle had to be able to operate from land or sea and have a radius of action of approximately 300 km (186 miles). The mission was day/night reconnaissance first and communications relay second. Speed was not specified, but it had to be able to maintain station in the high winds that would be experienced at high altitudes. The altitude requirement was not specified, but it was thought probably to be 30,000 ft (9.14 km) or higher.
- *Long range:* This class of UAVs was not defined by JPO. However, an endurance UAV is capable of flying in a long range (e.g., thousands of miles) mission.

This classification system has been superseded. However, some of the terminology and concepts, particularly the use of a mix of range and mission to define a class of UAVs, persist today and it is useful for anyone working in the field to have a general knowledge of the terminology that has become part of the jargon of the UAV community.

The following sections outline some of the more recent terminology used to classify UAVs. Any government-dictated classification scheme is likely to change over time to meet the changing needs of program managers, and the reader is advised to search the literature and other references if the current standard of government classification is needed. Operation and certification of small unmanned aircraft systems is regulated [5] by Federal Aviation Administration under Part 107.

2.2.3 Classification by Missions

Defining the missions for UAVs is a difficult task because (1) there are so many possibilities and (2) there have never been enough systems in the field to develop all of the possibilities. This is not to say that the subject has not been thought about, because there have been repeated efforts to come up with comprehensive lists as part of classification schemes. All such lists tend to become unique to the part of the UAV community that generates them, and all tend to become out of date as new mission concepts continually arise.

Two major divisions of missions for UAVs are civilian and military, but there is significant overlap between these two in the area of reconnaissance and surveillance, which a civilian might call search and surveillance or observation, which is the largest single application of UAVs in both the civilian and military worlds.

The development of UAVs has been led by the military and there are other areas long recognized as potential military missions that also have civilian equivalents. These include atmospheric sampling for radiation and/or chemical agents, providing relays for line-of-sight communications systems, and meteorological measurements.

An area of interest to both the military and civilian worlds is to provide a high-altitude platform capable of lingering indefinitely over some point on the Earth that can perform many of the functions of a satellite at lower cost and with the capability of landing for maintenance or upgrade and of being re-deployed to serve a different part of the world whenever needed.

Within the military arena, another division of missions has become prominent during the last decade. An increasing mission for military UAVs is the delivery of lethal weapons (i.e., precision strike). This mission has a number of significant distinctions from nonlethal missions in the areas of AV design and raises new issues related to the level of human control over the actions of the AV.

Of course, all missiles are "unmanned aerial vehicles." However, we consider systems that are designed to deliver an internal warhead to a target and destroy themselves while destroying that target as flying weapons (e.g., a missile), and distinguish them from vehicles that are intended to be recoverable and reused for many flights (i.e., UAV). As discussed later in this book, although there are areas in common between flying weapons and reusable aircraft, there are also many areas in which the design tradeoffs for weapons differ from those for the aircraft.

As of this writing, the primary form of active armed UAV is an unmanned platform, such as the MQ-1 Predator (see Figure 11.1) and MQ-9 Reaper (see Figure 1.5) carrying precision-guided munitions and the associated target acquisition and fire-control systems such as imaging sensors and laser designators. This is evolving to include the delivery of small guided bombs and other forms of dispensed munitions. These systems can be considered unmanned ground attack aircraft and unmanned combat aerial vehicle (UCAV). The future seems to hold unmanned fighters and bombers, either as supplements to manned aircraft or as substitutes.

DOD and Boeing have – in the past few years – developed and tested a number of UCAVs such as Boeing X-45, Northrop Grumman X-47, and Kratos XQ-58 Valkyrie. Moreover, some European countries are currently developing UCAVs. Under development are: BAE Corax (also known as Raven), EADS Barracuda (Germany), Dassault nEUROn (France), Elbit Hermes 450 (Israel), BAE Taranis (UK), Boeing X-45A, and EADS Surveyor (Multinational). Furthermore, Chengdu Wing Loong (Chinese title, GJ-2) – a long range UAV with a strike capability and a satellite data link – has been developed by China in 2018.

There is an ambiguous class of military missions in which the UAV does not carry or launch any weapons, but provides the guidance that allows the weapons to hit a target. This is accomplished using laser designators on the AV that "point out" the target to a laser-guided weapon launched from a manned aircraft or delivered by artillery. As we have seen, this mission was a primary driver for the resurgence of interest in UAVs in the US Army in the late 1970s. It remains a major mission for many of the smaller tactical UAVs in use by the military.

The classes of UAVs – Close-Range, Short-Range, Mid-Range, and Endurance – imply missions by virtue of their names, but the services often employ them in such unique ways that it is impossible to say that there is only one mission associated with each name. For example, the Air Force's airfield battle damage assessment mission and the Army's target designation mission both could utilize similar airframes (e.g., having the same weight and shape), but would require an entirely different range, endurance, speed, and payload capabilities. Some missions appear to be common to all the services such as reconnaissance, but the Army wants "close" reconnaissance to go out to 30 km, and the Marine Corps believes that 5 km is about right.

Among the core missions of UAVs for both military and civilian use are ISR, Intelligence, Surveillance, and Reconnaissance (search), which often are combined, but are different is important ways, as seen in the following definitions:

- *Intelligence*: The activity to obtain – by visual or other detection methods – information about what is present or happening at some point or in some area.
- *Surveillance*: The systematic observation of aerospace, surface or subsurface areas, places, persons, or things by visual, aural, electronic, photographic, or other means.
- *Reconnaissance*: The activity to identify a target (e.g., individual, car, or building) by visual or other detection methods at some point or in some area via comparing the obtained information with the reference data.

Thus, surveillance implies long endurance and, for the military, somewhat stealthy operations that will allow the UAV to remain overhead for long periods of time. Because of the interrelationship between intelligence, surveillance, and reconnaissance, the same assets are usually used to accomplish all three missions.

Insitu ScanEagle missions include ISR, as well as special services operations, escort operations, sea-lane and convoy protection, protection of high-value and secure installations, and high-speed wireless voice, video, and data communications relay.

These missions imply the detection and identification of stationary and moving targets both day and night – quite a formidable task, as we will see when discussing payloads and data links. The hardware requirements for the detection and identification capabilities impact almost every subsystem in the air vehicle as well as the ground station. Each UAV user may have requirements for the range from the UAV base to the area to be searched, the size of the area that must be searched, and the time on station required for surveillance, so intelligence/surveillance /reconnaissance missions and hardware can vary significantly.

There are both land- and air-based missions in both the military and civilian worlds. A landbased operational base may be fixed or may need to be transportable. If it is transportable, the level of mobility may vary from being able to be carried in a backpack to something that can be packed up and shipped in large trucks or on a train and then reassembled at a new site over a period of days or even of weeks. Each of these levels affects the tradeoffs between various approaches to AV size, launch, and recovery methods, and almost every other part of the system design.

Ship-based operations almost always add upper limits to AV size. If the ship is an aircraft carrier, the size restrictions are not too limiting, but may include a requirement to be able to remove or fold the long, thin wings that, as we will see later, are typical of long-endurance aircraft.

Associated with the military reconnaissance mission is target or artillery spotting. After a particular target is found, it can be fired upon while being designated with a laser to help guide a precision-guided munition. For conventional (unguided) artillery, the fire can be adjusted so that each succeeding round will come closer to, or hit, the target. Accurate artillery, naval gunfire, and close air support can be accomplished using UAVs in this manner. All of these missions can be conducted with the reconnaissance and surveillance payloads, except that a laser designator feature must be added if one is to control precision-guided munitions. This added feature raises the cost of the payload significantly.

An important mission in the military and intelligence area is Electronic Warfare (EW). Listening to an enemy transmission (communication or radar) and then either jamming it or analyzing its transmission characteristics falls under the category of EW.

Although developing and flight testing a number of experimental combat UAVs such as X-58 in the past 10 years, UCAVs are still not operational. As the latest development, in 2021, Singaporebased Kelley Aerospace [3] has developed and flight-tested the first supersonic UCAV, called Arrow. The plan is to reach speeds up to Mach 2.1.

In 2020, the Air Force awarded [3] contracts to Kratos, Boeing, Northrop Grumman, General Atomics, and Voly Defense Solutions authorizing the companies to compete for the Skyborg project, an effort to field an unmanned wingman cheap enough to sustain losses in combat but capable of supporting manned fighters in hostile environments.

The XQ-58 Valkyrie (Figure 2.1), with a wing span of 22 ft and stealth features, has 8 hardpoints (2 weapon bays with 4 in each) with a capacity of up to 550 lb (250 kg) bombs. The UCAV has a maximum speed of Mach 0.85, a range of 2,449 miles, and a service ceiling of 45,000 ft. The XQ-58 was designed to act as a "loyal wingman" that is controlled by a manned aircraft (a team of manned–unmanned aircraft) to accomplish tasks such as scouting and attracting enemy fire, if attacked.

Figure 2.1 The Kratos XQ-58 Valkyrie (Source: 88 Air Base Wing Public Affairs / Wikimedia Commons / Public Domain)

To summarize, the intelligence/surveillance/reconnaissance mission accounts for most of the UAV activity to date, and its sensors and data-links are the focus of much of today's development. Target spotting follows closely, with EW third. However, in terms of visibility and criticality, weapon delivery has become the most highly watched application and is a major focus of future development. Other missions will come into their own in time, with their way paved by success in the applications and missions now being actively carried out.

2.2.4 The Tier System

A set of definitions that has become pervasive in the UAV community stems from an attempt to define a hierarchy of UAV requirements in each of the US services. The levels in these hierarchies were called "tiers" and terms such as "tier II" are often used to classify a particular UAV or to describe a whole class of UAVs.

The tiers are different in each US service, which can lead to some confusion. These tiers are tabulated with brief descriptions in Tables 2.2 through 2.4 for US Air Force, Marine Corps, and Army respectively.

Table 2.2 US Air Force tiers

No.	Tier	Mission/Group	Example
1	N/A	Small/micro-UAV	
2	I	Low altitude, long endurance	
3	II	Medium altitude, long endurance (MALE)	MQ-1 Predator
4	II+	High altitude, long endurance (HALE) conventional UAV. Altitude: 60,000–65,000 ft (19,800 miles), less than 300 knots (560 km/h) airspeed, 3,000 nautical-miles (6,000 km) radius, 24 h time-on-station capability. Tier II is complementary to the Tier III aircraft.	RQ-4 Global Hawk
5	III−	HALE low-observable (LO) UAV. Same as the Tier II+ aircraft with the addition of LO.	RQ-3 DarkStar

2.3 Examples of UAVs by Size Group

Table 2.3 Marine Corps tiers

No.	Tier	Mission/Group	Example
1	N/A	Micro-UAV	Wasp
2	Tier I	Mini-UAV	Dragon Eye
3	Tier II	Small	RQ-2 Pioneer
4	Tier III	Medium	Shadow

Table 2.4 Army tiers

No.	Tier	Mission/Group	Example
1	Tier I	Small UAV	RQ-11A/B Raven
2	Tier II	Short-Range Tactical UAV	Role filled by the RQ-7A/B Shadow 200
3	Tier III	Medium-Range Tactical UAV	

The most recent classification of systems in use in the United States is related to missions although the old Tier system is still in existence. Eighteen missions relate to four general classes of UAVs – small, tactical, theater, and combat. It is quite specific to US military requirements and is not presented in this book.

2.3 Examples of UAVs by Size Group

We attempt here to provide a broad survey of the many types of UAVs that have been or are being designed, tested, and fielded throughout the world. The intent of this survey is to introduce those who are new to the UAV world to the wide variety of systems that have appeared over the few decades since the revival of interest in UAVs began in the 1980s.

There are a variety of guides to UAVs available and a great deal of information is posted on the Internet. We use *The Concise Global Industry Guide* [6] as a source for quantitative characteristics of current UAVs and a variety of open-source postings and our own personal files for information on systems no longer in production.

As a general organizing principle, we will start with the smallest UAVs and proceed to some that are the size of a corporate jet. The initial efforts on UAVs in the 1980s concentrated on AVs that had typical dimensions of 2 or 3 m (6.6–9.8 ft), partly driven by the need to carry sensors and electronics that at that time had not reached the advanced state of miniaturization that has since become possible. In more recent years, there has been a growing interest in extending the size range of UAVs down to insect-sized devices at one extreme and up to medium air transport sizes at the other end.

Some of the motivation for smaller UAVs is to make them man portable so that a soldier or a border guard can carry, launch, and control a model-airplane-sized UAV that allows him or her to take a look over the next hill or behind the buildings that are in front of him or her. Further miniaturization, to the size of a small bird or even an insect, is intended to allow a UAV to fly inside a building or perch unnoticed on a window sill or roof gutter and provide a look inside the building or into a narrow street.

The realm of small UAVs is one in which there is no competition from manned vehicles. It is unique to vehicles that take advantage of the micro-miniaturization of sensors and electronics to allow humans to view the world from a flying vehicle that could land and take off from the palm of their hand and can go places that are not accessible to anything on a human scale.

The motivation for larger UAVs is to provide long endurance at high altitudes with the ability to fly long distances from a base and then loiter over an area for many hours using a larger array of sensors to search for something or keep watch over some area. Increasingly, in the military arena, the larger UAVs also provide a capability to carry a large weapons payload a long distance and then deliver it to the destination area.

There now is increasing talk about performing missions such as heavy air transportation, bombers, and even passenger transportation with unmanned systems. Whatever may be the outcome of those discussions, it is likely that there eventually will be unmanned systems of all sizes.

In the following sections, we use intuitive-size classes that are not in any sense standardized but are convenient for this discussion. In the next six sections, general features of micro, mini, very small, small, medium, and large UAVs are presented.

2.3.1 Micro-UAVs

A new class of UAVs called micro-UAVs is of growing interest, which have the attributes of large insects and very small birds. This is a term for a class of UAVs that are, as of this writing, still largely in the conceptual or early stages of development. They are envisioned to range in size from a large insect to a model airplane with a one-inch wingspan. The advent of the micro-UAV produces a whole new series of challenges associated with scale factors, particularly micro-autopilot and control in atmospheric phenomena such as wind and gust.

Although there are still no active micro-UAVs in existence, DARPA have attempted a program to develop micro-UAVs including a flapping wing. The world's smallest and lightest micro-helicopter UAV was experimented and demonstrated by Japanese Epson Corporation – with a prop diameter of 130 mm and a mass of 8.9 g. Insight into the art of bird-size UAVs may be obtained from Hank Tennekes' book, *The Simple Science of Flight from Insects to Jumbo Jets* [7].

Assuming that payload and power-plant problems can be solved, the low wing loading of these types of air vehicles may prohibit operation in all except the most benign environmental conditions (e.g., no wind). Some of these problems, and solutions, will be discussed in Part Two of this book.

2.3.2 Mini-UAVs

This category stems from the old expendable definition and includes hand-launched as well as small UAVs that have some type of launcher (could be hand-launched). It is not officially defined by the JPO as a class of UAVs, but numerous demonstrations and experiments have been conducted over the past ten years.

These UAVs are exemplified by the electric-powered Raven and Bayraktar mini-UAVs in the selection of UAV examples earlier in this chapter. The Wasp UAV described among the previous examples is another example of a UAV at the upper limit of what might be called a "mini"-UAV. The ANAFI Parrot can be considered as a mini-UAV where it has a mass of about 0.5 kg, equipped with two 4K 21-megapixel cameras, and can fly for about 32 minutes.

2.3 Examples of UAVs by Size Group

Another example for this class is the HR quadcopter, which has four electric engines, dimensions of 7.1 in × 6.7 in × 1.5 in, and a takeoff mass/weight of about 120 g (4.23 oz). It is equipped with a 1080P HD Camera and can take aerial photos/videos, with the built-in WiFi module to broadcast a live video on a mobile device.

2.3.3 Very Small UAVs

For the purposes of this discussion, "very small UAVs" range from "micro" sized, which are about the size of a large insect, up to an AV with dimensions of the order of a 30–50 cm (12–20 in). There are two major types of small UAVs. One type uses flapping wings to fly like an insect or a bird and the other one uses a more or less conventional aircraft configuration, usually rotary wing for the micro size range. The choice of flapping wings or rotary wings often is influenced by the desire to be able to hover and land/perch on small surfaces to allow surveillance to continue without having to expend the energy to hover. Another advantage of flapping wings is covertness, as the UAV may look a lot like a bird or insect and be able to fly around very close to the subjects of its surveillance or perch in plan-view without giving away the fact that it is actually a sensor platform.

At the small end of this range and for flapping wings, there are many special issues related to the aerodynamics that allow the small UAVs to fly. However, in all cases the basic aerodynamic principles and equations apply, and one needs to understand them before proceeding to the special conditions related to very small size or flapping wings. Part Two of this book introduces the basic aerodynamics and some discussion to the issues for small size and flapping wings.

Examples of very small UAVs include the Chinese DJI mini 2, which is a quadcopter that can hover (i.e., vertically takeoff and land); the Israeli IAI Malat Mosquito, which is an oval flying wing with a single tractor propeller; the US Aurora Flight Sciences Skate, which is a rectangular flying wing with twin tractor engine/propeller combinations that can be tilted to provide "thrust vectoring" for control; and the Australian Cyber Technology CyberQuad Mini, which has four ducted fans in a square arrangement.

The Mosquito wing/fuselage is 35 cm (14.8 in) long and 35 cm (14.8 in) in total span. It uses an electric motor with batteries and has an endurance of 40 minutes, and claims a radius of action of about 1.2 km (0.75 mile). It is hand or bungee launched and can deploy a parachute for recovery.

The DJI Phantom 4 Pro quadcopter has four electric engines, dimensions of 9.9 in × 15.7 in × 6.75 in, and a takeoff weight of about 3 lb. It is equipped with an electro-optic camera and autopilot that allows fully autonomous waypoint navigation. It can be controlled via a smart phone as a ground control station using a Wi-Fi connection, by which the live videos are received.

The Skate wing/fuselage has a wingspan of about 60 cm (24 in) and length of 33 cm (13 in). It folds in half along its centerline for transport and storage. It has twin electric motors on the leading edge that can be tilted up or down and allow vertical takeoff and landing (VTOL) and transition to efficient horizontal flight. There are no control surfaces, with all control being accomplished by tilting the motor/propeller assemblies and controlling the speed of the two propellers. It can carry a payload of 227 g (8 oz) with a total takeoff weight of about 1.1 kg (2 lb).

The CyberQuad Mini has four ducted fans, each with a diameter of somewhat less than 20 cm (7.8 in), mounted so that the total outside dimensions that include the fan shrouds are about 42 cm by 42 cm (16.5 in). The total height including the payload and batteries, which are located in a fuselage at the center of the square, is 20 cm (7.8 in).

The CyberQuad Mini includes a low-light level electro-optic camera or thermal camera and a control system that allows fully autonomous waypoint navigation. The toy has two onboard cameras, one facing forward and one facing down, and is controlled much like a video game from a

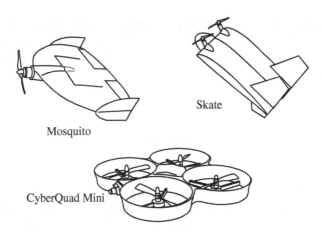

Figure 2.2 Very small UAVs

portable digital device such as a tablet computer or a smart phone. Drawings of these UAVs are shown in Figure 2.2.

2.3.4 Small UAVs

What we will call "small UAVs" have at least one dimension of greater than 50 cm (19.7 in) and go up to dimensions of a meter or two. Many of these UAVs have the configuration of a fixed-wing model airplane and are hand-launched by their operator by throwing them into the air much as we launch a toy glider. In the past ten years, the quadcopters for this class – with VTOL capability – have been operational and gained a significant market.

Examples of small UAVs include the US AeroVironment Raven and the Turkish Byraktar Mini, both conventional fixed-wing vehicles. There are also a number of rotary-wing and multi-copter UAVs in this size grouping, but they are basically scaled-down versions of the medium rotary-wing systems discussed in the following section and we do not offer an example in this group.

The RQ-11A Raven is an example of a UAV that is in the "model airplane" size range. It has a 1.4-m (4.6-ft) wingspan and is about 1 m (3.3 ft) long. It weighs only a little less than 4.4 lb (mass of 2 kg) and is launched by being thrown into the air by its operator in much the same way that a toy glider is put into flight. It uses electrical propulsion and can fly for nearly an hour and a half. The Raven and its control "station" can be carried around by its operator on his/her back. It carries gimbaled visible, near-infrared (NIR), and thermal imaging systems for reconnaissance as well as a "laser illuminator" to point out target to personnel on the ground. (Note that this is not a laser for guiding laser-guided weapons, but more like a laser pointer, operating in the NIR to point things out to people using image-intensifier night-vision devices.)

The latest model, the RQ-11B Raven, was added to the US Army's Small UAV (SUAV) program in a competition that started in 2005. Built by AeroVironment, the Raven B includes a number of improvements from the earlier Raven A, including improved sensors, a lighter Ground Control System, and the addition of the onboard laser illuminator. Endurance was improved, as was interoperability with battlefield communication networks.

The Bayraktar small UAV was developed by Baykar Makina, a Turkish company. It is a conventionally configured, electrically powered AV somewhat larger than the Raven, with a length of

2.3 Examples of UAVs by Size Group

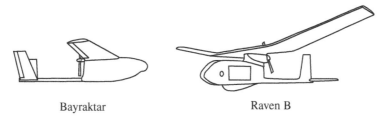

Figure 2.3 Small UAVs

Figure 2.4 Boeing-Insitu ScanEagle (Source: U.S. Navy / Wikimedia Commons / Public Domain)

1.2 m (3.86 ft), wingspan of 2 m (5.22 ft), and weight of 5 kg (105 lb) at takeoff. It is advertised to have a spread-spectrum, encrypted data link, which is a highly desirable, but unusual, feature in an off-the-shelf UAV. The data link has a range of 20 km (12.4 miles), which would limit the operations to that range, although it may depend on the local geography and where the ground antenna is located, as discussed in detail in Part Five of this book (Data Links).

The Bayraktar has a gimbaled day or night camera. It offers waypoint navigation with GPS or other radio navigation systems. Despite its slightly greater size and weight, it is launched much like the Raven. It can be recovered by a skidding landing on its belly or with an internal parachute. It is fielded with small army units and has been heavily used by the Turkish Army since it was fielded in about 2006. Drawings of these examples are shown in Figure 2.3.

The Boeing-Insitu ScanEagle (Figure 2.4) with a wing span of 3.11 m (10.2 ft) and a maximum takeoff weight of 58 lb (mass of 26.5 kg) is also classified as a small UAV. Insitu has produced about 3,000 EcanEagle air vehicles, which had a total of about 140,000 land and maritime sorties. The UAV carries a maximum of 5 kg of payload (EO/IR and ViDar sensors) for day/night and maritime surface search missions. The vehicle performance is: (1) maximum speed: 80 knots, (2) endurance: 18 hours, and (3) ceiling: 19,500 ft.

The ScanEagle can provide daytime and night-time intelligence, surveillance, and reconnaissance (ISR) – including high-resolution digital full motion video – in extreme environments.

2.3.5 Medium UAVs

We call a UAV "medium" if it is too large to be carried around by one person and still smaller than a light aircraft. (As with all of these informal size descriptions, we do not claim rigorousness in this definition. Some attempts at standardized and universal classifications of UAVs are described later in this chapter.)

The UAVs that sparked the present resurgence of interest, such as the Pioneer and Skyeye, are in the medium class. They have typical wingspans of the order of 5–10 m (16–32 ft) and carry payloads of from 100 to about 200 kg (220–440 lb). There are a large number of UAVs that fall into this size group. The Israeli–US Hunter (retired!!) and the UK Watchkeeper are more recent examples of medium-sized, fixed-wing UAVs.

There are also a large number of rotary-wing UAVs in this size class. A series of conventional helicopters with rotor diameters of the order of 2 m (6.4 ft) have been developed in the United Kingdom by Advanced UAV Technologies. There are also a number of ducted-fan systems configured much like the CyberQuad Mini but having dimensions measured in meters instead of centimeters. Finally, we mention the US Boeing Eagle Eye, which is a medium-sized VTOL system that is notable for using tilt-wing technology.

The RQ-2 Pioneer is an example of an AV that is smaller than a light manned aircraft, but larger than what we normally think of as a model airplane. It was for many years the workhorse of the stable of US tactical UAVs. Originally designed by the Israelis and built by AAI in the United States, it was purchased by the US Navy in 1985. The Pioneer provided real-time reconnaissance and intelligence for ground commanders. High-quality day and night imagery for artillery and naval gunfire adjust and damage assessment were its prime operational missions.

The 205-kg (452-lb), 5.2-m (17-ft) wingspan AV had a conventional aircraft configuration. It cruised at 200 km/h and carried a 220-kg (485-lb) payload. Maximum altitude was 15,000 ft (4.6 km) and endurance was 5.5 hours. The ground control station could be housed in a shelter on a High Mobility Multipurpose Wheeled Vehicle (HMMWV) or truck. The fiberglass air vehicle had a 26-hp engine and was shipboard capable. It had piston and rotary engine options.

The Pioneer could be launched from a pneumatic or rocket launcher or by conventional wheeled takeoff from a prepared runway. Recovery was accomplished by conventional wheeled landing with arrestment or into a net. Shipboard recovery used a net system.

The BAE Systems Skyeye R4E UAV system was fielded in the 1980s and is roughly contemporary with the Pioneer, with which it has some common features, but the air vehicle is significantly larger in size, which allows expanded overall capability. It uses launchers similar to the Pioneer but does not have a net-recovery capability. It uses a ground control station similar in principle to that of the Pioneer. It is still in service with Egypt and Morocco.

The Skyeye air vehicle is constructed of lightweight composite materials and was easy to assemble and disassemble for ground transport because of its modular construction. It has a 7.32-m (24-ft) wingspan and is 4.1 m (13.4 ft) long. It is powered by a 52-hp rotary engine (Teledyne Continental Motors) providing high reliability and low vibration. The maximum launch weight is 570 kg (1,257 lb) and it can fly for 8–10 h and at altitudes up to 4,600 m (14,803 ft). The maximum payload weight is about 80 kg (176 lb).

Perhaps the most unique feature of the Skyeye when it was fielded was the various ways in which it could be recovered. The Skyeye has no landing gear to provide large radar echoes or obstruct the view of the payload. The avoidance of a nose wheel is particularly significant as a nose gear often obstructs the view of a payload camera looking directly forward, precluding landing based on the view through the eyes of the camera. However, it can land on a semi-prepared surface by means of a retractable skid located behind the payload. This requires the landing to be controlled by observing the air vehicle externally during its final approach. This is particularly dangerous during night operations.

2.3 Examples of UAVs by Size Group

The landing rollout, or perhaps more accurately the "skid-out," is about 100 m (322 ft). The Skyeye also carries either a parafoil or a parachute as an alternative recovery system. The parafoil essentially is a soft wing that is deployed in the recovery area to allow the air vehicle to land much more slowly. The parafoil recovery can be effective for landing on moving platforms such as ships or barges. The parachute can be used as an alternative means of landing or as an emergency device. However, using the parachute leaves one at the mercy of the vagaries of the wind, and it primarily is intended for emergency recoveries. All of these recovery approaches are now offered in various fixed-wing UAVs, but having all of them as options in one system still would be unusual.

The RQ-5A Hunter was the first UAS to replace the terminated Aquila system as the standard "Short Range" UAS for the US Army. The Hunter does not require a recovery net or launcher, which significantly simplifies the overall minimum deployable configuration and eliminates the launcher required by the Skyeye. Under the appropriate conditions, it can take off and land on a road or runway. It utilizes an arresting cable system when landing, with a parachute recovery for emergencies. It is not capable of net recovery because it has a tractor ("puller") propeller that would be damaged or broken or would damage any net that was used to catch it. It also has a rocket-assisted takeoff option to allow the launch to occur when no suitable road or runway is available.

The Hunter is constructed of lightweight composite materials, which afford ease of repair. It has a 10.2-m (32.8-ft) wingspan and is 6.9 m (22.2 ft) long. It is powered by two four-stroke, two-cylinder (v-type), air-cooled Moroguzzi engines, which utilize fuel injection and individual computer control. The engines are mounted in-line – tractor and pusher – giving the air vehicle twin engine reliability without the problem of unsymmetrical control when operating with a single engine. The air vehicle weighs approximately 885 kg (1,951 lb) at takeoff (maximum), has an endurance of about 12 h, and a cruise speed of 120 knots.

The Hermes 450/Watchkeeper is an all-weather, intelligence, surveillance, target acquisition and reconnaissance UAV. Its dimensions are similar to the Hunter. The Watchkeeper is manufactured in the United Kingdom by a joint venture of the French company Thales and the Israeli company Elbit Systems. It has a weight of 450 kg (992 lb) including a payload capacity of 150 kg (331 lb). The Watchkeeper was in service in Afghanistan with British forces late in 2011.

A series of rotary-wing UAVs called the AT10, AT20, AT100, AT200, AT300, and AT1000 have been developed by the UK firm Advanced UAV Technology. They are all conventionally configured helicopters with a single main rotor and a tail boom with a tail rotor for yaw stability and control. The rotor diameters vary from 1.7 m (5.5 ft) in the AT10 to about 2.3 m (7.4 ft) for the AT1000. Speed and ceiling increase as one moves up the series, as does the payload capacity and payload options. All are intended to be launched by vertical takeoff and all claim the ability for autonomous landings on moving vehicles.

The Northrop Grumman MQ-8B Fire Scout is an example of a conventionally configured VTOL UAV. It looks much like a typical light helicopter. It has a length of 9.2 m (30 ft) (with the blades folded so that they do not add to the total length), height of 2.9 m (9.5 ft), and a rotor diameter of 8.4 m (27.5 ft). It is powered by a 420 hp turbine engine. The Fire Scout is roughly the same size as an OH-58 Kiowa light observation helicopter, which has a two-man crew and two passenger seats.

The Kiowa has a maximum payload of about 630 kg (1,389 lb), compared to the 270 kg (595 lb) maximum payload of the Fire Scout, but if one takes out the weight of the crew and other things associated with the crew, the net payload capability of the Fire Scout is similar to that of the manned helicopter. The Fire Scout is being tested by the US Army and Navy for a variety of missions that are similar to those performed by manned helicopters of a similar size.

The tilt-rotor Bell Eagle Eye was developed during the 1990s. It uses "tilt-wing" technology, which means that the propellers are located on the leading edge of the wing and can be pointed up for takeoff and landing and then rotated forward for level flight. This allows a tilt-wing aircraft to utilize wing-generated lift for cruising, which is more efficient than rotor-generated lift, but still to operate like a helicopter for VTOL capability.

The Eagle Eye has a length of 5.2 m (16.7 ft) and weighs about 1,300 kg (2,626 lb). It can cruise at up to about 345 km/h (knots) and at altitudes up to 6,000 m (19,308 ft). Some of these UAVs are shown in Figure 2.5.

The Bayraktar TB2 – as a precision strike MALE UAV – is developed by Baykar Makina, a Turkish company. It has a length of 6.5 m (21 ft), a wingspan of 12 m (39 ft), and a maximum takeoff weight of 1430 lb. The UAV is equipped with a 100 hp piston engine, carries a 330 lb payload (EO/IR/LD imaging and targeting sensor), and features four hardpoints for laser-guided smart munition (missiles/rockets). The performance includes: (1) maximum speed of 120 knots, (2) range of 150 km, (3) service ceiling of 27,000 feet, and (4) endurance of 27 hours. As of July 2021, the TB2 UAV has completed 300,000 flight-hours globally.

2.3.6 Large UAVs

Our informal size groupings are not finely divided, and we will discuss all UAVs that are larger than a typical light manned aircraft in the group called "large." This includes, in particular, a

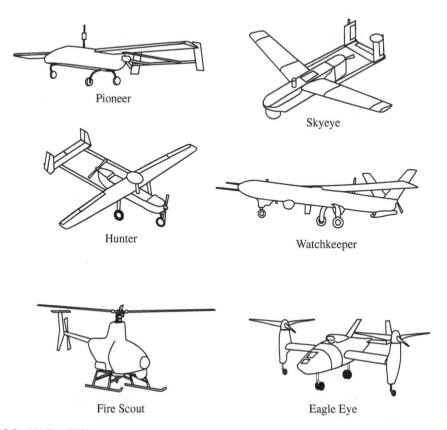

Figure 2.5 Medium UAVs

2.3 Examples of UAVs by Size Group

group of UAVs that can fly long distances from their bases and loiter for extended periods to perform surveillance functions. They also are large enough to carry weapons in significant quantities. The lower range of such systems includes the US General Atomics Predator A, which has a significant range and endurance, but can carry only two missiles of the weight presently being used.

The limitation to two missiles is serious as it means that after firing the two missiles that are onboard, the UAV either has lost the ability to deliver weapons or must be flown back to its base to be rearmed. For this reason, a second generation of UAVs designed for missions similar to that of the Predator, including a Predator B model (i.e., Reaper), is now appearing that is larger and able to carry many more weapons on a single sortie.

The Cassidian Harfang is an example of a system much like the Predator A and the Talarion, also by Cassidian, is an example of the emerging successors to the Predator A.

At the high end of this size group, an example is an even larger UAV designed for very long range and endurance and capable of flying anywhere in the world on its own, the US Northrop Grumman Global Hawk (see Figure 1.4).

There are a number of specialized military and intelligence systems for which information available to the public is very limited. An example of this is the US Lockheed Martin RQ-170 Sentinel. The RQ-170 Sentinel is reported to be a stealthy AV manufactured by Lockheed Martin, but limited performance data are available. It is a flying wing configuration much like the Northrop B-2 Spirit bomber and is in the medium-to-large size class, with a wingspan of around 20 m (65 ft) and a length of 4.5 m. This UAV is equipped with a single turbofan engine with 41.26 kN of thrust. In 2011, Iran stated that its Army's electronic warfare unit had downed an RQ-170 that violated Iranian airspace along its eastern border through overriding its controls, and had captured the lightly damaged UAV.

The MQ-1 Predator A is larger than a light single-engine private aircraft and provides medium altitude, real-time surveillance using high-resolution video, infrared imaging, and synthetic aperture radar. It has a wingspan of 17 m (55 ft) and a length of 8 m (26 ft). It adds a significantly higher ceiling (7,620 m or 24,521 ft) and longer endurance (40 h) to the capabilities of the smaller UAVs. GPS and inertial systems provide navigation, and remote control can be via satellite. Cruise speed is 220 km/h (119 knots) and the air vehicle can remain on station for 24 h, 925 km (575 mi) from the operating base. It can carry an internal payload of 200 kg (441 lb) plus an external payload (hung under the wings) of 136 kg (300 lb).

The Harfang UAV is produced by Cassidian, which is a subsidiary of the French company EADS. It is about the same size as the Predator and is designed for similar missions. The configuration is different, using a twin-boom tail structure. There are a variety of possible payloads. Its stated performance is similar to that of the Predator, but it has a shorter endurance of 24 h. It takes off and lands conventionally on wheels on a runway. Remote control can be via satellite.

Talarion is under development by Cassidian as a second-generation successor to the Predator/Harfang class of UAVs. It uses two turbojet engines and can carry up to 800 kg (1,764 lb) of internal payload and 1,000 kg (2,205 lb) of external payload with a ceiling of over 15,000 m (49,215 ft) and speeds around 550 km/h (297 knots).

The RQ-4 Global Hawk (see Figure 1.4) is manufactured by Northrop Grumman Aerospace Systems. It flies at high altitude and utilizes radar, electro-optical, and infrared sensors for surveillance applications. It uses a turbofan engine and appears to have a shape that reduces its radar signature, but is not a "stealth" aircraft. It is 14.5 m (47 ft) long with a 40 m (129 ft) wingspan and has a maximum weight at takeoff of 1,460 kg (3,219 lb). It can loiter at 575 km/h (310 knots) and has an endurance of 32 h. It has a full set of potential payloads and it appears that it is routinely

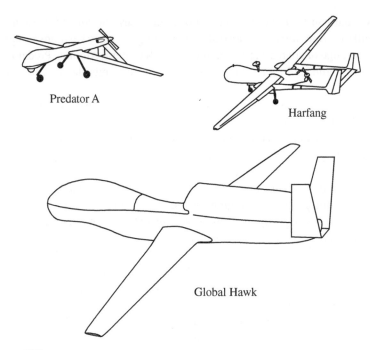

Figure 2.6 Large UAVs

controlled via satellite links. More information about the Global Hawk is presented in Section 5 of Chapter 1.

Some of these large UAVs are illustrated in Figure 2.6.

2.4 Expendable UAVs

Expendable UAVs are not designed to return after accomplishing their mission. In the military world, this often means that they contain an internal warhead and are intended to be crashed into a target destroying it and themselves. This type of expendable is discussed further in Chapter 11 and we make the argument there that it is not really a UAV, but rather a missile of some sort. There is a considerable area of overlap between guided missiles and UAVs, as illustrated by the fact that the first "UAVs" of the aviation era were mostly guided weapons. An alternative definition of an expendable is that it can (and should) be recovered if possible, but can have a very high loss rate.

The electric-motor-powered Raven, described in Section 2.3.4, is an example of an expendable, but recoverable, UAV. It is hand launched and uses a hand-carried ground control station. The Raven is used to conduct reconnaissance missions out to about 5 km and is recoverable, but if it does not return or crashes during landing, the loss is considered acceptable.

However, in the past few years, a new type of lethal UAVs (as loitering munition) has emerged. This precision-guided loitering munition is a relatively new concept and technology and generates a low (or very low) collateral damage. As example is Switchblade 300 (Figure 2.7), designed and manufactured by AeroVironment. This UAV – which was first fielded by the US Army and Marine Corp in 2011 – utilizes a small munition to break apart prior to impact and has a low acoustic, thermal, and radar signature. Switchblade 300 is interoperable with current Puma, Raven, and Wasp UAV control stations.

Figure 2.7 Switchblade UAV (Source: Business Wire, Inc.)

This UAV – with a mass of 2.7 kg (weight of 6 lb), a length of 2 ft, and a pusher electric motor – carries a small munition with a range of 10 km; it can be in the air for about 10 minutes. The operating altitude is 500 ft AGL (above ground level) and 15,000 ft MSL (mean sea level), with a cruise speed of 63 mph and a dash of 98 mph. Its tubular fuselage contains the battery, motor, avionics, and a payload (gimbaled nose camera). The UAV is folded inside a tube with wing sections unfolding once it gets airborne. The two wing sections are spring loaded, low-mounted, and sweep out upon launching.

Switchblade – armed with a 40-mm grenade-like warhead – is meant to strike targets beyond the range – or shielded from the line of sight – of the rifles, machine guns, and other weapons of an infantry platoon or squad. Switchblade was fielded on an emergency basis in Afghanistan beginning in 2012.

There are a number of similar UAVs developed by other companies. Examples are Lockheed Martin Terminator, Mistral HERO-30SF, Textron Systems BattleHawk, SAAB AT4 CS, Booz Allen Hamilton Vampire, Prox Dynamics Black Hornet, and Area-I Air-Launched, Tube-Integrated, Unmanned System, or ALTIUS-600.

Questions

1) What do HALE, MALE, UCAV, FLOT, AGL, MSL, and EW stand for?
2) Write classes of UAVs based on manufacturing location.
3) Write classes of UAVs based on user.
4) Write classes of UAVs based on mission.

5) Write classes of UAVs based on size.
6) Write classes of UAVs based on wing configuration.
7) Write classes of UAVs based on number of uses.
8) Write classes of UAVs based on altitude/range/endurance.
9) Which FAR part governs small UAVs?
10) What group of aircraft are often referred to as "model airplane"?
11) What is the definition of very low-cost close-range required by the Marine Corps?
12) What is the definition of close-range in this chapter?
13) What is the definition of short-range in this chapter?
14) What is the definition of mid-range in this chapter?
15) What is the loiter capability of an endurance UAV?
16) How do we distinguish missiles from "unmanned aerial vehicles"?
17) Define intelligence as the mission for a UAV.
18) Define surveillance as the mission for a UAV.
19) Define reconnaissance as the mission for a UAV.
20) What is the mission for tier I UAVs in the US Air Force? Provide an example.
21) What is the mission for tier II UAVs in the US Air Force? Provide an example.
22) What is the mission for tier II+ UAVs in the US Air Force? Provide an example.
23) What is the mission for tier III− UAVs in the US Air Force? Provide an example.
24) What is the weight group for tier I UAVs in the US Marine Corps? Provide an example.
25) What is the weight group for tier II UAVs in the US Marine Corps? Provide an example.
26) Provide an example for tier III UAVs in the US Marine Corps.
27) What is the weight group for tier I UAVs in the US Army? Provide an example.
28) What is the mission for tier II UAVs in the US Army? Provide an example.
29) What is the mission for tier III UAVs in the US Army? Provide an example.
30) What mission accounts for most of the UAV activity to date?
31) What is tier in UAV classification?
32) List size classes of UAVs that are presented in this chapter.
33) What is the typical size range for very small UAVs?
34) Briefly describe the characteristics of DJI Phantom 4 Pro.
35) Briefly describe the characteristics of RQ-11A Raven.
36) What is the range of data link for Bayraktar UAV?
37) Write typical wingspan and payload weight for medium UAVs.
38) Write three examples of air vehicles for the class of medium UAVs.
39) What were prime operational missions of RQ-2 Pioneer?
40) Briefly describe the characteristics of BAE Systems Skyeye R4E UAV.
41) Write four UAVs that are equipped with a rotary or piston engine.
42) What is the most unique feature of the Skyeye, when it was fielded? Discuss.
43) Write two disadvantages of a nose gear when the camera payload is located under the fuselage nose?
44) Which UAS was the first one to replace the terminated Aquila system as the standard "Short Range" UAS for the US Army?
45) Write names of three UAVs that were made of composite materials.
46) Briefly describe the characteristics of RQ-5A Hunter.
47) What was the mission of Hermes 450/Watchkeeper?
48) Write length, height, and engine power of Northrop Grumman MQ-8B Fire Scout.
49) Write the name of a UAV with a tilt-rotor configuration.

Questions 45

50) What are the maximum cruise speed and ceiling of Bell Eagle Eye?
51) Write two example UAVs (one for the lower range and one for the high end) of the large UAV group.
52) Write: (1) wingspan, (2) length, and (3) total payload weight of the MQ-1 Predator A.
53) Write cruise speed and endurance of the MQ-1 Predator A.
54) What can an MQ-1 Predator A provide as its mission?
55) What are the payload sensors for surveillance applications in the RQ-4 Global Hawk?
56) What is an expendable UAV?
57) What is the type and mission of the Switchblade 300?
58) Write features of the wing and fuselage of the Switchblade 300.
59) What are the cruise speed and ceiling of the Switchblade 300?
60) What are two payloads of the Switchblade 300?
61) Write the maximum speed, endurance, and ceiling of ScanEagle.
62) For ScanEagle, what are: (1) the maximum takeoff weight and (2) weight of payload?
63) What are sensors of Boeing-Insitu ScanEagle?
64) According to FAR Parts 48 and 107, may you operate an sUAS at night in the US? Why?
65) What UAV is defined as a small unmanned aircraft by FAR?
66) Name four expendable UAVs and their manufacturers.
67) List missions of Insitu ScanEagle.
68) Describe the performance of the XQ-58 Valkyrie.
69) What is the capacity of bombs in the XQ-58 Valkyrie?
70) Write: (1) wingspan, (2) length, and (3) total payload weight of the Bayraktar TB2.
71) Write the maximum speed and endurance of the Bayraktar TB2.
72) What can a Bayraktar TB2 provide as its mission?
73) What is the payload of Bayraktar TB2?

Part II

The Air Vehicle

This part introduces the subsystem at the heart of any UAS, the air vehicle. The section begins with a basic discussion of the aerodynamics (Chapter 3), followed by illustrations of how to analyze the key areas of air-vehicle performance (Chapter 4) and flight stability and control (Chapter 5). The various means of propulsion (Chapter 6) commonly used by UAVs are explored, including an introduction to the subject of rotary wing (e.g., quadcopter) and ducted fan concepts. Finally, some structural and load (Chapter 7) topics of importance to UAV engineers are described.

Introduction to UAV Systems, Fifth Edition. Paul Gerin Fahlstrom, Thomas James Gleason, and Mohammad H. Sadraey.
© 2022 John Wiley & Sons, Inc. Published 2022 by John Wiley & Sons, Inc.
Companion website: www.wiley.com/go/fahlstrom/uavsystems5e

3

Aerodynamics

3.1 Overview

Aerodynamics is the science that involves the study of the behavior (i.e., dynamics) of air when confronting a moving object (e.g., air vehicle). The UAV has a number of components that are characterized by their aerodynamic outputs (e.g., lift), two of which are wing and tail. A wing/tail is considered as a lifting surface in which the lift is produced due to the pressure difference between lower and upper surfaces. In contrast, surfaces such as aileron, rudder, and elevator are referred to as the control surfaces. Lifting surfaces are generally fixed while control surfaces are always moving up/down or left/right. Both lifting and control surfaces are functions of their aerodynamic features.

The primary forces that act on an air vehicle are thrust, lift, drag, and gravity (or weight). They are shown in Figure 3.1. The UAV components that have a direct contact with moving air contribute to aerodynamic features. A number of elements have considerable contributions on an air vehicle's aerodynamic features. They are mainly: (1) wing, (2) tail, (3) fuselage, (4) engine cowling, (5) landing gear, and (6) payload. Three primary aerodynamic components of an air vehicle are: (1) wing, (2) tail, and (3) fuselage.

The primary aerodynamic function of the wing and tail is to generate sufficient lift force or simply lift (L). However, they have two other undesirable products, namely drag (D) and nose-down pitching moment (M). The fuselage is not fundamentally considered as an aerodynamic component based on its function. However, fuselage has a considerable role in creating drag, while it generates a little lift.

In this chapter, aerodynamic forces (mainly lift and drag), airfoil, pressure distribution, friction drag, induced drag, drag polar, air vehicle drag, boundary layer, and aerodynamic efficiency will be briefly covered.

3.2 Aerodynamic Forces

The primary function of the UAV aerodynamic components (e.g., wing and tail) is to generate sufficient lift force or simply lift (L). However, they have two other aerodynamic products, namely drag force or drag (D) and nose-down (Figure 3.2) pitching moment (M). While a UAV designer is

---

Introduction to UAV Systems, Fifth Edition. Paul Gerin Fahlstrom, Thomas James Gleason, and Mohammad H. Sadraey.
© 2022 John Wiley & Sons, Inc. Published 2022 by John Wiley & Sons, Inc.
Companion website: www.wiley.com/go/fahlstrom/uavsystems5e

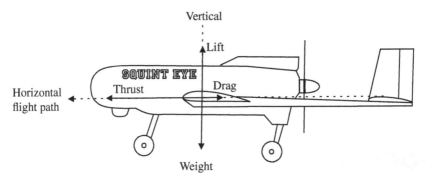

Figure 3.1 Forces on an air vehicle during a level flight

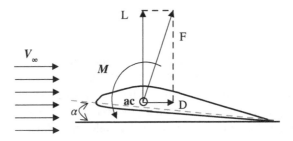

Figure 3.2 Aerodynamic lift, drag, and pitching moment

looking to maximize the lift, the other two (drag and pitching moment) must be minimized. In this chapter, lift and drag are presented, while in Chapter 5, the pitching moment will be introduced.

The aerodynamic forces of lift and drag [8] are functions of the following factors: (1) aircraft configuration, aircraft/wing angle of attack (α), (3) aircraft geometry, (4) airspeed (V), (5) air density (ρ), (6) Reynolds number of the flow, and (7) air viscosity:

$$L = \frac{1}{2}\rho V^2 S C_L \tag{3.1}$$

$$D = \frac{1}{2}\rho V^2 S C_D \tag{3.2}$$

where ρ is air density, V is velocity, S is the wing planform area, and C_L and C_D are the lift and drag coefficients respectively. The calculations of lift and drag coefficients will be presented in the coming sections.

The lift force or simply lift is always defined as the component of the aerodynamic force perpendicular to the relative wind. The drag is always defined as the component of the aerodynamic force parallel to the relative wind (V_∞). In other words, lift is always perpendicular and drag is always parallel to the relative wind. Figure 3.2 shows an airfoil section and the directions of lift and drag.

In reality, the aerodynamic force is located at the center of pressure (cp), which is moving with the variations of angle of attack (α). However, the aerodynamic center (ac) which is frequently selected to be the center of lift, is located nearly at the quarter chord (i.e., 1/4 of C). The pitching moment is the bi-product of moving the location of aerodynamic force from cp to ac. The moment can be taken with respect to any point, but traditionally is taken about a point 25% rearward of the wing leading edge, known as the quarter chord. The aerodynamic center has a desired property – the variation of moment coefficient with respect to the angle of attack is zero (i.e., moment coefficient remains constant).

3.3 Mach Number

It is customary to compare the speed of an air vehicle (V) with the speed of sound (a). The Mach number is defined as the ratio of airspeed over the speed of sound:

$$M = \frac{V}{a} \tag{3.3}$$

The speed of sound at the sea level standard condition is 340 m/s. M is used to define four different flight regimes for airspeed: (1) subsonic, (2) transonic, (3) supersonic, and (4) hypersonic.

When the flight speed is less than the speed of sound – where $M < 1$ – it is defined as subsonic speed. When $0.8 \leq M \leq 1.2$, the flight regime is loosely defined as transonic. If the flight speed is less than the speed of sound, but M is sufficiently near 1, the airflow expansion over the top surfaces of the wing/tail/fuselage may result in locally supersonic regions. A flowfield where $M > 1$ everywhere is defined as supersonic. At supersonic speeds, a shock wave (e.g., normal, oblique, and bow) is created by nature to adjust the flow properties (e.g., air pressure and temperature).

The flow regime for $M > 5$ is given a special label, hypersonic flow. For values of $M > 5$, the shock wave is very close to the surface, and the flowfield between the shock and the body (the shock layer) becomes very hot indeed, hot enough to dissociate or even ionize the gas. The aerodynamic characteristics of an air vehicle is strongly a function of Mach number, some of which will be discussed in this chapter.

3.4 Airfoil

The most important element of a lifting surface (e.g., wing/tail) for aerodynamic purposes is its cross-section or airfoil. Any section of the wing cut by a plane parallel to the aircraft xz plane is called an airfoil. It usually looks like a positive cambered section where the thicker part is in front of the airfoil. A typical airfoil section is shown in Figure 3.3, where several geometric parameters are illustrated. The leading edge is often curved (for a subsonic air vehicle) and the trailing ledge is sharp.

If the mean camber line is a straight line, the airfoil is referred to as symmetric airfoil; otherwise it is called a cambered airfoil. The camber of airfoil is usually positive.

Two of the most important parameters of an airfoil are camber and the thickness-to-chord ratio. The wing/tail is a three-dimensional component, while the airfoil is a two-dimensional (2d) section. Because of the airfoil section, two other outputs of the airfoil, and consequently the wing/tail, are drag and pitching moment.

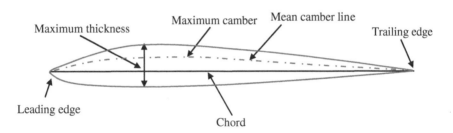

Figure 3.3 Airfoil geometric parameters

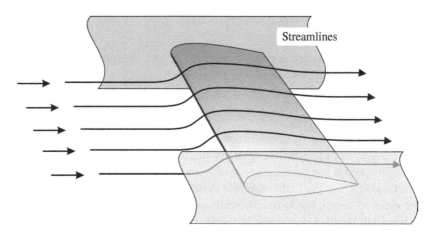

Figure 3.4 Infinite span wing

Any particular airfoil cross-sectional shape has a characteristic set of curves for the coefficients of lift, drag, and moment that depend on the angle of attack and Reynolds number. These are determined from wind tunnel tests and are designated by lowercase subscripts.

Basic aerodynamic data are usually measured from a wing that extends from wall to wall in the wind tunnel, as shown in Figure 3.4. Extending the wing from wall to wall prevents spanwise airflow and results in a true two-dimensional pattern of air pressure. This concept is called the infinite-span wing, because a wing with an infinite span could not have air flowing around its tips, creating spanwise flow and disturbing the two-dimensional pressure pattern that is a necessary starting point for describing the aerodynamic forces on a wing. A real airplane wing has a finite span, and often tapers and twists. The analysis of aerodynamic forces begins with the two-dimensional coefficients, which then are adjusted to account for the three-dimensional nature of the real wing.

Figure 3.4 also illustrates a few of the infinite number of streamlines around a wing. A streamline is a curve in the flowfield that is tangent to the local velocity vector at every point along the curve. Upstream of the wing, the flow is uniform with a constant velocity.

In the 1950s, airfoils were classified by the National Advisory Committee for Aeronautics (NACA), the forerunner of the present NASA, and were cataloged using a four/five digits code. The details of NACA airfoils have been presented in a book published by Abbott and Von Donehoff [9]. The NACA airfoils are one of the most common and one of the oldest airfoil families.

Airfoil cross-sections and their two-dimensional coefficients were classified in a standard system maintained by the NACA (predecessor of the present National Aeronautics and Space Administration (NASA)) and identified by a NACA numbering system, which is described in most aerodynamic textbooks. Figures 3.5 to 3.7 show the data contained in the summary charts of the NASA database for NACA airfoil 23021 as an example of the information available on many airfoil designs.

Figure 3.5 shows the profile of a cross-section of this airfoil which has a thickness-to-chord ratio of 21%. The x (horizontal) and y (vertical) coordinates of the surface are plotted as x/c and y/c, where c is the chord of the airfoil, its total length from nose to tail.

Two-dimensional lift and moment coefficients for this airfoil are plotted as a function of angle of attack in Figure 3.6. Two-dimensional drag and moment coefficients for this airfoil are plotted as a function of the lift coefficient in Figure 3.7. Moments and how they are specified are further discussed in Section 3.5. The moment coefficient in the plot is around an axis located at the quarter-cord, as mentioned previously.

3.4 Airfoil

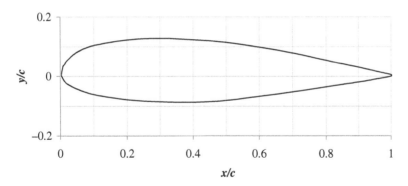

Figure 3.5 NACA 23021 airfoil profile

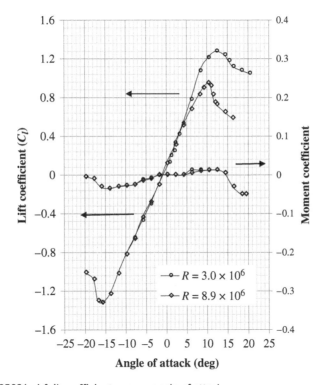

Figure 3.6 NACA 23021 airfoil coefficients versus angle of attack

Figures 3.6 and 3.7 show two curves for each coefficient. Each curve is for a specified Reynolds number (R). The NASA database contains data for more than two Reynolds numbers, but Figures 3.6 and 3.7 reproduced only $R = 3.0 \times 10^6$ and $R = 8.9 \times 10^6$. The two moment curves lie nearly on top of each other and are hardly distinguishable.

The NASA LRN 1015 (NASA TM 102840) airfoil is used on the Northrop Grumman RQ-4 Global Hawk (see Figure 1.4) wing. The airfoil maximum thickness is 15.2% at 40% chord, and its maximum camber is 4.9% at 44% chord.

The airfoils used on the MQ-1 Predator (see Figure 11.1) wing is a proprietary material of General Atomics. However, it is estimated that the airfoil maximum thickness is 10% at 26.6% chord, and

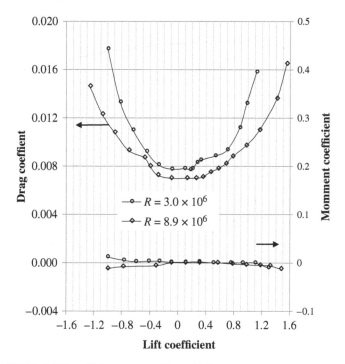

Figure 3.7 NACA 23021 airfoil coefficients versus lift coefficient

its maximum camber is 3.4% at 45.1% chord. The airfoil has a high camber, which provides the desired aerodynamic performance at high Reynolds numbers.

With the progress of the science of aerodynamics, there is a variety of techniques and tools to accomplish the time-consuming job of calculating lift and drag. A variety of tools and software based on aerodynamics and numerical methods have been developed in the past decades. A number of CFD[1] Software packages based on Navier-Stokes equations, the vortex lattice method, and thin airfoil theory are available in the market. The application of such software packages, which is expensive and time-consuming, at this early stage of wing design seems un-necessary.

3.5 Pressure Distribution

The aerodynamic forces on an object in the airflow (e.g., wing) can be calculated from pressure distribution around the object. The lift of a wing/tail is produced due to the pressure difference between the lower and upper surfaces. An airfoil-shaped body moved through the air will vary the static pressure on the top surface and on the bottom surface of the airfoil.

In a positive cambered airfoil (at a positive angle of attack), the upper surface static pressure in less than ambient pressure, while the lower surface static pressure is higher than ambient pressure. This is due to the higher airspeed at the upper surface and the lower speed at the lower surface of the airfoil (see Figure 3.8). As the airfoil angle of attack increases, the pressure difference between the upper and lower surfaces will be higher.

The wing three-dimensional (3d) lift coefficient (C_L) is directly a function of the airfoil two-dimensional (2d) lift coefficient, c_l. In turn, the variable c_l is a function of variations of pressure

---

[1] Computational Fluid Dynamics.

3.5 Pressure Distribution

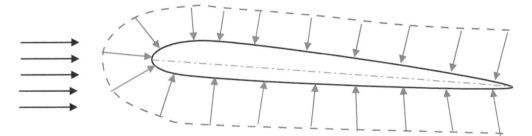

Figure 3.8 Pressure distribution for an airfoil section

coefficients on the upper and lower surfaces. For the case of a small angle of attack (less than 5 degrees), the 2d lift coefficient is calculated from

$$c_l = \frac{1}{c}\int_0^c \left(C_{pl} - C_{pu}\right) dx \tag{3.4}$$

where C_{pl} and C_{pu} are pressure coefficients at the lower and upper surfaces respectively, and c is the airfoil chord. Thus, the lift is produced due to the pressure difference between the lower and upper surfaces of the wing/tail. References such as [8, 10, and 11] provide techniques to determine the pressure distribution around any lifting surface such as the wing. The variations of lift coefficient versus angle of attack are often linear below the stall angle. The UAVs are usually flying at an angle of attack below the stall angle (about 15 degrees).

The Northrop Grumman X-47B (Figure 3.9) an experimental unmanned combat air vehicle (UCAV) has provisions for various payloads such as EO/IR/SAR/GMTI. Its aerodynamic design

Figure 3.9 Northrop Grumman X-47B UCAV

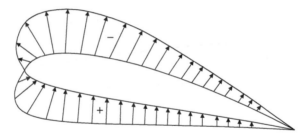

Figure 3.10 Net pressure distribution over an airfoil

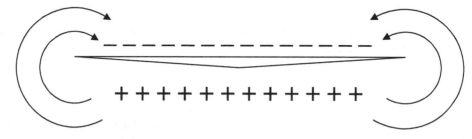

Figure 3.11 Spanwise pressure distribution around a 3d wing

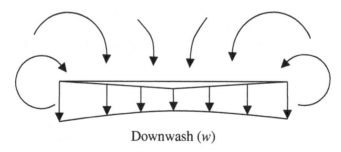

Downwash (w)

Figure 3.12 Downwash

allows for a maximum speed of Mach 0.9+ and a range of 3,900 km. This UCAV with a blended wing-body tailless configuration, a maximum takeoff weight of 44,567 lb (mass of 20,215 kg), and a wing span of 62.1 ft (18.9 m) is equipped with a single turbofan engine.

Consider the net pressure distribution about an airfoil, as shown in Figure 3.10. It is apparent that a wing would have positive pressure on its underside and negative (in a relative sense) pressure on the top. This is shown in Figure 3.11 as plus signs on the bottom and minus signs on the top as viewed from the front or leading edge of the wing.

Such a condition would allow air to spill over from the higher pressure on the bottom surface to the lower pressure on the top, causing it to swirl or form a vortex. The downward velocity or downwash onto the top of the wing created by the swirl would be greatest at the tips and reduced toward the wing center, as shown in Figure 3.12.

Aerodynamics textbooks (e.g., Reference [8]) are a good source to consult for information about mathematical techniques for calculating the pressure distribution over the wing and for determining the flow variables.

3.6 Drag Polar

Another important concept concerning the three-dimensional air vehicle is what is known as the aircraft drag "polar," a term introduced by Eiffel many years ago, which is a curve of C_L plotted against C_D. A typical airplane drag polar curve is illustrated in Figure 3.13.

The drag polar will later be shown to be parabolic in shape and define the minimum drag (or zero-lift drag), C_{Do}, or drag that is not attributable to the generation of lift. A line drawn from the origin and tangent to the polar gives the minimum lift-to-drag ratio that can be obtained. It will also be shown later that the reciprocal of this ratio is the tangent of the power-off glide angle of an air vehicle. The drag created by lift or induced drag is also indicated on the drag polar.

The drag coefficient is the sum of two terms: (1) zero-lift drag coefficient (C_{Do}) and (2) induced drag coefficient (C_{Di}). The first part is mainly a function of friction between air and the aircraft body (i.e., skin friction), but the second term is a function of local air pressure, which is represented by the lift coefficient. Pressure drag is mainly produced by flow separation. The sum of the pressure drag and skin friction (friction drag – primarily due to laminar flow) on a wing is called profile drag. This drag exists solely because of the viscosity of the fluid and the boundary layer phenomena.

The drag coefficient is a function of several parameters, particularly UAV configuration. A mathematical expression for the variation of the drag coefficient as a function of the lift coefficient is

$$C_D = C_{D_o} + K C_L^2 \tag{3.5}$$

This equation is sometimes referred to as aircraft "drag polar." The variable K is referred to as the induced drag correction factor. It is obtained from

$$K = \frac{1}{\pi \cdot e \cdot AR} \tag{3.6}$$

where e is the Oswald span efficiency factor and AR is the wing aspect ratio. The aspect ratio is defined as the ratio of wingspan over wing mean aerodynamic chord (b/C). It is also equal to wingspan squared divided by wing area or b^2/S. The variable AR is further discussed in this chapter.

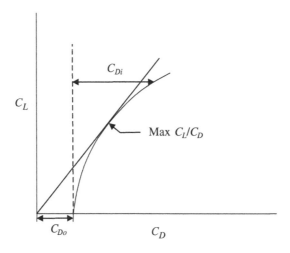

Figure 3.13 Airplane drag polar

3.7 The Real Wing and Airplane

A real three-dimensional conventional aircraft normally is mainly composed of a wing, a fuselage, and a tail. The wing geometry has a shape, looking at it from the top, called the planform. It often has twist, sweepback, and dihedral (angle with the horizontal looking at it from the front) and is composed of two-dimensional airfoil sections. The details of how to convert from the "infinite wing" coefficients to the coefficients of a real wing or of an entire aircraft is beyond the scope of this book, but the following discussion offers some insight into the things that must be considered in that conversion.

A full analysis for lift and drag must consider not only the contribution of the wing but also by the tail and fuselage and must account for varying airfoil cross-section characteristics and twist along the span. Determining the three-dimensional moment coefficient also is a complex procedure that must take into account the contributions from all parts of the aircraft.

A crude estimate (given without proof) of the three-dimensional wing lift coefficient, indicated by an uppercase subscript (C_L), in terms of the "infinite wing" lift coefficient is

$$C_L = \frac{C_l}{\left(1 + \dfrac{2}{AR}\right)} \tag{3.7}$$

where C_l is also the two-dimensional airfoil lift coefficient. From this point onward, we will use uppercase subscripts, and assume that we are using coefficients that apply to the 3d wing and aircraft.

3.8 Induced Drag

Drag of the three-dimensional airplane wing plays a particularly important role in airplane design because of the influence of drag on performance and its relationship to the size and shape of the wing planform.

The most important element of drag introduced by a wing – at high angles of attack – is the "induced drag," which is drag that is inseparably related to the lift provided by the wing. For this reason, the source of induced drag and the derivation of an equation that relates its magnitude to the lift of the wing will be described in some detail, although only in its simplest form.

The lowest induced drag is generated when the lift distribution over the wing is elliptical (Figure 3.14), which provides a constant downwash along the span, as shown in Figure 3.14. Aerodynamicists in the past – including Ludwig Prandtl – believed that a wing whose planform is elliptical would have an elliptical lift distribution. But further research has not proven this idea. The notion of a constant downwash velocity (w) along the span will be the starting point for the development of the effect of three-dimensional drag.

Considering the geometry of the flow with downwash, as shown in Figure 3.15, it can be seen that the downward velocity component for the airflow over the wing (w) results in a local "relative wind" flow that is deflected downward. This is shown at the bottom, where w is added to the velocity of the air mass passing over the wing (V) to determine the effective local relative wind (V_{eff}) over

3.8 Induced Drag

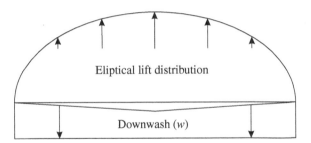

Figure 3.14 Elliptical lift distribution

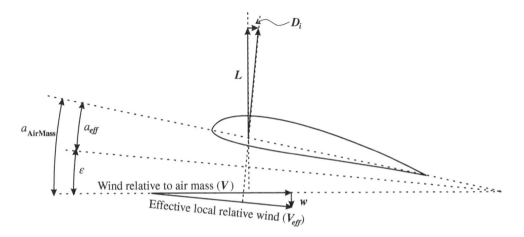

Figure 3.15 Induced drag diagram

the wing. Therefore, the wing "sees" an angle of attack that is less than it would have had there been no downwash.

The lift (L) is perpendicular to V and the net force on the wing is perpendicular to V_{eff}. The difference between these two vectors, which is parallel to the velocity of the wing through the air mass, but opposed to it in direction, is the induced drag (D_i). This reduction in the angle of attack is

$$\varepsilon = \tan^{-1}\left(\frac{w}{V}\right) \tag{3.8}$$

Then, the induced drag coefficient (C_{Di}) is given by

$$C_{Di} = \frac{C_L^2}{\pi AR} \tag{3.9}$$

This expression reveals to us that air vehicles with short stubby wings (small AR) will have relatively high-induced drag and therefore suffer in range and endurance. Air vehicles that are required to stay aloft for long periods of time and/or have limited power, as, for instance, most electric-motor-driven UAVs, will have long (high AR) thin wings.

3.9 Boundary Layer

A fundamental axiom of fluid dynamics and aerodynamics is the notion that a fluid flowing over a surface has a very thin layer adjacent to the surface that sticks to it and therefore has a zero velocity. The next layer (or lamina) adjacent to the first has a very small velocity differential, relative to the first layer, whose magnitude depends on the viscosity of the fluid. The more viscous the fluid, the lower the velocity differential between each succeeding layer. At some distance δ, measured perpendicular to the surface, the velocity is equal to the free-stream velocity of the fluid. The distance δ is defined as the thickness of the boundary layer (BL).

The boundary layer – at subsonic speeds – is often composed of three regions beginning at the leading edge of a surface: (1) the laminar region where each layer or lamina slips over the adjacent layer in an orderly manner, creating a well-defined shear force in the fluid, (2) a transition region, and (3) a turbulent region, where the particles of fluid mix with each other in a random way, creating turbulence and eddies. The transition region is where the laminar region begins to become turbulent. The shear force in the laminar region and the swirls and eddies in the turbulent region both create drag, but with different physical processes. The cross-section of a typical boundary layer might look like Figure 3.16.

Figure 3.17 illustrates a typical boundary layer (BL) over a wing/tail airfoil, where two laminar and turbulent portions are distinguished. As we move along the flow, the thickness of the BL is increased, and the flow becomes more and more turbulent.

The shearing stress that the fluid exerts on the surface is called skin friction and is an important component of the overall drag. The two distinct regions in the boundary layer (laminar and turbulent) depend on the velocity of the fluid, the surface roughness, the fluid density, and the fluid viscosity. These factors, with the exception of the surface roughness, were combined by Osborne

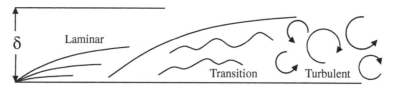

Figure 3.16 Typical boundary layer over a flat surface

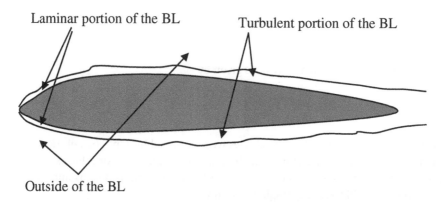

Figure 3.17 Typical boundary layer over a wing/tail airfoil

3.9 Boundary Layer   **61**

Reynolds in 1883 into a formula that has become known as the Reynolds number, which mathematically is expressed as

$$R = \rho V \left(\frac{l}{\mu} \right) \tag{3.10}$$

where ρ is fluid (here, air) density, V is air velocity, μ is air viscosity, and l is a characteristic length.

In aeronautical work, the characteristic length is usually taken as the chord of a wing surface. The Reynolds number is an important indicator of whether the boundary layer is in a laminar or turbulent condition. Laminar flow creates considerably less drag than turbulent, but nevertheless causes difficulties with small surfaces, as we shall learn later. Typical Reynolds numbers for various air vehicles – including a bird and an insect – are shown in Table 3.1.

Laminar flow causes drag by virtue of the friction between layers and is particularly sensitive to the surface condition. Normally, laminar flow results in less drag and is desirable. The drag of the turbulent boundary layer is caused by a completely different mechanism (e.g., vortex) that depends on knowledge of viscous flow.

In any flow, two fundamental laws are always applicable: (1) energy conservation law and (2) mass conservation law. For an incompressible flow ($M < 0.3$), the energy conservation law indicates that for an ideal fluid (no friction), the sum of the static pressure (P) and the dynamic pressure (q), where $q = \frac{1}{2}\rho V^2$, is constant:

$$P + \frac{1}{2}\rho V^2 = \text{const.} \tag{3.11}$$

Applying this principle to flow in a duct (e.g., convergent–divergent duct), with the first half representing the first part of an airplane wing, the distribution of pressure and velocity in a boundary layer can be analyzed. The flow inside a duct is very similar to a flow over and under a wing.

In an incompressible flow where the air density remains constant along the flow, this equation – for two arbitrary points (1 and 2) – is expanded as

$$P_1 + \frac{1}{2}\rho V_1^2 = P_2 + \frac{1}{2}\rho V_2^2 \tag{3.12}$$

This equation is referred to as Bernoulli's equation. As the fluid (assumed to be incompressible) moves through the duct or over a wing, its velocity increases (because of the law of conservation of mass) and, as a consequence of Bernoulli's equation, its pressure decreases, causing what is known

Table 3.1 Typical Reynolds numbers

No.	Air Vehicle Type	Reynolds Number
1	Large subsonic UAVs	5,000,000
2	Small UAVs	400,000
3	Mini-UAVs – Quadcopters	50,000
4	A Seagull	100,000
5	A Gliding Butterfly	7,000

as a favorable pressure gradient. The pressure gradient is favorable because it helps push the fluid in the boundary layer on its way.

After reaching a maximum velocity (usually at the maximum thickness point of an airfoil or the minimum diameter of a duct), the fluid begins to slow. If the slope of the divergent section is high, the flow consequently forms an unfavorable pressure gradient (i.e., hinders the boundary layer flow) as seen by the velocity profiles in Figure 3.18.

Small characteristic lengths and low speeds result in low Reynolds numbers and consequently laminar flow, which is normally a favorable condition. A point is reached in this situation where the unfavorable pressure gradient actually stops the flow within the boundary layer and eventually reverses it. The flow stoppage and reversal results in the formation of turbulence, vortices, and in general a random mixing of the fluid particles.

At this point, the boundary layer detaches or separates from the surface and creates a turbulent wake. This phenomenon is called separation, and the drag associated with it is called vortex drag. Whether the boundary layer is turbulent or laminar depends on the Reynolds number, as does the friction coefficient, as shown in Figure 3.19.

It would seem that laminar flow is always desired (for less pressure drag), and usually it is, but it can become a problem when dealing with very small UAVs that fly at low speeds. The favorable and unfavorable pressure gradients previously described also exist at very low speeds, making it possible for the laminar boundary layer to separate and reattach itself. This keeps the surface essentially in the laminar flow region, but creates a bubble of fluid within the boundary layer. This is called laminar separation and is a characteristic of the wings of very small, low-speed airplanes (e.g., small model airplanes and very small UAVs).

The bubble can move about on the surface of the wing, depending on the angle of attack, speed, and surface roughness. It can grow in size and then can burst in an unexpected manner.

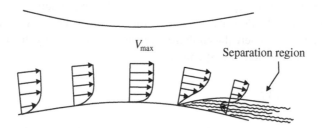

Figure 3.18 Boundary layer velocity profile inside a convergent–divergent duct

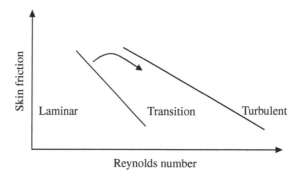

Figure 3.19 Skin friction versus Reynolds number

The movement and bursting of the bubble disrupts the pressure distribution on the surface of the wing and can cause serious and sometimes uncontrollable air-vehicle motion. This has not been a problem with larger, higher speed airplanes because most of the wings of these airplanes are in turbulent flow boundary layers due to the high Reynolds number at which they operate.

Specially designed airfoils are required for small lifting surfaces to maintain laminar flow, or the use of "trip" devices (known as turbulators) to create turbulent flow. In either case, the laminar separation bubble is either eliminated or stabilized by these airfoils. Laminar separation occurs with Reynolds numbers of about 75,000. Control surfaces, such as the elevator and aileron, are particularly susceptible to laminar separation.

3.10 Friction Drag

The friction drag mainly includes all types of drag that do not depend on production of the lift. Every aerodynamic component of aircraft (i.e., the components that are in direct contact with flow) generates friction drag. Typical components are the wing, horizontal tail, vertical tail, fuselage, landing gear, antenna, engine nacelle, camera, and strut. The zero-lift drag is primarily a function of the external shape of the components.

Since the performance analysis is based on aircraft drag, the accuracy of aircraft performance analysis relies heavily on the calculation accuracy of friction drag (i.e., C_{Do}). The C_{Do} of an aircraft is simply the summation of C_{Do} of all contributing components:

$$C_{D_o} = C_{D_{o_f}} + C_{D_{o_w}} + C_{D_{o_{ht}}} + C_{D_{o_{vt}}} + C_{D_{o_{LG}}} + C_{D_{o_N}} + C_{D_{o_S}} + C_{D_{o_p}} + \ldots \tag{3.13}$$

where C_{Dof}, C_{Dow}, C_{Doht}, C_{Dovt}, C_{DoLG}, C_{DoN}, C_{DoS}, and C_{DoP} are respectively representing fuselage, wing, horizontal tail, vertical tail, landing gear, nacelle, strut, and external payload contributions in aircraft C_{Do}. The three dots at the end of Equation (3.13) illustrate that there are other components that are not shown here. They include non-significant components such as the antenna and pitot tube. In the majority of fixed-wing conventional air vehicles, wing and fuselage each contribute about 30%-40% (a total of about 60%–80%) to aircraft C_{Do}.

Reference [12] provides a build-up technique to calculate the contribution of each component to C_{Do} of an aircraft. The majority of the equations are based on flight test data and wind tunnel test experiments, so the technique is mainly relying on empirical formulas.

3.11 Total Air-Vehicle Drag

The total resistance to the motion of a subsonic air-vehicle wing is made up of two components: the drag due to lift (induced drag) and the profile drag, which in turn is composed of the friction drag and the pressure drag (due to flow separation). For the overall air vehicle, the drag of all the non-lift parts (e.g., fuselage, landing gear, and payload) are lumped together and called parasite (or parasitic) drag. If the various drag components are expressed in terms of drag coefficients, then simply multiplying their sum by the dynamic pressure (q) and a characteristic area (usually the wing area, S) results in the total drag:

$$D = \frac{1}{2} \rho V^2 S \left(C_{Do} + C_{Di} \right) \tag{3.14}$$

where C_{Do} is the sum of all the profile drag coefficients (it is also referred to as the zero-lift drag coefficient) and C_{Di} is the induced drag coefficient, whose quadratic form results in the parabolic shape of the polar curve (see Figure 3.13).

The ScanEagle UAV – developed by Boing Insitu – is composed of four field-replaceable major modules/components: (1) nose (including payload sensors), (2) fuselage, (3) Wing, and (4) prop-driven engine. The UAV has a cylindrical fuselage of 2 m long with a mid-mounted swept-back wing with winglets, endplate vertical tail and movable rudders. The nose carries a pitot tube, which is fitted with an anti-precipitation system for cold weather operation. The air vehicle is fitted with a pusher piston engine (0.97 kW) with a two-blade propeller. All of these external components contribute to the total vehicle drag.

Moreover, the internal components include avionics, autopilot, communication system, fuel system, and mechanical/electric systems. The nose houses a gimballed and inertially stabilized turret which is fitted with EO/IR cameras. The air vehicle is not fitted with landing gear. The vehicle carries a maximum of 4.3 kg of fuel.

3.12 Flapping Wings

There is interest in UAVs that use flapping wings to fly like a bird. The details of the physics and aerodynamics of flight using flapping wings are beyond our scope, but the basic aerodynamics can be appreciated based on the same mechanisms for generating aerodynamic forces that we have outlined for fixed wings. The following discussion is based largely on *Nature's Flyers: Birds, Insects, and the Biomechanics of Flight* [13].

The flapping of the wings of birds is not a pure up and down or rowing backstroke as commonly thought. The wings of a flying bird move up and down as they are flapped, but they also move forward due to the bird's velocity through the air mass. Figure 3.20 shows the resulting velocity and force triangles when the wing is moving downward. The net velocity of the wing through the air mass is the sum of the forward velocity of the bird's body (V) and the downward velocity of the wing, driven by the muscles of the bird (w), which varies over the length of the wing, being greatest at the wing tip. The resulting total velocity through the air mass is forward and down, which means that the relative wind over the wing is to the rear and up.

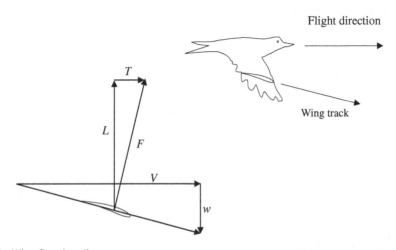

Figure 3.20 Wing flapping diagram

3.12 Flapping Wings

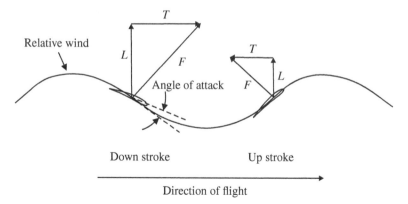

Figure 3.21 Flight of a bird

The net aerodynamic force generated by that relative wind (F) is perpendicular to the relative wind and can be resolved into two components, lift (L) upward and thrust (T) forward (Figure 3.21).

The velocity and force triangles vary along the length of the wing because w is approximately zero at the root of the wing, where it joins the body of the bird and has a maximum value at the tip of the wing, so that the net force, F, is nearly vertical at the root of the wing and tilted furthest forward at the tip. As a result, it sometimes is said that the root of the bird's wing produces mostly lift and the tip produces mostly thrust. This is dissimilar to a fixed-wing air vehicle, where the lift at the wingtip is almost zero, while at the wing root, it is often the maximum.

It is also possible for the bird to introduce a variable twist in the wing over its length, which could maintain the same angle of attack as w increases and the relative wind becomes tilted more upward near the tip. This twist can also be used to create an optimum angle of attack that varies over the length of the wing. This can be used to increase the thrust available from the wing tip.

Figure 3.22 shows how flapping the wing up and down can provide net lift and net positive thrust. The direction of the relative wind is tangent to the curved line that varies over the up and down strokes. To maximize the average lift and thrust, the angle of attack is "selected" by the bird to be large during the down stroke, which creates a large net aerodynamic force. This results in a large lift and large positive thrust. During the up stroke, the angle of attack is reduced, leading to a smaller net aerodynamic force. This means that even though the thrust is now negative, the average thrust over a complete cycle is positive. The lift remains positive, although smaller than during the up stroke.

The bird can make the negative thrust during the up stroke even smaller by bending its wings during the up stroke, as shown in Figure 3.22. This largely eliminates the forces induced by the

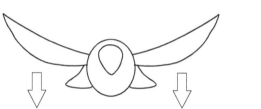
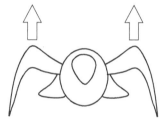

Figure 3.22 Wing articulation

outer portions of the wings, which are the most important contributors to thrust, while preserving much of the lift produced near the wing roots.

This simplified description of how flapping wings can allow a bird to fly is as far as we are going to go in this introductory text. There are some significant differences between how birds fly and how insects fly, and not all birds fly in exactly the same way. In the early days of heavier-than-air flight, there were many attempts to use flapping wings to lift a human passenger. All were unsuccessful. As interest has increased in recent years in small, even tiny, UAVs, the biomechanics of bird and insect flight are being closely re-examined and recently have been successfully emulated by machines.

3.13 Aerodynamic Efficiency

One of the important parameters in a cruising flight is the aerodynamic efficiency. This parameter is directly a function of the aircraft lift-to-drag ratio. One of the objectives in aerodynamic design of an air vehicle is to produce the maximum lift with a minimum drag. The aerodynamic efficiency of an air vehicle relies heavily on the section of its wing/tail; that is "*airfoil.*"

Using the equations for lift and drag, one can readily prove the following relation for the maximum lift-to-drag ratio:

$$\left(\frac{L}{D}\right)_{max} = \left(\frac{C_L}{C_D}\right)_{max} \tag{3.15}$$

The lift versus drag curve is further discussed in Section 3.5. Reference [12] has derived the following expression for the maximum lift-to-drag coefficients ratio:

$$\left(\frac{C_L}{C_D}\right)_{max} = \frac{1}{2\sqrt{KC_{D_o}}} \tag{3.16}$$

With this relationship, one is able to evaluate the maximum lift-to-drag ratio of any aircraft. The only necessary information is the aircraft zero lift drag coefficient (C_{Do}) and the induced drag correction factor (K). Graphically, the maximum L/D corresponds to the slope of the tangent to the C_D-C_L figure (drag polar), which is shown in Figures 3.7 and 3.13. Figure 3.7 shows the two-dimensional drag coefficient (C_d) and the moment coefficient as a function of the lift coefficient (C_l) for the NACA 23021 airfoil.

Aerodynamic efficiency (η_E) is defined as the ratio of the difference between lift and drag over the lift:

$$\eta_E = \frac{L - D}{L} \tag{3.17}$$

In a cruising flight, lift is equal to the aircraft weight ($L = W$). Thus,

$$\eta_E = \frac{W - D}{W} \tag{3.18}$$

The maximum aerodynamic efficiency is

$$\eta_{Emax} = 1 - \frac{1}{(L/D)_{max}} \tag{3.19}$$

Equations (3.19) and (3.16) reveal that the following factors play an important role in air vehicle aerodynamic efficiency: (1) aircraft zero lift drag coefficient (C_{Do}) and (2) induced drag correction factor (K). Furthermore, due to Equation (3.6), the Oswald span efficiency factor (e) and the wing aspect ratio (AR) are also impacting the aerodynamic efficiency.

As a conclusion, in order to increase the air vehicle maximum aerodynamic efficiency, one must: (1) decrease the aircraft zero lift drag coefficient and (2) increase the wing aspect ratio. High aspect ratio wings (long and slender) are conducive to good range and endurance. Short stubby (low AR) wings generate low aerodynamic efficiency (i.e., penalize the length of time-on-target during reconnaissance missions), but are good for highly maneuverable fighters.

The wing aspect ratio for the MALE UAV General Atomics MQ-1 Predator (see Figure 11.1) wing is about 22.5, while for the HALE UAV Northrop Grumman RQ-4 Global Hawk (see Figure 1.4) wing is about 25. The wing aspect ratio for the small UAV AeroVironment RQ-11 Raven (see Figure 18.11) is 3.7.

Example 3.1

A MALE UAV with a turbofan engine has a zero-lift drag coefficient of 0.021 and an induced drag factor of 0.04. Determine the maximum aerodynamic efficiency of the air vehicle.

Solution:

$$\left(\frac{C_L}{C_D}\right)_{max} = \frac{1}{2\sqrt{KC_{D_o}}} = \frac{1}{2\sqrt{0.04 \times 0.021}} = 17.2 \qquad \text{from (3.16)}$$

$$\eta_{E\,max} = 1 - \frac{1}{\left(L/D\right)_{max}} = 1 - \frac{1}{17.2} = 0.942 = 94.2\% \qquad \text{from (3.19)}$$

Questions

1) Define aerodynamics.
2) Name primary forces that act on an air vehicle.
3) Name elements that make considerable contributions on air vehicle's aerodynamic features.
4) What is the primary aerodynamic function of a wing?
5) Name two aerodynamic forces.
6) The aerodynamic forces of lift and drag are functions of a number of factors. What are they?
7) Define lift.
8) Draw the side-view of a wing at an angle of attack. In the figure, illustrate: (1) total aerodynamic force, (2) lift, (3) drag, and (4) name and the location of the forces.
9) Define Mach number.
10) List the flight regimes, when the airspeed is compared with the speed of sound. Briefly define each one.
11) What are the two most important parameters of an airfoil?
12) Provide at least five geometric parameters of an airfoil.
13) Briefly compare the patterns of air pressure over a two-dimensional airfoil and a three-dimensional wing.

14) What do NACA and NASA stand for?

15) Define streamline.

16) What is the thickness-to-chord ratio of the NACA airfoil 23021?

17) What is the lift coefficient of the NACA 23021 airfoil at an angle of attack of 10 degrees when $R = 3 \times 10^6$?

18) What is the maximum lift coefficient of the NACA 23021 airfoil when $R = 3 \times 10^6$?

19) What is the minimum drag coefficient of the NACA 23021 airfoil when $R = 8.9 \times 10^6$?

20) What is the drag coefficient of the NACA 23021 airfoil when the lift coefficient is 0.2 at $R = 8.9 \times 10^6$?

21) What airfoil is used on the wing of the Northrop Grumman RQ-4 Global Hawk?

22) What airfoil is used on the wing of the General Atomics MQ-1 Predator?

23) Compare the static pressure of the top and bottom surfaces for a positive cambered airfoil at a positive angle of attack. Draw a figure.

24) Why is a vortex generated at each wing tip?

25) Briefly describe how the lift is generated on a wing.

26) Draw a typical drag polar and discuss its characteristics.

27) The drag coefficient is the sum of two terms. What are they?

28) Write the drag polar equation.

29) Define aspect ratio.

30) Discuss how to convert the "infinite wing" (i.e., airfoil) lift coefficient to the lift coefficient of a real wing.

31) What is the most important element of drag introduced by a wing at high angles of attack?

32) What is the significance of an elliptical lift distribution over the wing? Discuss.

33) What can be concluded from Equation (3.9) about induced drag?

34) Discuss characteristics of the boundary layer of a fluid flowing over a surface.

35) Name three regions of the boundary layer.

36) How is mathematically Reynolds number expressed?

37) What is the typical Reynolds number for small UAVs?

38) What is the typical Reynolds number for large subsonic UAVs?

39) What is the typical Reynolds number for Quadcopters?

40) What is the relation between static pressure and dynamic pressure for an incompressible flow?

41) What can be concluded from Bernoulli's equation for variations of pressure and flow speed?

42) Discuss when a flow separation happens inside a duct. What is the consequence of flow separation?

43) What do C_{Do} and C_{Di} stand for?

44) What are two major contributors to C_{Do} of a fixed-wing conventional air vehicle?

45) What are two components of the net aerodynamic force generated by the relative wind in a flapping wing? Show them in a figure.

46) Briefly compare the lift at the wingtip of: (1) a flapping wing and (2) a fixed-wing of a conventional air vehicle.

47) Briefly compare the aerodynamic outcome at the wing tip and the wing root of a flapping wing.

48) Compare the angle of attack variation which is "selected" by a bird during an up and down stroke.

49) What factors impact the air vehicle maximum aerodynamic efficiency?

50) A HALE UAV with a wing area of 74 m^2 and the following feature is cruising with an airspeed of 400 km/h at an altitude that the air density is 0.3 kg/m^3:

$$K = 0.043, \quad C_{Do} = 0.026, \quad C_L = 0.25$$

Determine the air vehicle drag at this flight condition.

Questions 69

51) A MALE UAV with a turbofan engine has a zero-lift drag coefficient of 0.024 and an induced drag factor of 0.05. Determine the maximum aerodynamic efficiency of the air vehicle.

52) A mini-UAV with a wing area of 1.2 m^2 is cruising with an airspeed of 20 m/s at an altitude where the air density is 1.1 kg/m^3. The UAV has the following features:

$$C_{Di} = 0.005, \ C_{Do} = 0.026$$

Determine the air vehicle drag for this flight condition.

53) A small UAV with a wing area of 6.5 m^2 is cruising with an airspeed of 50 m/s at an altitude where the air density is 0.92 kg/m^3. The UAV has the following features:

$$C_{Di} = 0.008, \ C_{Do} = 0.028$$

Determine the air vehicle drag for this flight condition.

54) What is the wing aspect ratio for the General Atomics MQ-1 Predator?

55) What is the wing aspect ratio for the Northrop Grumman RQ-4 Global Hawk?

56) What is the wing aspect ratio for the AeroVironment RQ-11 Raven?

57) Describe aerodynamic features of the ScanEagle.

4

Performance

4.1 Overview

The objective of Chapters 4 and 5 is to give the reader the ability to analyze air vehicle flight dynamics. The subject of flight dynamics is about the motion (i.e., behavior) of air vehicles that fly in the atmosphere. It encompasses two broad areas: (1) flight performance and (2) flight stability and control. The first area – which includes topics such as cruise/maximum speed, range, endurance, ceiling, climb/descent/glide, launch/recovery, takeoff/landing, and maneuverability – is presented in this chapter. Fundamentals, definitions, concept, and analysis of flight stability and control will be presented in Chapter 5.

The basic aerodynamics equations are a powerful way to calculate key performance characteristics in terms of the design characteristics and mission profiles of the AV. They are the tools with which an aeronautical engineer designs any aircraft. This chapter provides examples of their power by showing how very basic equations can be put together to estimate important characteristics, such as range and endurance. The power of the basic equations extends to all areas of the design.

Calculations of this sort are central to the whole design process. The examples show how the basic configuration of the AV (weight and drag), altitude, and flight modes drive fuel consumption, which, in turn, drives range and endurance. These and other similar equations are used to tradeoff between AV weight, fuel load, mission payload weight, wing size and design, and all of the other factors that go into a complete AV design.

This chapter illustrates how the basic governing equations – including aerodynamics and propulsion – can be used to predict the performance of an aircraft and shows how that performance is related to the key elements of the aircraft design. An UAV is deployed to perform a flight mission; main phases in a typical flight mission is demonstrated in Figure 4.1.

The flight mission for a UAV begins with a launch/takeoff, and then climbs to a cruising altitude. Next, the UAV will cruise to its destination (i.e., target location) for performing the primary function using payload (e.g., aerial photography, aerial filming, and surveillance using a camera). After the mission is done, it is programmed to cruise back, descend, and be recovered/land. In this chapter, the analysis of air vehicle performance during launch/takeoff, climb, cruise (maximum speed, range, and endurance), descend (glide, when no engine is used), and landing/recovery is discussed. As an illustration of the power of the basic equations, expressions for two of the most important capabilities of a UAV, range and endurance, are derived.

Introduction to UAV Systems, Fifth Edition. Paul Gerin Fahlstrom, Thomas James Gleason, and Mohammad H. Sadraey.
© 2022 John Wiley & Sons, Inc. Published 2022 by John Wiley & Sons, Inc.
Companion website: www.wiley.com/go/fahlstrom/uavsystems5e

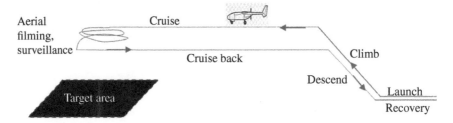

Figure 4.1 Main phases in a typical flight mission

This textbook is not intended to be a basis for even an introductory course in aerodynamics and flight dynamics, but simplified derivations of these expressions are provided in the following sections, both to illustrate the power of even a simple mathematical description of the dynamics of flight and to provide some useful equations for estimating the key performance characteristics of UAVs.

4.2 Cruising Flight

In a cruising flight, both airspeed and altitude are kept at constant values, which requires a fixed engine thrust. The primary forces that act on an air vehicle are: (1) thrust (T), (2) lift (L), (3) drag (D), and (4) gravity (or weight, W). They are shown in Figure 4.2 for a cruising level flight (x and z axes are also illustrated). The expressions for these quantities are functions of aerodynamics and propulsion that govern the performance of an air vehicle.

The air vehicle is at an equilibrium point, so the sum of the forces along x and z are equal to zero:

$$T = D \tag{4.1}$$

$$W = L \tag{4.2}$$

Plugging aerodynamic forces equations (i.e., lift and drag) from Chapter 3, we will have:

$$T = \frac{1}{2}\rho V^2 S C_D \tag{4.3}$$

$$mg = \frac{1}{2}\rho V^2 S C_L \tag{4.4}$$

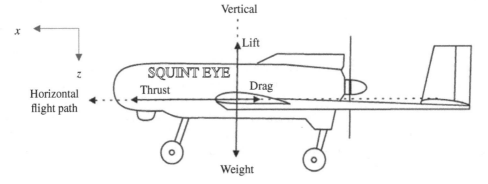

Figure 4.2 Forces on an air vehicle in a level flight

where S denotes wing area, V airspeed, and ρ air density. Plugging equivalents for lift and drag coefficients (i.e., C_L and C_D) from Chapter 3, one can derive an equation [12] for cruising speed as

$$T_C = \frac{1}{2}\rho V_C^2 S C_{Do} + \frac{2KW^2}{\rho V_C^2 S} \tag{4.5}$$

This is a nonlinear algebraic equation; it does not have any closed-form solution. Equation (4.5) gives the cruise speed (V_C) at any altitude due to a given engine thrust (T_C) at that altitude. It is remembered that K is the induced drag factor, C_{Do} is the aircraft zero lift drag coefficient, and ρ denotes the air density.

As a parenthetical comment, Equation (4.5) also explains why cruise airspeed decreases as a function of altitude (see Figure 6.16 and Equations (6.20) and (6.21)). It should be noted that, as discussed in Chapter 6, engine thrust for a prop-driven aircraft is a function of engine power. Thus, to employ Equation (4.5) for a prop-driven engine, Equation (6.17) must be simultaneously used.

A question of interest is: what is the minimum speed at which a fixed-wing airplane can still fly? This is important for understanding landing, take-off, launch from a catapult, and arrested recovery. To find the minimum velocity at which the airplane can fly, we set lift equal to weight in Equation (4.2) to balance the vertical forces (along the z-axis) and solve for velocity. If the maximum lift coefficient, C_{Lmax}, is known then the minimum velocity can be seen to be directly proportional to the square root of the wing loading W/S. Needless to say, an airplane with a large wing area and low weight can fly slower than a heavy, small-winged airplane. The equation for minimum velocity is

$$V_{min} = V_s = \sqrt{\frac{2W}{\rho S C_{L_{max}}}} \tag{4.6}$$

This velocity is referred to as the stall speed (V_s). When the aircraft speed is reduced to its lowest steady level value (stall speed), it implies that the lift coefficient has been reached to its maximum value (C_{Lmax}). The maximum lift coefficient for a large number of air vehicles is about 1.5 to 2.5.

Please note that the velocity (V) in the equations above is airspeed (i.e., the relative speed between the air vehicle and the air), not ground speed. There are a number of units for airspeed including m/s, km/h, and mile/h. However, in British units, the unit of knot – which is equivalent to a nautical mile per hour - is the most common. Recall, 1 knot = 0.514 m/s. Moreover, most regulations regarding airspeed in Federal Aviation Regulations (FAR) are in knots. For instance, based on FAR Part 107, the ground speed of the small unmanned aircraft may not exceed 87 knots (100 miles per hour).

The cruising speed for the Northrop Grumman RQ-4 Global Hawk with a turbofan engine is 310 knots (570 km/h), while for the General Atomics MQ-9 Reaper with a turboprop engine is 169 knots. The AeroVironment RQ-11 Raven with an electric engine has a cruise speed of 30 km/h (16.2 knots).

4.3 Range

The range of a UAV is an important performance characteristic, particularly for MALE and HALE UAVs (e.g., Northrop Grumman RQ-4 Global Hawk). Range is defined as the distance flown by consuming all fuel (for air breathing engines) or battery energy (for electric engines). Missions such as border monitoring or tracking a moving target requires a long range.

74   *4 Performance*

It is relatively easy to calculate in a reasonable approximation. The range is dependent on a number of basic aircraft parameters and strongly interacts with the weight of the mission payload, because fuel can be exchanged for payload within limits set by the ability of the air vehicle to operate with varying center of gravity conditions. The fundamental relationship for calculating range and endurance is the decrease in weight of the air vehicle caused by the consumption of fuel.

For a propeller-driven aircraft, this relationship is expressed in terms of the specific fuel consumption, c, which is the rate of fuel consumption per unit of power produced at the engine shaft, P_e. With this definition, we can see that

$$-\frac{dW}{dt} = cP_e \tag{4.7}$$

For a jet aircraft a different measure of specific fuel consumption is used, the thrust specific fuel consumption, c_t, which is the rate of fuel consumption per unit thrust produced by the jet engine. Thus:

$$-\frac{dW}{dt} = c_t T \tag{4.8}$$

It is worth spending a moment discussing the units of c and c_t. They are equal to the weight of fuel burned per unit time per unit of power or thrust produced by the engine. For c in English units, this would be pounds per second per foot-pound per second (or pounds per hour per horsepower). The pound per second cancels out and the units of c are 1/ft (or in metric units, 1/m). For c_t, the units are pounds per second per pound of thrust, so the net units of c_t are 1/s. Since power equals thrust × velocity, we can express c_t in terms of c:

$$c_t = \frac{cV}{\eta} \tag{4.9}$$

The typical value of specific fuel consumption for a propeller-driven non-electric-engine aircraft is about 0.5 lb/(hp.h). In the following two sections, the range for propeller-driven and jet-propelled aircraft is presented.

4.3.1   Range for a Non-Electric-Engine Propeller-Driven Aircraft

For a propeller-driven aircraft, we start from Equation (4.7). In a level flight, $T = D$, or $TV = DV$. The term TV is the power delivered to the air vehicle by the propulsion system. It is called power available (P_A). Moreover, DV is equal to the power required to maintain a level flight, which is called power required (P_R). Since $P_A = \eta P_e$ and $P_A = P_R = DV$ for level flight, we can rewrite that equation as

$$-\frac{dW}{dt} = \left(\frac{c}{\eta}\right) DV \tag{4.10}$$

Because $L/D = W/D$, $D = W/(L/D)$, one can substitute for D into Equation (4.10) and solve for V dt, which becomes

$$V\,dt = -\left(\frac{\eta}{c}\right)\left(\frac{L}{D}\right)\frac{dW}{W} \tag{4.11}$$

4.3 Range

Assuming that both L/D and η/c are constant, the range R can be determined by integrating $V\,dt$ over the total flight. The result [12] is

$$R = \left(\frac{\eta}{c}\right)\left(\frac{L}{D}\right)\ln\left(\frac{W_0}{W_1}\right) \tag{4.12}$$

where W_0 is the empty weight of the aircraft (all fuel expended) and W_1 is the weight at takeoff. This equation was derived early in the history of aeronautics and is known as the Breguet range equation. The weight of fuel as a fraction of the takeoff weight is then given by

$$\frac{W_{\text{fuel}}}{W_{\text{To}}} = \frac{W_1 - W_0}{W_1} = 1 - \frac{W_0}{W_1} \tag{4.13}$$

For non-electric-engine propeller-driven aircraft, Equation (4.12) provides a value for the range directly, based on some basic parameters of the aircraft (η, c, and fuel capacity) and L/D. Examination of the equation indicates that the range of the aircraft is increased by having higher propeller efficiency, lower specific fuel consumption, and large fuel capacity (a large difference between W_1 and W_0). All of this is quite intuitive.

The more interesting result of examining this equation is that it shows that for the maximum range, we must fly with a speed ($V_{R\text{max}}$) at the maximum value of L/D. This can be shown to occur at a velocity given by

$$V_{R\text{max}-\text{prop}} = V_{(L/D)\text{max}} = \sqrt{\frac{2mg}{\rho S\sqrt{\dfrac{C_{D_o}}{K}}}} \tag{4.14}$$

and with a maximum value of L/D of

$$\left(\frac{L}{D}\right)_{\text{max}} = \sqrt{\frac{\pi ARe}{4C_{\text{Do}}}} \tag{4.15}$$

A similar equation for the maximum value of the lift-to-drag ratio is presented in Chapter 3 as Equation (3.16).

Example 4.1

A UAV has a zero-lift drag coefficient of 0.03, a wing aspect ratio of 8, and an Oswald efficiency factor of 0.9. Determine the maximum value of L/D.

Solution:

$$\left(\frac{L}{D}\right)_{\text{max}} = \sqrt{\frac{\pi ARe}{4C_{\text{Do}}}} = \sqrt{\frac{3.14 \times 8 \times 0.9}{4 \times 0.03}} = 13.7$$

From Equation (4.15), we see that if we are looking for a long range, we need to have a large aspect ratio, so we will need to have long, narrow wings. A range chart for a typical propeller-driven aircraft is shown in Figure 4.3.

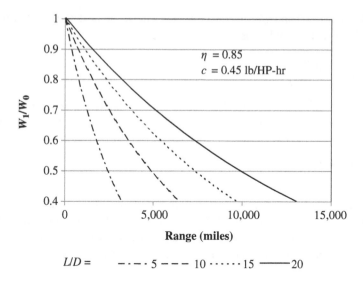

Figure 4.3 Range versus weight ratio for propeller-driven aircraft

It is common in aeronautical engineering to use rather mixed systems of units, such as horsepower for engine power combined with pounds for thrust and miles per hour instead of feet per second. It is left as an exercise for the reader to derive forms of this equation that will give the correct answer, when using some of the more common systems of mixed units.

Since we assumed that $L = W$ in deriving the simple range equation for propeller-driven aircraft, it applies only to a flight in which those conditions are maintained from start to finish. Since the weight of the air vehicle continually decreases as fuel is burned, it is necessary to decrease the lift over time to maintain the condition $L = W$.

In a cruising flight, this goal can be accomplished by one of the following three approaches: (1) decreasing the velocity, or (2) increasing the altitude, or (3) decreasing the aircraft angle of attack over time. For a cruising flight, we have three options for the flight program: (1) flight with constant altitude, constant angle of attack, and decreasing velocity, or (2) flight with constant velocity, constant angle of attack, and increasing altitude, or (3) flight with constant velocity, constant altitude, and decreasing angle of attack.

Therefore, Equation (4.12) only applies to (1) flight with constant altitude and decreasing velocity or (2) flight with constant velocity and increasing altitude. The second flight program is referred to as the cruise-climb flight. For the range equations for other flight programs, study Reference [12].

Nonetheless, it is a useful way to make a quick estimate of the range of an air vehicle based on the weight of fuel available relative to the weight of the air vehicle without fuel.

4.3.2 Range for a Jet-Propelled Aircraft

For a jet-propelled aircraft the situation is somewhat different. To develop a form of the equation that is specifically for a jet aircraft, we start from Equation (4.8) and find that

$$V \mathrm{d}t = -\left(\frac{1}{c_t}\right)\left(\frac{L}{D}\right) V \frac{\mathrm{d}W}{W} \tag{4.16}$$

4.3 Range

Proceeding as before, and assuming that V and the lift-to-drag ratio (i.e., L/D) are constant over the flight (keep both the airspeed and the lift coefficient (i.e., the angle of attack) constant as the weight of the aircraft decreases), we can integrate and get a simplified range equation for a jet-propelled aircraft of

$$R = \frac{V}{c_t} \frac{L}{D} \ln\left(\frac{W_0}{W_1}\right) \tag{4.17}$$

This equation is known as the Breguet range equation for jet aircraft. We can see that the maximum range will occur if the flight is made at the maximum value of (VL/D). We know that $L = W$ for level flight, and can write

$$L = W = \frac{1}{2}\rho V^2 S C_L$$

Using the fact that – in a cruising flight – $L/D = C_L/C_D$, one can derive the following expression for VL/D:

$$V\frac{L}{D} = \sqrt{\frac{2W}{\rho S C_L}} \frac{C_L}{C_D} \tag{4.18}$$

This cannot be substituted directly into Equation (4.17), because that equation was derived by integrating over W and the expression for VL/D involves W. Substituting Equation (4.18) into Equation (4.16) and integrating under the assumption that ρ, C_L, S, and $C_L^{1/2}/C_D$ are all constant, we find another form of the range equation for jet-propelled aircraft:

$$R = \frac{1}{c_t}\sqrt{\frac{2}{\rho S}}\frac{C_L^{1/2}}{C_D}\left(W_0^{1/2} - W_1^{1/2}\right) \tag{4.19}$$

Similar to a propeller-driven aircraft, for a jet aircraft, we want low specific fuel consumption (in this case, thrust specific fuel consumption) and large fuel capacity. In addition, for the jet-propelled aircraft we would like to have the minimum possible air density, so we would prefer to fly at high altitudes.

In deriving this equation for the range of a jet-propelled aircraft, we assumed that air density (ρ) and lift coefficient (C_L) were constant. To maintain ρ constant, the altitude must remain constant. To maintain $L = W$, the velocity must be decreased as the weight of the aircraft decreases.

One goal in a cruising flight is to maximize the range. In order to maximize the range for a constant-speed cruising flight, we have two options: (1) reduce the lift coefficient and (2) increase the cruising altitude. For both cases, this requires the velocity to have values such that $C_L^{1/2}/C_D$ is maximized. Reference [12] has shown that the maximum range for these cases occurs at a velocity $V_{R\max}$ given by

$$V_{R_{\max-jet}} = \sqrt{\frac{2W}{\rho S \sqrt{\frac{C_{D_o}}{3K}}}} \tag{4.20}$$

As for a propeller-driven aircraft, a long range for a jet-propelled aircraft requires a high aspect ratio wing.

The range for the Northrop Grumman RQ-4 Global Hawk with a turbofan engine is 22,780 km, while for the General Atomics MQ-9 Reaper with a turboprop engine, it is 1,900 km. The AeroVironment RQ-11 Raven with an electric engine has a range of 10 km.

4.4 Endurance

Endurance is defined as the flight duration by consuming all fuel (for air breathing engines) or battery energy (for electric engines). The endurance is important for the missions where the payloads (e.g., camera) are employed for functions such as aerial filming or a tower inspection. Long endurance is a primary performance goal for MALE and HALE UAVs (e.g., General Atomics MQ-1 Predator and Northrop Grumman RQ-4 Global Hawk). In this section, endurance for propeller-driven and jet-propelled aircraft is discussed.

4.4.1 Endurance for a Non-Electric-Engine Propeller-Driven Aircraft

Endurance is the time that the aircraft can remain airborne before running out of fuel. To estimate the endurance of a propeller-driven aircraft, we start from Equation (4.11):

$$V dt = -\left(\frac{\eta}{c}\right)\left(\frac{L}{D}\right)\frac{dW}{W} \tag{4.21}$$

Using the fact that $W = L$ in a level flight, we can derive V from Equation (4.3). Then, substituting V in the left-hand side and C_L/C_D for L/D in the right-hand side, we find for dt:

$$dt = -\left(\frac{\eta}{c}\right)\sqrt{\frac{\rho S}{2}}\left(\frac{C_L^{3/2}}{C_D}\right)\left(\frac{dW}{W^{3/2}}\right) \tag{4.22}$$

If we assume, as before, that everything but the weight of the aircraft is constant over time, we can integrate this and find that the endurance (E) is given by

$$E = \left(\frac{\eta}{c}\right)\sqrt{2\rho S}\frac{C_L^{3/2}}{C_D}\left(W_1^{-1/2} - W_0^{-1/2}\right) \tag{4.23}$$

Examination of this equation shows that maximum endurance occurs for much the same aircraft parameters as are needed for long range: high propeller efficiency, low specific fuel consumption, and large fuel capacity. However, in contrast to the desire to operate at high altitudes (low ρ) for long range, Equation (4.23) indicates that for long endurance, one would prefer to fly at sea level, where ρ has its highest value. Some representative curves for endurance of an example small UAV are shown in Figure 4.4.

For maximum endurance, it is necessary for the aircraft to operate at the maximum value of $C_L^{3/2}/C_D$. Reference [12] has shown that the maximum endurance for these cases occurs at a velocity V_{Emax} given by

$$V_{E_{\max-prop}} = \sqrt{\frac{2W}{\rho S \sqrt{\dfrac{C_{D_o}}{3K}}}} \tag{4.24}$$

The velocity for maximum endurance shown above is not the same velocity that results in the maximum range, shown in Equation (4.14). They differ by a factor of the fourth root of 3, or 1.32, with the velocity for maximum range (at maximum L/D) being 1.32 times the velocity for maximum endurance (at maximum $C_L^{3/2}/C_D$). This difference is not surprising, since endurance is time and range is velocity $\times$ time. As for long range, long endurance calls for a high aspect ratio wing.

4.4 Endurance

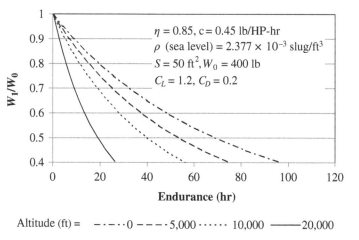

Figure 4.4 Endurance versus weight ratio for propeller-driven aircraft

4.4.2 Endurance for a Jet-Propelled Aircraft

For a jet-propelled aircraft, we start from Equation (4.8) and use the facts that, in a level flight, $T = D$ and $L = W$. Because $L/D = W/D$, $D = W/(L/D)$, one can substitute for D into Equation (4.8) and solve for dt, which becomes

$$dt = -\frac{L}{D}\frac{1}{c_t}\frac{dW}{W} \tag{4.25}$$

If we assume that L/D and c_t are constant over time, we can integrate this to produce a simple expression for the endurance of the jet-propelled aircraft:

$$E = \frac{1}{c_t}\frac{L}{D}\ln\frac{W_0}{W_1} \tag{4.26}$$

For maximum endurance, the flight must be performed at the velocity that produces maximum L/D. This already has been considered in connection with the range of a propeller-driven aircraft and the velocity that meets this condition is given by Equation (4.14) as

$$V_{E\max-\text{jet}} = V_{(L/D)\max} = \sqrt{\frac{2mg}{\rho S \sqrt{\frac{C_{D_o}}{K}}}} \tag{4.27}$$

This velocity varies as the weight of the aircraft decreases. The maximum value of L/D is given by Equation (4.15).

With little more than these basic equations, it is possible to derive expressions that provide reasonable approximations for the range and endurance of propeller and jet-propelled aircraft.

The endurance for the Northrop Grumman RQ-4 Global Hawk with a turbofan engine is 32+ hours, while for the General Atomics MQ-9 Reaper with a turboprop engine, it is 14 hours fully loaded. The AeroVironment RQ-11 Raven with an electric engine has a range of 60–90 minutes.

4.5 Climbing Flight

Consider an airplane – which is climbing in a steady, linear flight – is in equilibrium with all the forces acting on it, as shown in Figure 4.5. The climb angle (θ) is defined as the angle between the flight path and horizontal. The equations of motion for this condition can be written as:

$$\text{Lift} = L = W\cos(\theta) \quad \text{(along } z\text{-axis)} \tag{4.28}$$

where W is aircraft weight and

$$\text{Thrust} = T = D + W\sin(\theta) \quad \text{(along } x\text{-axis)} \tag{4.29}$$

where D is drag. Multiplying the second equation by velocity V results in

$$TV = DV + WV\sin\theta \tag{4.30}$$

In dynamics, mechanical power is defined as the product of force applied on an object and the object's velocity. Here, TV is the power delivered to the air vehicle by the propulsion system (either jet or pro-driven). It is called available power (P_A) and DV is equal to the power required to maintain a level flight, which is called required power (P_R). Thus, we obtain

$$P_A = P_R + WV\sin(\theta) \tag{4.31}$$

Since the vertical component of the airspeed, $V\sin(\theta)$ is equivalent to the rate of climb, dh/dt, Equation (4.31) can be rewritten as

$$W\frac{dh}{dt} = P_A - P_R \tag{4.32}$$

In a prop-driven aircraft, the power available can be obtained from the power delivered at the shaft of the engine (P_e) and propeller efficiency (η), expressed as

$$P_A = P_e\eta \tag{4.33}$$

Power is commonly expressed in horsepower, but the basic units of power are foot-pound per second in English units and watt (or Newton-meter per second) in metric units, and the equations here use those basic units. A detailed discussion for the relation between engine power, engine thrust, and propeller efficiency is provided in Chapter 6.

Solving Equation (4.32) for dh/dt (i.e., dividing by W), the rate of climb, ROC, is easily obtained:

$$ROC = \frac{dh}{dt} = \frac{P_A - P_R}{W} \tag{4.34}$$

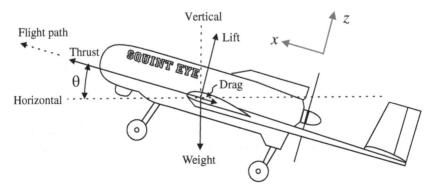

Figure 4.5 Force diagram in a climbing flight

4.5 Climbing Flight

The units of the rate of climb will be the same as those of V (e.g., m/s, and km/h, mile/h). However, in British units, the unit of foot per minute (ft/min or fpm) is most common. Moreover, most regulations regarding rate of climb in Federal Aviation Regulations (FAR) are in the unit of ft/min.

Since P_R is equal to drag × velocity (for both jet and prop-driven aircraft), our previous discussion of the components of drag as a function of velocity is applicable, and both P_A and P_R can be plotted against velocity, as shown in Figure 4.6. From Equation (4.34), the maximum rate of climb takes place at a velocity that has the maximum distance between the two curves. This is also the point where the slopes or derivatives of the two curves are equal. Therefore, the velocity for the maximum rate of climb can be read off the chart or calculated. The maximum and minimum airspeeds also can be obtained by directly reading the chart.

Drag and power are, of course, dependent on air density (among other parameters) and, therefore, flight altitude affects both curves. From the "power versus velocity" chart, we can readily conclude that, at the maximum speed (V_{max}), the available power is equal to the required power. This implies that there is no excess power to climb.

Figure 4.7 shows typical power-available and power-required curves for several altitudes. One can see that as the altitude increases, the distance between the curves, as well as the points where they intersect, become increasingly closer together, until the airplane is no longer capable of flight (i.e., there is no place where P_A is greater than P_R).

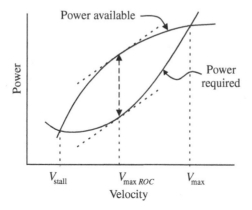

Figure 4.6 Power versus velocity

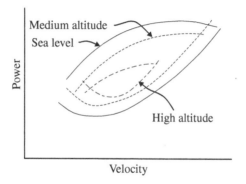

Figure 4.7 Power versus velocity for several altitudes

Example 4.2

A MALE UAV with a mass of 4,000 kg is equipped with a turboprop engine. The UAV -which is climbing with an airspeed of 90 knots – is producing 6,300 N of drag. In this flight condition, the engine is generating 700 kW of shaft power, while the propeller efficiency is 72%. Determine the rate of climb.

Solution:

The engine available power is

$$P_A = P_e \eta = 700 \times 0.72 = 504 \text{ kW} \qquad \text{from (4.33)}$$

The required power for this flight condition is

$$P_R = DV = 6,300 \times 90 \times 0.514 = 291,690 \text{ W} = 291.7 \text{ kW}$$

$$ROC = \frac{P_A - P_R}{W} = \frac{504,000 - 291,690}{4,000 \times 9.81} = 5.412 \frac{\text{m}}{\text{s}} = 1065.4 \frac{\text{ft}}{\text{min}} \qquad \text{from (4.34)}$$

4.6 Gliding Flight

A gliding flight is a free flight where the engine thrust/power is zero, so the UAV is descending due to its weight. UAVs – which are recovered via break-apart landing (e.g., AeroVironment RQ-11 Raven) or a vertical net – need to glide to the landing point. The goal for this glide is to fly with a glide angle and glide speed, such that to minimize the UAV deceleration at the impact with the ground/net.

The ability of the air vehicle to glide is measured by its sink rate and is easy to determine using the equations of motion or balance (Equations (4.28) and (4.29)). With the power off ($T = 0$), the glide governing equations of motion become

$$W \sin(\theta) = -D \qquad (4.35)$$

$$W \cos\theta = L \qquad (4.36)$$

Dividing the first equation by the second, the tangent of the glide angle is determined:

$$\tan\theta = \frac{-D}{L} = \frac{-C_D}{C_L} \qquad (4.37)$$

The tangent of the glide angle is the reciprocal of L/D. The sink rate is stated as a positive number, so if the velocity is known, the sink/glide rate becomes

$$\text{Sink rate} = V\frac{D}{W} = V \tan(\theta) \qquad (4.38)$$

The units of the sink rate will be the same as those of V (e.g., m/s, km/h, knot). Most UAVs do not use a lot of power during the approach and, because of their high L/D values, approach at rather shallow angles (say about 4-8 degrees). This means that copious amounts of runway or cleared space is necessary, even for net recovery. One way to overcome this deficiency, so that small fields can be used for recovery, is to reduce the L/D during the approach with the use of drag producing devices, such as flaps. These devices lead to additional system complexity and cost. Another way is by deploying a parafoil, which inherently has a low L/D.

4.7 Launch

The launch process is basically a linear accelerated motion, where the UAV is accelerating along the ramp until it reaches a safe launch speed. The UAV is frequently pushed forward by an external force added to the UAV engine thrust. The external force is provided by the launcher via a specific mechanism. The UAV engine must be at its maximum power/thrust, since the UAV must continue the flight after being launched. As provided in Chapter 17, there are multiple techniques for launching a UAV, including conventional takeoff and employing a mechanical ramp launcher. In this section, only the performance of a UAV when catapulted via a ramp launcher is presented. For an air vehicle performance in a conventional takeoff operation, the reader may study References [12, 14, 15].

Figure 4.8 shows the launch forces and parameters for a UAV launch operation on a launcher with a launch angle of θ. The launch angle is recommended to be close to the UAV maximum climb angle. The applied force, F, is the sum of an external force and the engine thrust. The x-axis is selected to be along the UAV motion path, and the z-axis is perpendicular to the x-axis. Thus, the UAV weight has two components, $W \cos(\theta)$ along the z-axis and $W \sin(\theta)$ along the x-axis (i.e., opposite to the x-axis).

From Newton's second law, along the launcher ramp, the sum of the forces along the x-axis creates the acceleration

$$\sum F_x = ma \qquad (4.39)$$

where m denotes the UAV mass and a represents the linear acceleration. Three main forces are active during a launch operation: (1) applied force (F), (2) UAV weight (W), and friction force (F_f) between the UAV cart and the launcher rails. However, if you would like to be more accurate, include UAV lift (L) and drag (D) forces. However, due to the low speed, both forces may be neglected.

The UAV-Ramp friction force (F_f) is

$$F_f = \mu N \qquad (4.40)$$

where μ is the friction coefficient between UAV legs (or cart) and the rails and N denotes the normal force on the rails. Typical values for the friction coefficient between two metallic surfaces

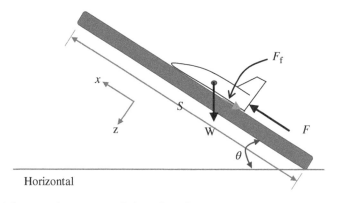

Figure 4.8 Launch forces and parameters during a launch

range from 0.1 to 0.2. Ignoring the UAV lift, the normal force is the component of the UAV weight perpendicular to the rails:

$$N = W\cos(\theta) \tag{4.41}$$

From the theory of dynamics, when a moving object with an initial velocity of V_1 accelerates to a new velocity of V_2, the distance (S) covered is governed by the following equation:

$$V_2^2 - V_1^2 = 2aS \tag{4.42}$$

Here, for a launcher, the distance S is the launcher stroke. The initial velocity at launch is often zero ($V_1 = 0$). The launch force must be strong enough so that the UAV at the end of the launch achieves a velocity that is at least 10% greater than the UAV stall speed.

4.8 Recovery

The recovery operation is basically a linear accelerated motion (with a negative acceleration), where the UAV is decelerating along a pre-programmed flight path until it stops. The UAV engine often is at idle condition, to allow the UAV to glide. As provided in Chapter 18, there are multiple techniques for recovery of a UAV, including conventional landing and employing a vertical net. In this section, only performance of a UAV when recovered via a vertical net or a hanging cable is presented. For air vehicle performance in a conventional landing operation, the reader may study Reference [12]. Figure 4.9 illustrates the recovery of a UAV using a vertical net in three steps: (1) UAV flying into a net, (2) UAV impacts the net and the net is extended to absorb the shock, and (3) UAV is tangling/hanging.

From Newton's second law, the sum of the forces along the flight path creates a deceleration:

$$\Sigma F = ma \tag{4.43}$$

where m denotes the UAV mass and a represents the linear deceleration. The main force active during a recovery operation is the force that the UAV exerts on the net/cable during impact. Note that we ignored lift and weight, since they are almost at equilibrium prior to recovery. We also neglected drag force for simplicity. From Equation (4.39), plugging zero for the final velocity, V_2, we obtain:

$$-V_1^2 = 2aS \tag{4.44}$$

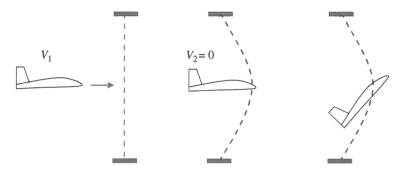

Figure 4.9 Recovery of a UAV using a vertical net

Questions 85

where S is the distance travelled from the impact point until it stops. The UAV velocity at the time of impact (i.e., V_1) is about 10% greater than the stall speed. It is a common practice to make the deceleration unitless by dividing it by g (i.e., 9.81 m/s^2).

The net/cable must have two important features: (1) be strong enough to handle the impact force without any failure and (2) be flexible enough to extend it such that it provides a low deceleration (say, less than 3g) to the UAV structure.

Example 4.3

A UAV with a mass of 25 kg and a velocity of 6 m/s is recovered via a vertical net. During recovery, the net is extended 0.8 m.

(a) Determine the deceleration in terms of g. (b) What force is applied on the net?

Solution:
(a)

$$-V_1^2 = 2aS \rightarrow a = \frac{-V_1^2}{2S} = \frac{-\left(6^2\right)}{2\times 0.8} = -22.5\frac{\text{m}}{\text{s}^2} \qquad \text{from (4.44)}$$

Dividing this deceleration by 1 g yields

$$a = \frac{-22.5}{9.81} = -2.3\,g$$

(b)

$$\Sigma F = ma = 25\times\left(-22.5\right) = -562.5\,N \qquad \text{from (4.43)}$$

Questions

1) What do FAR, ROC, W/S, and L/D stand for?
2) What is the subject of flight dynamics about?
3) Write two broad areas of flight dynamics.
4) Draw a figure to demonstrate main phases in a typical flight mission of a UAV. Name main phases.
5) What primary forces are active during a level cruising flight? Show them in a figure.
6) Write two basic governing equations (along two axes) for a level cruising flight.
7) What is the minimum speed at which a fixed-wing airplane can still fly?
8) What are typical values for the maximum lift coefficient.
9) Write cruise speed for: (a) the Northrop Grumman RQ-4 Global Hawk, (b) the General Atomics MQ-9 Reaper, and (c) the AeroVironment RQ-11 Raven.
10) Based on FAR Part 107, what is the maximum ground speed that a small unmanned aircraft may have?
11) Define range.

12) Write range for: (a) the Northrop Grumman RQ-4 Global Hawk, (b) the General Atomics MQ-9 Reaper, and (c) the AeroVironment RQ-11 Raven.
13) Write primary factors that impact range of a fixed-wing non-electric-engine propeller-driven aircraft.
14) Write primary factors that impact the range of a fixed-wing jet-driven aircraft.
15) Write a Breguet range equation for a fixed-wing jet-driven aircraft.
16) Write a Breguet range equation for a fixed-wing non-electric-engine propeller-driven aircraft.
17) Define endurance.
18) Write primary factors that impact endurance of a fixed-wing non-electric-engine propeller-driven aircraft.
19) Write primary factors that impact endurance of a fixed-wing jet-driven aircraft.
20) What is the typical value of specific fuel consumption for a propeller-driven non-electric-engine aircraft?
21) Write endurance for: (a) the Northrop Grumman RQ-4 Global Hawk, (b) the General Atomics MQ-9 Reaper, and (c) the AeroVironment RQ-11 Raven.
22) Define specific fuel consumption for: (a) a propeller-driven aircraft and (b) a jet-propelled aircraft.
23) What is the unit of specific fuel consumption for: (a) a non-electric-engine propeller-driven aircraft and (b) a jet-propelled aircraft?
24) A jet-propelled UAV is desired to have the maximum range in a mission. What should be the velocity for this goal?
25) A prop-driven UAV is desired to have the maximum range in a mission. What should be the velocity for this goal?
26) Write three approaches to maintain $L = W$ during a long cruising flight.
27) What is a "cruise-climb" flight? Explain.
28) What is the best altitude to fly at which the endurance is maximized for a non-electric-engine propeller-driven aircraft?
29) Write four ways to increase range for a non-electric-engine propeller-driven aircraft.
30) Write four ways to increase range for a jet-driven aircraft.
31) Write four ways to increase endurance for a non-electric-engine propeller-driven aircraft.
32) Write four ways to increase endurance for a jet-driven aircraft.
33) For a non-electric-engine propeller-driven aircraft, compare the velocity for maximum endurance with the velocity for maximum range.
34) Define climb angle.
35) Define available power and required power.
36) Write two basic governing equations (along two axes) for a climbing flight.
37) Write the relation between propeller engine power (delivered at the shaft) and available power for a propeller-driven aircraft.
38) Write typical units for rate of climb.
39) From a "power versus velocity" chart, discuss when the maximum rate of climb takes place.
40) What is the relation between the available power and the required power at the maximum speed (V_{max})?
41) Define gliding flight.
42) Write two basic governing equations (along two axes) for a gliding flight.
43) Briefly describe the launch process.
44) Briefly describe the recovery process.
45) What forces are mainly active during a launch operation?

Questions

46) What forces are mainly active during a recovery operation?
47) Briefly compare launch with recovery from two aspects: (1) velocity and (2) acceleration.
48) What are the two important features of a net/cable for recovery of a UAV?
49) What are typical values for the friction coefficient between two metallic surfaces?
50) A MALE UAV has a zero-lift drag coefficient of 0.026, a wing aspect ratio of 12, and an Oswald efficiency factor of 0.84. Determine the maximum value of L/D.
51) A HALE UAV has a zero-lift drag coefficient of 0.021, a wing aspect ratio of 16, and an Oswald efficiency factor of 0.92. Determine the maximum value of L/D.
52) The turboprop engine of a MALE UAV is generating 900 hp of shaft power. If the propeller efficiency is 0.75, determine the available power.
53) An electric engine of a small UAV is generating 400 W of shaft power. If the propeller efficiency is 0.72, determine the available power.
54) A MALE UAV with a mass of 3,200 kg is equipped with a turboprop engine. The UAV – which is climbing with an airspeed of 110 knots – is producing 7,400 N of drag. In this flight condition, the engine is generating 800 kW of shaft power, while the propeller efficiency is 72%. Determine the rate of climb.
55) A small UAV with a mass of 2 kg is equipped with an electric engine. The UAV – which is climbing with an airspeed of 12 m/s – is producing 6 N of drag. In this flight condition, the engine is generating 180 W of shaft power, while the propeller efficiency is 68%. Determine the rate of climb.
56) A small fixed-wing UAV with a mass of 15 kg and a velocity of 9 m/s is recovered via a vertical net. During recovery, the net is extended 1.4 m. (a) Determine the deceleration in terms of g. (b) What force is applied on the net?
57) A small fixed-wing UAV with a mass of 7 kg and a velocity of 5 m/s is recovered via a vertical net. During recovery, the net is extended 0.6 m. (a) Determine the deceleration in terms of g. (b) What force is applied on the net?

5

Flight Stability and Control

5.1 Overview

The UAV flight safety is a function of several parameters including trim, stability, and control. An air vehicle must be stable if it is to remain in flight. Moreover, it must be controllable as well as trimmable. Each UAV must meet the trim, stability, and control requirements that are developed by the customer and the federal government. Three concepts of trim, stability, and control are so inter-related that one must make sure all are met concurrently. For instance, the horizontal tail is designed based on the trim requirements, but was later revised based on stability and control requirements.

It is necessary to define suitable flight dynamic quantities to describe the flight safety requirements. This chapter is devoted to present the definition, fundamental parameters, basic governing equations, and design requirements for trim, stability, and control. In this chapter, we also discuss the basic concepts of stability and the associated area of use of automatic control systems (autopilots) to fly the air vehicle. The equations provided in this chapter describe the behavior (i.e., dynamics) of the air vehicle to any input (either pilot command or undesired input such as gust).

Trim is defined as the aircraft condition when the sum of all forces along each axis (x, y, and z) and the sum of all moments about the aircraft center of gravity about all axes are zero. Each flight condition with such a feature is referred to as the trim point. An aircraft must be trimmable at all flight conditions inside the flight envelope.

Stability is the tendency of an aircraft to return to its initial equilibrium/trim flight condition, if disturbed. This feature allows the aircraft to remain in its present state of rest or motion despite small disturbances. An aircraft must be stable at all flight conditions within the flight envelope.

Control is defined as an operation to change the flight condition from the current trim point to a new trim point. An aircraft must be controllable at all flight conditions within the flight envelope. The ground pilot must make sure not to push an air vehicle out of the flight envelope.

The flight regime of an aircraft usually includes all permissible combinations of airspeeds, altitudes, weights, centers of gravity, and configurations. This regime is primarily shaped by aerodynamics, propulsion, structure, and dynamics of the aircraft. The borders of this flight regime are referred to as the flight envelope. As the flight is within the boundaries of the published flight envelope, the safety of the flight is guaranteed by the aircraft designer and manufacturer.

Pilots are always trained and warned through flight instruction manuals not to fly out of the flight envelope, since the aircraft is either not stable, not controllable, or not structurally strong

---

Introduction to UAV Systems, Fifth Edition. Paul Gerin Fahlstrom, Thomas James Gleason, and Mohammad H. Sadraey.
© 2022 John Wiley & Sons, Inc. Published 2022 by John Wiley & Sons, Inc.
Companion website: www.wiley.com/go/fahlstrom/uavsystems5e

enough outside the boundaries of the flight envelope. A mishap or a crash can be expected if an aircraft is flown outside the flight envelope.

The flight envelope has various types, each of which is usually the allowable variations of one flight parameter versus another flight parameter. These envelopes are calculated and plotted by flight dynamics engineers and employed by ground pilots.

5.2 Trim

Trim is one of the basic requirements of a UAV safe flight. The terms trim, balance, and equilibrium are used interchangeably in the literature. Trim is defined as the aircraft condition where the sum of all forces along each axis (x, y, and z) and the sum of all moments about each axis are zero. Non-aerodynamic moments (e.g., thrust) are measured with respect to the aircraft center of gravity.

Each flight condition with such features is referred to as the trim point. Trim point is the condition at which all flight parameters of the UAV have steady state values (e.g., airspeed (V), altitude (h), and angle of attack (α)). There are an infinite number of trim points for an air vehicle.

An aircraft must be trimmable at all flight conditions inside the *flight envelope*. The flight envelope encompasses a number of figures; each figure is represented by allowable variations of flight parameters (e.g., UAV weight, center of gravity, altitude, airspeed, and load factor) and control parameters (e.g., elevator, aileron, rudder deflections, and engine throttle). Each flight/control parameter has a minimum value and a maximum value. These values are always published by the UAV designer/manufacturer and must be available to the UAV pilot. The flight safety is guaranteed if the UAV is flown inside the flight envelope. Figure 5.1 illustrates a typical flight envelope; it exhibits allowable variations of center of gravity along the x-axis and UAV weight. This figure is often called the "weight and balance" diagram.

When an UAV is at trim, the aircraft will not rotate about its center of gravity (cg) and will either keep moving in a desired direction (with a constant speed) or will move in a desired circular motion (with a constant angular speed). When the summations of all forces (along all three axes) and moments (about all three axes) are zero, the UAV is in the trim condition, where

$$\sum F = 0 \tag{5.1}$$

$$\sum M = 0 \tag{5.2}$$

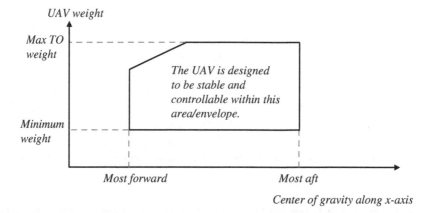

Figure 5.1 A typical flight envelope

5.2 Trim

Any moving object (here, UAV) has six degrees of freedom (6-DOF): (1) three linear motions along the x, y, and z axes and (2) three angular motions about the x, y, and z axes (roll, pitch, and yaw respectively). The air vehicle trim/equilibrium must be maintained along and about three axes (x, y, and z): (1) lateral axis (x), (2) longitudinal axis (y), and (3) directional axis (z). These three trim conditions are introduced in the following three sections.

5.2.1 Longitudinal Trim

When the summation of all forces along the x-axis (such as drag and thrust) is zero and the summation of all forces in the z direction (such as lift and weight) is zero, and the summation of all moments – including the aerodynamic pitching moment – about the y-axis is zero, the aircraft is said to have a longitudinal trim:

$$\sum F_x = 0 \tag{5.3}$$

$$\sum F_y = 0 \tag{5.4}$$

$$\sum M_{cg} = 0 \tag{5.5}$$

In another term, the sum of all forces in the xz plane should be zero. In a fixed-wing UAV, the horizontal tail (including the elevator) is mainly responsible for maintaining longitudinal trim and making the summations zero, by generating a necessary horizontal tail lift and contributing to the summation of moments about the y-axis. The horizontal tail can be placed at the rear fuselage or close to the fuselage nose. The first one is called the conventional tail or aft tail, while the second one is referred to as the first tail, foreplane, or canard.

The longitudinal trim in a conventional fixed-wing aircraft is provided mainly through the horizontal tail. To support the longitudinal trimability of the aircraft, conventional aircraft employ an elevator, which is part of the horizontal tail.

Consider the side view of a fixed-wing air vehicle in Figure 5.2 that is shown in longitudinal trim. The figure depicts the aircraft when the aircraft center of gravity (cg) is behind the wing-fuselage aerodynamic center (ac_{wf}). There are a number of moments about the y-axis (i.e., CG or cg) that must be balanced by the moment of the horizontal tail's lift: (1) the wing-fuselage aerodynamic pitching moment, (2) the moment of lift about the aircraft's center of gravity, (3) the engine thrust moment, and (4) the UAV drag moment.

By the application of the trim equation, and sum the moments about the center of gravity, we obtain to the following equation:

$$\sum M_{cg} = 0 \Rightarrow M_{owf} + M_L + M_{L_h} + M_D + M_T = 0 \tag{5.6}$$

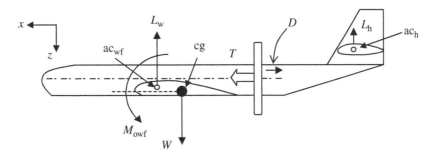

Figure 5.2 Forces and moments in a fixed-wing UAV in the xz plane

In deriving this equation, a few minor moments – including the moment tail drag, D_t – are neglected for simplicity. There are mainly five longitudinal moments about the aircraft center of gravity: (1) wing-fuselage aerodynamic pitching moment (M_{owf} or M_{ac}, simply M), (2) wing-fuselage lift moment (M_{Lwf}), (3) UAV drag moment (M_D), (4) horizontal tail lift moment (M_{Lwf}), and (5) engine thrust's longitudinal moment (M_T).

As discussed in Chapter 3, the aerodynamic pitching moment is the third natural outcome of any lifting surface (see Figure 3.2). To follow the format of lift and drag, this moment is modelled as the product of dynamic pressure, a reference area, and a coefficient, plus wing chord, c. This distance is conventionally chosen as the moment arm:

$$M = \frac{1}{2}\rho V^2 S C C_m \tag{5.7}$$

where C_m is the aircraft pitching moment coefficient and C denotes the wing mean aerodynamic chord. The three most important contributors to this aerodynamic moment are horizontal tail, wing, and fuselage.

Knowledge of the pitching moment is critical to the understanding of longitudinal stability and control. The aircraft pitching moment coefficient (C_m) is a function of a number of factors including the pitching moment coefficient of the wing airfoil cross-section (c_m). Both of these aerodynamic coefficients (C_m and c_m) are of primary interest to the UAV designer.

The two-dimensional moment coefficient (c_m) is simply extracted from airfoil graphs. For instance, the value of c_m for NACA airfoil 23021 (see Figure 3.6) at low angles of attack (from 0 to 5 degrees) is about zero, while at higher angles of attack (from 5 to 12 degrees) it is about 0.01. Determining the three-dimensional moment coefficient (C_m) is a complex procedure that must take into account the contributions from all parts of the aircraft. Reference [16] provides technique, equations, and figures to determine C_m for various aircraft configurations.

Figure 5.3 is a simplified longitudinal moment balance diagram of the aerodynamic forces acting on the aircraft. Summing these moments about the aircraft center of gravity (CG) results in

$$M_{CG} = L x_a + D z_a + M_{ac} - T z_e - L_t x_t \quad \left(\text{if } D_t = 0\right) \tag{5.8}$$

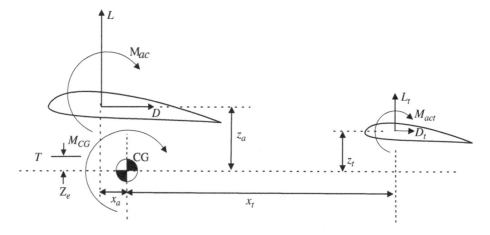

Figure 5.3 Longitudinal moment balance diagram

where T and Tz_e are the engine thrust and thrust pitching moment respectively. Note that the last two terms ($L_t x_t$ and Tz_e) have a negative sign, since they are counterclockwise. Dividing Equation (5.8) by $\frac{1}{2}\rho V^2 SC$ (see Equation (5.7)), the non-dimensional three-dimensional pitching moment coefficient about the CG is obtained:

$$C_{M_{CG}} = C_L\left(\frac{x_a}{c}\right) + C_D\left(\frac{z_a}{c}\right) + C_{m-ac} - \frac{Tz_e}{qSc} - C_{Lt}\left(\frac{S_t}{S}\right)\left(\frac{x_t}{c}\right) \tag{5.9}$$

where S_t denotes the area of the horizontal tail surface and S the area of the wing. Pitching moment, the torque about the aircraft center of gravity, has a profound effect on the pitch control and longitudinal stability of the air vehicle. A negative pitching moment coefficient is required to maintain stability and is obtained primarily from the tail (the last term in the equation). Any flight condition where the $C_{M_{CG}}$ is equal to zero is assumed as a longitudinal trim point.

5.2.2 Directional Trim

When the summation of all forces along the y-axis (e.g., centrifugal force and side force) is zero; and the summation of all moments including aerodynamic yawing moment about the z-axis (i.e., N) is zero, the aircraft is said to have the directional trim:

$$\sum F_y = 0 \tag{5.10}$$

$$\sum N_{cg} = 0 \tag{5.11}$$

In a fixed-wing UAV, the vertical tail (including the rudder) is mainly responsible to maintain directional trim and to make the summations zero, by generating a necessary vertical tail lift (to the left or right). The aerodynamic yawing moment is modelled as:

$$N = \frac{1}{2}\rho V^2 SbC_n \tag{5.12}$$

where C_n is the aircraft yawing moment coefficient and b denotes the wing span. The most important contributor to this aerodynamic moment is the vertical tail (through the rudder).

5.2.3 Lateral Trim

When the summation of all forces in the z direction (such as lift, weight) is zero and the summation of all moments including the aerodynamic rolling moment about the x-axis is zero, the aircraft is said to have a lateral trim:

$$\sum F_z = 0 \tag{5.13}$$

$$\sum L_{cg} = 0 \tag{5.14}$$

The aerodynamic rolling moment is modelled as

$$L = \frac{1}{2}\rho V^2 SbC_l \tag{5.15}$$

where C_l is the aircraft rolling moment coefficient. In a fixed-wing UAV, the wing (through the aileron) is mainly responsible for maintaining lateral trim and making the summation of moment

5.2.4 Summary

Trim is the equilibrium/balance of forces and moments in a flight operation. All three longitudinal, lateral, and directional trims are required throughout a flight operation. Both stability and control are directly related and are impacted by trim. Before a pilot controls (makes any changes to) the UAV, the air vehicle must already be in trim condition. The flight stability can be interpreted as "marinating a trim condition." Longitudinal trim is mainly created by the horizontal tail and directional trim is primarily created by the vertical tail.

In a fixed-wing UAV, longitudinal trim (in the x–z plane) is almost independent of lateral (in the y–z plane) and directional (in the x–y plane) trim. Thus, longitudinal trim can be mathematically modeled and analyzed without any refence to lateral and directional trim. However, lateral and directional trim are highly coupled. Any change in the lateral trim will impact the directional trim, and any change in the directional trim will impact the lateral trim. Therefore, lateral–directional trim is usually maintained simultaneously.

5.3 Stability

Stability refers to the tendency of an object (here, UAV) to oppose any disturbance, and to return to its present state of rest or motion, if disturbed. We categorize the concept of stability in two modes and can be treated in three axes independently. In terms of modes, there are: (1) static stability and (2) dynamic stability. Static stability refers to the tendency of an object (here, UAV) to oppose any disturbance, while dynamic stability is to return to the present state of rest or motion, if disturbed.

Stability implies that the forces acting on the airplane (thrust, weight, and aerodynamic forces) are in directions, and will gain new values, that tend to restore the airframe to its original equilibrium position after it has been disturbed (by a wind gust or other forces). The stability is primarily provided by the air vehicle configuration. When the air vehicle configuration is changed, its stability characteristics will vary. In general, fixed-wing air vehicles are often stable (e.g., AeroVironment RQ-11 Raven), while VTOL UAVs including quadcopters (e.g., DJI Phantom) are all always unstable. In VTOL UAVs, the stability is often artificially provided by the automatic flight control systems.

For example, a pendulum hanging down on a hinge can be impacted and will, at most, rock back and forth a few times and then settle back to its original state. This occurs because the forces on the pendulum generated by a small tilt oppose the increase in tilt and tend to return the pendulum to its original state. This is called "positive stability" or simply "stability".

In theory, one could balance the same pendulum upright on the hinge (i.e., inverted pendulum). However, even a very slight disturbance would cause it to fall on its side, because as soon as the center of gravity ceased to be directly above the point where the pendulum on the hinge contacted the level surface, the net forces on the pendulum would increase the tilt, which would increase the forces, and so on. This is referred to as "negative stability" or instability.

With positive stability, an increasing disturbance in the state of the object generates increasing restoring moments/forces, while with negative stability, an increasing disturbance in the state generates increasing disturbing moments/forces that can "blow up" into a catastrophe.

For military and highly maneuverable UAVs (e.g., Boeing X-5 and BAE Raven), negative stability can be acceptable and even desirable. This is illustrated by a bicycle, which has only a small tolerance of disturbance before it will fall over unless the rider shifts his/her weight slightly to

5.3 Stability 95

correct the effects of the disturbance. Negative stability often coincides with very high maneuverability and can be desirable in some situations. However, negative stability requires a control system that can function at a high enough bandwidth to correct any small disturbances before the disturbing feedback forces become uncontrollable.

For aircraft, it has only been in the last few decades that automatic control systems have become capable enough and reliable enough that designers have been willing to design aircraft that have negative stability when operating within their design flight envelope.

If the air vehicle is not statically stable, the smallest disturbance will cause ever-increasing deviations from the original flight state. A statically stable airplane will have the "tendency" to return to its original position after a disturbance, but it may overshoot, turn around, go in the opposite direction, overshoot again, and eventually oscillate to destruction. In this case, the airplane would be statically stable but dynamically unstable. If the oscillations are damped and eventually die out, then the air vehicle is said to be dynamically stable.

In terms of the axis, we have: (1) lateral stability (about the x-axis), (2) longitudinal stability (about the y-axis), and (3) directional stability (about the z-axis). An air vehicle is desired to be stable in both modes and about all three axes (x, y, and z). The pitch axis is the most critical, and stability about it is called longitudinal stability. Some instability can be tolerated about the roll and yaw axes, which are combined in most analysis and called lateral stability. The concept of stability in two modes and three axes are introduced in the following sections.

5.3.1 Longitudinal Static Stability

Longitudinal stability refers to the tendency of a UAV to oppose any longitudinal disturbance, and to return to its present longitudinal state of rest or motion, if disturbed in the xz plane. Longitudinal static stability is defined as the tendency of an UAV to oppose any longitudinal disturbance. In a fixed-wing air vehicle, the longitudinal stability is mainly provided by the horizontal tail.

To have longitudinal static stability, an aircraft needs to develop a restoring pitching moment, when it is displaced from its equilibrium point. The factors that affect longitudinal stability can be determined by referring to Figure 5.3, which shows the balance of forces on the air vehicle. A restoring pitching moment is the one UAV input to oppose any longitudinal disturbance.

The pitching moment coefficient is a good representative of the pitching moment (Equation (5.7)). The pitching moment coefficient is a function of the lift coefficient, and this fact is used to evaluate the static stability of an airplane.

The most important flight parameters in the longitudinal plane are: (1) angle of attack (α), (2) airspeed (V), (3) altitude (h), (4) pitch angle (θ), and (5) climb angle (γ). In a longitudinally stati-cally stable aircraft, these parameters are expected to be maintained if the UA is disturbed. All these parameters are inter-related, but they can be represented by the most sensitive one, which is the angle of attack.

When the UAV angle of attack is increased, a negative (i.e., nose down) pitching moment is desired to oppose the disturbance and restore the trim point. In addition, when the UAV angle of attack is decreased, a positive (i.e., nose up) pitching moment is desired. This stability requirement can be visualized as the rate of change of the pitching moment coefficient (C_m) with respect to the angle of attack (α), which is modelled as a new parameter C_{m_α}. Thus, the requirement for longitu-dinal static stability is to have a negative value for this parameter:

$$C_{m_\alpha} = \frac{dC_m}{d\alpha} < 0 \tag{5.16}$$

The higher the negative value, the more the aircraft will be longitudinally statically stable. Typical values for C_{m_α} of a longitudinally statically stable aircraft can range from −1 to −3 1/rad. There are more than 30 parameters like this, called "stability derivatives," but they are specialized variables that influence the dynamic characteristics of the air vehicle and their discussion is beyond the scope of this text.

Since the angle of attack is directly a function of the lift coefficient, we can see the same feature (i.e., negative slope) in analyzing longitudinal static stability. If the UAV pitching moment coefficient is plotted against the lift coefficient, Figure 5.4 results. The UAV which is represented in this figure has a zero C_m at a lift coefficient of 1. This is just one trim point.

Using this plot, one can reason as follows: if a disturbance (e.g., a gust hits under the fuselage nose that increases C_L) causes the nose to rise and the restoring moment causes the nose to fall (e.g., the change in pitching moment is negative), the air vehicle will tend to restore itself to its original condition. If the pitching moment causes the nose to rise further after a nose-up disturbance, then the airplane will continue to pitch up and is statically unstable.

Mathematically, a stable system must have a pitching moment versus lift coefficient curve with a negative slope, as shown above. The question is, how can the air vehicle be made to behave in this manner? Each of the terms in the longitudinal equilibrium equation (Equation (5.9)) contributes to the pitching moment in either a negative or a positive sense. The contribution of the horizontal tail (the last term in the equation) is of major importance because of its minus sign and the large value that can be obtained by designing in a large x_t and S_t. A large tail area located a long distance aft of the air-vehicle center of gravity is a powerful stabilizer.

A plot of the pitching moment coefficient versus the lift coefficient for each of the contributors is shown in Figure 5.5. Reference [16] has shown that the slope of C_m versus C_L is a function of the relation between locations of the aircraft center of gravity (cg) and the aircraft neutral point (np or NP). It is obtained as

$$\frac{C_m}{C_L} = x_{cg} - x_{np} \tag{5.17}$$

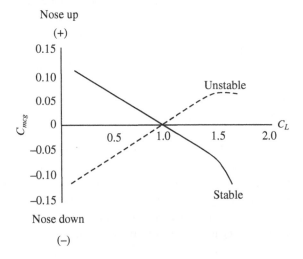

Figure 5.4 Pitching moment coefficient versus lift coefficient

5.3 Stability

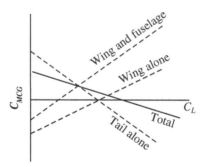

Figure 5.5 Contributors to pitching moment

From this relation, for any flying object, whether an arrow or a complete aircraft, the center of gravity (cg) must be ahead of the aircraft aerodynamic center (also called the neutral point) for the pitching moment–lift curve to remain negative, a condition for static longitudinal stability, as previously mentioned. Adding surfaces that produce lift and drag, such as a horizontal stabilizer, behind the CG has the effect of moving the overall airplane neutral point rearward, increasing stability. Techniques to determine C_{m_a} and the neutral point are beyond the scope of this book, but are presented in Reference [16].

The distance between the NP and CG has a profound effect on the stability of the air vehicle. Air vehicles that have a small distance between the CG and the NP are less stable than those with large separations. It is necessary for the CG to be forward of the NP for longitudinal stability. The horizontal tail is an important lifting/control surface for both stability and the ability to control the vehicle. A larger distance of the tail aft of the wing results in greater control and stability. This is generally true but placing horizontal surfaces ahead of the CG (such as canards) can result in snappy control, but at the expense of stability.

5.3.2 Directional Static Stability

Directional (or weathercock) stability refers to the tendency of a UAV to oppose any directional disturbance (e.g., a gust to the left fuselage nose), and to return to its present directional state of rest or motion, if disturbed in the *xy* plane. Directional static stability is defined as the tendency of an UAV to oppose any directional disturbance (e.g., yawing disturbance). In a fixed-wing air vehicle, the directional stability is mainly provided by the vertical tail. To have directional static stability, an aircraft needs to develop a restoring yawing moment (N) when it is displaced from its directional equilibrium point.

The most important flight parameter in the directional plane is the sideslip angle (β). In a cruising flight, the sideslip angle is usually zero. Moreover, in a turning flight, a small sideslip angle is generated to keep the turn coordinated. The zero sideslip angle in cruise and a small value for sideslip angle in a turn are part of the directional trim condition.

If a positive sideslip angle is generated by a directional disturbance (a gust to the right of the fuselage nose or a gust to the left of the vertical tail), a positive (i.e., nose to the left or vertical tail to the right) yawing moment (see Figure 5.6) is desired to oppose the disturbance and restore the trim point (i.e., $\beta = 0$). In addition, when a negative sideslip angle is generated by a directional disturbance, a negative (i.e., nose to the right) yawing moment is desired to oppose the disturbance and restore the trim point (i.e., $\beta = 0$). The yawing moment coefficient (C_n) is a good representative of the yawing moment (Equation (5.12)).

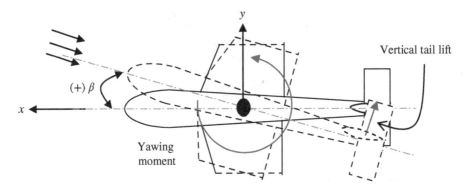

Figure 5.6 Directional stability (top view)

This static stability requirement can be visualized as the rate of change of the yawing moment coefficient with respect to the sideslip angle (β), which is modelled as a new parameter (stability derivative), C_{n_β}. Thus, the requirement for *directional static stability* is just to have a positive value for this parameter:

$$C_{n_\beta} = \frac{dC_n}{d\beta} > 0 \tag{5.18}$$

The higher the value, the more the aircraft will be directionally statically stable. C_{n_β} is the variation of the aircraft yawing moment coefficient with the dimensionless rate of change of angle of sideslip. A typical value for C_{n_β} of a directionally statically stable aircraft can range from +0.1 to +0.4 1/rad. The C_{n_β} is another non-dimensional stability derivative, it is often referred to as the directional static stability derivative.

Yaw (or directional) stability is easy to obtain by incorporating the proper amount of vertical fin or stabilizer area with a desired yaw moment arm. The mathematics of yaw analysis is similar to the pitch case, except that the wing contributes almost nothing to directional stability. The fuselage and vertical tail surfaces are the two major contributors.

Unlike the case of pitch, the yawing moment coefficient versus yaw angle must have a *positive* slope for stability. The reasoning is identical to that previously used in the pitch case; the vertical fin must be able to create a restoring moment that minimizes the yaw angle caused by a side force disturbance. A typical yawing moment coefficient versus sideslip angle is shown in Figure 5.7. Techniques to determine C_{n_β} and corresponding equations are beyond the scope of this book, but are presented in Reference [16].

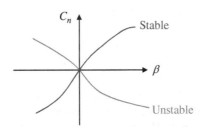

Figure 5.7 Directional stability

5.3 Stability

Note that the directional stability is not the same as the heading stability, with no aircraft features heading stability. An aircraft does not have any memory of its heading angle, which is defined as the difference between the current flight direction and a reference line (e.g., North direction). If a wind gust changes the heading of an aircraft, the aircraft does not have a tendency to return to its initial direction. To keep an aircraft on a desired heading angle, an automatic control system such as autopilot is required.

5.3.3 Lateral Static Stability

Lateral stability refers to the tendency of an UAV to oppose any lateral disturbance (e.g., a gust under a wing tip), and to return to its present lateral state of rest or motion, if disturbed in the yz plane. Lateral static stability is defined as to the tendency of a UAV to oppose any lateral disturbance. Lateral stability is not as critical as directional and longitudinal stability because a slight roll is not as destructive as pitch, which, if uncontrolled for very long, can cause the air vehicle to stall and discontinue flying. In a fixed-wing air vehicle, the lateral stability is mainly provided by the wing dihedral angle (Γ).

The most important flight parameter in the lateral plane (zy) is the bank angle (ϕ). In a cruising flight, the bank angle is usually zero. Moreover, in a turning flight, a desired bank angle is generated to keep the turn coordinated. The zero bank angle in cruise, or a desired value for bank angle in a turn, is part of the lateral trim condition.

Bank angle is defined as the difference between the aircraft xy plane and horizontal. The wing dihedral angle is the angle that a wing section makes with the fuselage (i.e., y axis) as you view from the front (i.e., in the yz plane).

When the aircraft bank angle is disturbed in a level flight, the vehicle begins to roll, which triggers two other undesired outputs: (1) loss of altitude and (2) sideslip angle. To have lateral static stability, an aircraft needs to develop a restoring rolling moment (L) when it is displaced from its lateral equilibrium point. Note that the restoring rolling moment will not resolve the loss of altitude, nor the disturbance in the bank angle, but it will only restore the initial zero sideslip angle (β).

If a positive sideslip angle is generated by a negative rolling disturbance (see Figure 5.8), a positive (i.e., left wing up) rolling moment is desired to oppose the disturbance and restore the trim point (i.e., $\beta = 0$). In addition, when a negative sideslip angle is generated by a positive rolling disturbance, a negative (i.e., right wing up) rolling moment is desired to oppose the disturbance and restore the trim point (i.e., $\beta = 0$). The rolling moment coefficient (C_l) is a good representative of the rolling moment (Equation (5.15)).

The rolling moment due to such sideslip is mainly created by the dihedral angle. An exaggerated picture, Figure 5.8, will help to visualize the concept. The static lateral stability requirement can be

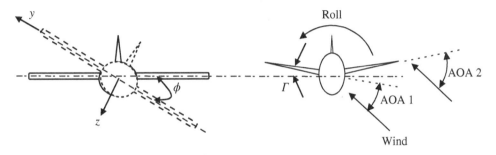

Figure 5.8 Lateral stability (front view)

visualized as the rate of change of the rolling moment coefficient (C_l) with respect to the sideslip angle (β), which is modelled as a new parameter C_{l_β}. Thus, the requirement for *lateral static stability* is to have a negative value for this parameter:

$$C_{l_\beta} = \frac{dC_l}{d\beta} < 0 \tag{5.19}$$

The higher the negative value, the more laterally statically stable will the aircraft be. A typical value for C_{l_β} of a laterally statically stable aircraft can range from– 0.05 to– 0.4 1/rad. The C_{l_β} is another non-dimensional stability derivative and is often referred to as the aircraft "dihedral effect" or "lateral static stability derivative." Techniques to determine C_{l_β} and corresponding equations are beyond the scope of this book, but are presented in Reference [16].

If the wings have a positive dihedral angle, wind impinging on a sideslipping air vehicle will create a greater induced angle of attack (AOA) on the downwind wing than the upwind wing. This causes the downwind wing to have greater lift than the upwind wing and causes it to roll in a direction so as to reduce the sideslip and hence create stability. The vertical location of the wing on the fuselage also creates roll stability, but the dominant factor is wing dihedral. The induced angle of attack is equal to the ratio of a normal component of the side velocity to the airspeed due to sideslip (i.e., $\Delta\alpha = \tan^{-1}(V_n/V)) \approx V_n/V$.

Note that the lateral stability is not the same as the roll stability, as an aircraft does not often feature a roll stability. An aircraft does not often have any memory of its bank angle, but it has a memory of its sideslip angle. If a wind gust changes the bank angle of an aircraft, the aircraft does not have a tendency to return to its bank angle. To keep an aircraft on a desired bank angle, an automatic control system such as autopilot is required.

Indeed, the first autopilot was developed in 1912 by Sperry Corporation. Its objective was to make the wing level, so it was called a wing leveler. This is a very helpful flight aid to a remote pilot in a long cruising flight, so he/she can spend a long-time cruise with his/her hands off the control sticks, trusting the wing leveler, as the air vehicle can maintain a level flight. The details of an automatic flight control system and autopilot are further discussed in Chapter 9.

5.3.4 Dynamic Stability

Dynamic stability is defined as the return to the present state of rest or motion if the vehicle is disturbed. To obtain dynamic stability, restoring forces/moments must have the capability of absorbing energy from the system. Dynamic stability is created by forces/moments that are proportional to the rate (velocity) of motion of the various surfaces, such as wing, tail, and fuselage, with the proportionality constant called a stability derivative. Three most important dynamic non-dimensional stability derivatives are: (1) C_{m_q} for the longitudinal plane, (2) C_{n_r} for the directional plane, and (3) C_{l_p} for the lateral plane. The variables p, q, and r are roll rate, pitch rate, and yaw rate respectively.

The stability derivative, when multiplied by the angular velocity of the vehicle, results in a force that usually reduces the angular velocity of the vehicle (i.e., absorbs energy). This phenomenon is called damping, which is a kind of friction. Because of the natural occurrence of friction in real systems, dynamic stability is usually, but not always, present if the system is statically stable.

Dynamic stability can be augmented by artificial means, such as an autopilot using feedback, which are used to control the air vehicle. A very popular stability augmenter for a long cruising flight is the yaw damper, which augments the directional stability.

However, dynamic instability may arise if the feedback in the control system is improperly designed or compensated and adds energy to the system that is out of phase with the forces that are acting to correct divergent motion. This can lead to an amplification of the instability instead of damping.

An analysis of dynamic stability is much more complex than static stability and requires a lot of calculations. In summary, one must derive values for all stability derivatives and then write longitudinal and lateral/directional characteristic equations (usually each of fourth order). If the roots of these equations all have negative real parts, the air vehicle is said to be dynamically stable. Two aircraft response modes in analyzing longitudinal dynamic stability are: (1) short-period mode and (2) long-period or Phugoid mode. Three aircraft response modes used in analyzing lateral/directional dynamic stability are: (1) roll mode, (2) spiral mode, and (3) Dutch roll. If all these response modes are damped, the air vehicle is dynamically stable.

The Dutch roll is a second-order oscillatory mode, which is a combination of yawing and side slipping. This mode is represented by a pair of complex conjugate roots. The spiral mode is a first order non-oscillatory mode that is the change in yawing angle with time. It has a long time constant. This mode is represented by a real root. The rooll mode is a first order non-oscillatory mode that is the change in bank angle with time. This mode is represented by a real root and has a short time constant.

5.4 Control

5.4.1 Aerodynamic Control

Control is defined as changing the flight condition from one trim condition (say cruise with an initial angle of attack) to another trim condition (say cruise with a new angle of attack). An aircraft must be controllable at all flight conditions within the flight envelope. In a fixed-wing aircraft, there are four longitudinal/lateral/directional control tools: (1) Elevator, (2) Aileron, (3) Rudder, and (4) Engine throttle. The first three control surfaces are components for aerodynamic control, since they create aerodynamic forces and aerodynamic moments. The forces acting on an aircraft in flight consist of aerodynamic, thrust, and gravitational forces. Any change in any of these forces will remove the trim condition and can be employed as a method to control the aircraft. For instance, while fuel is burning during flight, the control is needed to maintain the flight status (e.g., keep altitude or angle of attack).

Three aerodynamic forces are: (1) Lift, (2) Drag, and (3) Side force. The aerodynamic control of an air vehicle about the pitch, roll, and yaw axes is accomplished by three control surfaces (Figure 5.9) deflections: (1) Elevator (δ_E), (2) Aileron (δ_A), and (3) Rudder (δ_R) respectively. For this

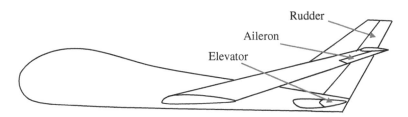

Figure 5.9 Control surfaces of a fixed-wing UAV

objective, aerodynamic pitching, rolling, and yawing moments are created by a combination of lifting surfaces (mainly wing and tails) and control surface deflections.

An UAV has six degrees of freedom (i.e., three linear motions along x, y, and z and three angular motions about x, y, and z). Thus, there are normally six outputs in an actual flight; six examples are three linear velocities (u, v, w) and three angular rates (p, q, r).

In a manned aircraft, when performed by an onboard pilot flying "heads up," it involves a complex subconscious synthesis of the horizon seen outside the aircraft, the feel of the controls and aircraft, and, literally, the feel of the "seat of the pants." This is the perceived net direction of the force/moment on the pilot's body due to the combination of gravity and the accelerations of the aircraft.

For a UAV, the "feel" of the system via the feedback of the airframe and control surfaces is essentially nonexistent. However, artificial feel could be designed into the ground controls to give the controller some sensation of flying, but most UAVs have autopilots and electronic controls without artificial feel. The forces/moments created by the control surfaces are not fed back to the operator. Nevertheless, they must be analyzed so as to determine the proper response of the airframe and to determine the size of the actuators.

All of the various flight conditions that a UAV may encounter must be investigated so that the control surfaces can be designed to the proper size and location. This usually requires a determination of the balance of forces and an integration of Newton's laws for air vehicle moments. A complete dynamic analysis of the motion of the air vehicle caused by deflection of the control surfaces is a much more complex problem, requiring the use of programming, code, and simulation.

In this section, concept, fundamentals, and parameters in the air vehicle control are reviewed. In Chapter 9, techniques, devices, sensors for controlling fixed-wing air vehicle and its payload (including automatic flight control system and autopilot) are presented. Chapter 20 is devoted to fundamentals and tools for control of rotary wing UAVs including quadcopters.

5.4.2 Pitch Control

Any rotational motion control in the x–z plane is called longitudinal control (e.g., pitch about the y-axis, plunging, climbing, cruising, pulling up, and descending). Any change in lift, drag, and pitching moment have the major influence on this motion. The pitch control is assumed as a longitudinal control. Three UAV flight angles controlled in the longitudinal control are: (1) angle of attack, (2) climb angle, and (3) pitch angle. Other flight parameters to be controlled in longitudinal motion include: (1) airspeed, (2) altitude, (3) rate of climb, and (4) Rate of descent. There are primarily two tools in longitudinal control: (1) Elevator and (2) Engine throttle.

In a fixed-wing conventional UAV (e.g., AeroVironment RQ-11 Raven and IAI RQ-5 Hunter), the elevator (see Figure 5.9) is part of the horizontal tail. The elevator located at the trailing edge – is very similar to a flap, but it is deflected either up or down.

The pitching moment (Equation (5.7)) in a UAV is generated by: (1) changing the lift coefficient of the horizontal tail via deflecting the elevator (see Figure 5.10) and (2) changing the throttle setting, which consequently varies the engine thrust. In a pull-up and push-over maneuver, the elevator deflection also determines the acceleration (in g) the airplane can generate and consequently the radius of vertical turn. The aerodynamic pitching moment (M) is primarily a function of the horizontal tail lift (L_{ht}) and the tail arm. Any angular motion that follows the right-hand rule is assumed positive. Thus, this is a negative pitch, since the y-axis is going to the page.

There are special flight conditions that require a specific motion or aerodynamic force/moments from the control surfaces. For instance, the ability of the elevator to pitch the airframe (elevator

5.4 Control

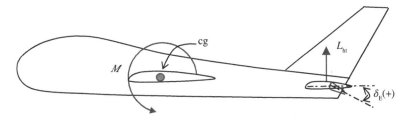

Figure 5.10 Longitudinal control via elevator deflection

control power) depends on the size, shape, moment arm, and air velocity over it. During landing, when the air vehicle is usually flying very slowly, it is necessary that enough elevator effectiveness is available to keep the nose high, so the vehicle generates a sufficient lift.

During a catapult launch, the airspeed also is low (near stall) and if the vehicle is disturbed, there must be enough control to maintain its attitude until the appropriate airspeed is obtained. During a conventional takeoff, the elevator must also be of a size and location to lift the nose wheel off the ground, while the main gear is still on the runway.

A typical variation of aircraft pitching moment coefficient (C_m) versus lift coefficient (C_L) as a function of elevator deflection is shown in Figure 5.11. The slope of C_m/C_L is constant for all flight conditions and can be determined by using Equation (5.17). The negative slope indicates that the aircraft neutral point is behind the aircraft center of gravity (i.e., $x_{cg} < x_{np}$), which implies aircraft static longitudinal stability. When C_m is not zero, it implies that the elevator has created a pitching moment for longitudinal control.

When C_m is zero, the aircraft is at longitudinal trim. For instance, as seen in Figure 5.11, when $C_L = 0.4$ (say $\alpha = 4$ degrees), the elevator is deflected about 4 degrees. In this case, when the elevator deflection is increased to +5 degrees, the aircraft will receive a negative pitching moment. Thus, the aircraft will pitch down, until the lift coefficient decreases to a new value of about 0.3 (say $\alpha = 3$ degrees).

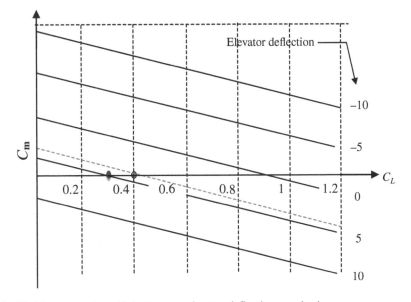

Figure 5.11 Pitching moment coefficient versus elevator deflection – revised

5.4.3 Directional Control

The rotational motion control about the z-axis and any motion along the y-axis is called directional or yaw control (e.g., yaw about the z-axis, side-slipping, and skidding). Any change in side-force and yawing moment (Equation (5.12)) has a major influence on this control. The directional control (in the x–y plane) is performed through a directional control surface or rudder. The yaw control is assumed to be a directional control. In a fixed-wing conventional UAV, the rudder (see Figure 5.9) is part of the vertical tail (e.g., AeroVironment RQ-11 Raven and IAI RQ-5 Hunter). The rudder – located at the trailing edge – is very similar to a flap, but it is deflected either left or right.

In some fixed-wing UAVs (e.g., General Atomics MQ-9 Reaper, AAI RQ-7 Shadow, and Northrop Grumman RQ-4 Global Hawk), the rudder and elevator are combined, and referred to as the ruddervator. In these UAVs, the traditional horizontal and vertical tails are in the V-tail unconventional arrangement. The ruddervator will function as an elevator when both pieces are deflected to opposite sides, while it will function as a rudder when both pieces are deflected to the same sides.

When the rudder is deflected, a vertical tail lift is created, which consequently generates a yawing moment (see Figure 5.12). However, there is a strong coupling between lateral and directional state variables. Any lateral moment (L) will produce a directional motion (along y). Any directional moment (N) will produce a lateral motion (e.g., ϕ).

Hence, any aileron deflection not only generates a rolling moment, but also creates a yawing moment. Moreover, any rudder deflection not only generates a yawing moment, but also creates a rolling moment. In other words, any deflection in aileron (δ_A) or rudder (δ_R) will generate state variables of β, ϕ, ψ, P, R. The reason lies behind the existence of a lateral moment arm and a directional moment arm at the same time.

A change in the aircraft direction and heading is obtained via a turn using the rudder as well the aileron, since a level turn is a combination of lateral and directional motions. The solution to adjust the relation between outputs of the aileron and rudder is to follow a specific technique. To have a nice turn, the pilot needs to employ both the rudder and aileron; such motion is referred to as the coordinated turn. The governing equations of a coordinated turn is beyond the scope of this book, but can be found in references such as [16].

A coordinated turn is when: (1) the side force is zero and (2) the centrifugal force is equal to the horizontal component of the aircraft lift. This turn features: (1) no slipping and no skidding, (2) a constant radius of turn, (3) a constant turn rate, and (4) fuel is distributed symmetrically.

In a coordinated turn for a fixed-wing conventional UAV, all three control surfaces are simultaneously employed: (1) the aileron to roll for creating the bank angle, (2) the rudder to yaw for creating a small sideslip angle, and (3) an elevator to pitch for increasing the angle of attack in order to maintain the altitude.

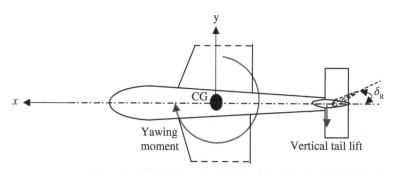

Figure 5.12 Directional control via rudder deflection (top view)

5.4.4 Lateral Control

Any rotational motion control about the x-axis is called lateral control (e.g., roll). Any change in the wing lift distribution and rolling moment (Equation (5.15)) will have a major influence on this motion. In a fixed-wing UAV (e.g., General Atomics MQ-9 Reaper) with a conventional configuration, the lateral motion is executed using aileron deflection. The rolling control is assumed to be a lateral control.

In a fixed-wing conventional UAV, the aileron (see Figure 5.9) is part of the wing (located at the outboard and trailing edge). An aileron is very similar to a flap, but is deflected either up or down. It has two pieces, one on each section of the wing, and they are deflected differentially.

Ailerons – which are located on the outboard of the wing – are designed to create a roll, which is necessary to turn, which requires a change in the bank angle. As previously discussed, a combination of rudder and horizontal stabilizer control is used to obtain a balanced flight in a turn. The aileron deflection also contributes to the load factor and acceleration (g) the airplane can generate and consequently the radius of turn.

When the ailerons are deflected, the wing lift distribution is differentially changed. The lift – on the section (say left, see Figure 5.13) that the aileron has been deflected downward – is increased. Moreover, the lift – on the section (say right) that the aileron has been deflected upward – is decreased. This change in the wing lift distribution will consequently generate a rolling moment. Based on the right-hand rule, this is a positive roll, since the x-axis is coming out of the page.

Since the lift is always perpendicular to the wing, when the aircraft is banking in a turn, the lift is tilted at an angle to the vertical and only the vertical component is available to oppose the downward force due to the weight of the aircraft. As a result of this, the AOA must be increased to increase the total lift, until its vertical component balances the weight. If this is not done, the aircraft will lose altitude during the turn.

To increase the AOA, an up-elevator is applied by the autopilot or ground pilot. Thus, for a proper turn at constant altitude, the rudder is deflected to yaw the aircraft and the "stick" is moved to the side and pulled back to bank the aircraft and increase the AOA. This is to have a "coordinate turn." An unmanned system must behave in the same way that a piloted system would when turning and the "coordination" must be built into its flight control system.

Adverse yaw is a turn control concept that one hears much about. The British term "adverse aileron drag" is a more accurate name and makes the concept easier to understand. When the ailerons are deflected so as to create roll, they create drag that tends to yaw the air vehicle in the opposite direction of the turn caused by rolling. This results in an unbalanced condition that is alleviated by the application of rudder deflection to counteract the yaw. This, of course, is one of the primary reasons for having a rudder. A pilot can sense this unbalance, but a UAV must automate the counter action in its flight control system.

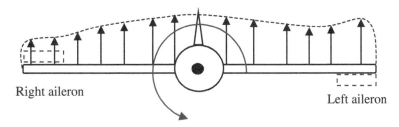

Figure 5.13 Aileron deflection to create a roll (front view)

Questions

1) What do DOF, CG, NP, and AOA stand for?
2) Briefly define trim, control, and stability.
3) What is the flight envelope? Draw a figure to show an example.
4) Define trim point.
5) How many degrees of freedom are there in any moving object (here, UAV)? What are they?
6) Define longitudinal trim.
7) Define lateral trim.
8) Define directional trim.
9) Explain how the longitudinal trim in a conventional fixed-wing aircraft is provided.
10) Explain how the lateral trim in a conventional fixed-wing aircraft is provided.
11) Explain how the directional trim in a conventional fixed-wing aircraft is provided.
12) List moments that are often contributing to the longitudinal trim in a conventional fixed-wing aircraft.
13) Which parameters are contributing to the aerodynamic pitching moment?
14) Which parameters are contributing to the aerodynamic rolling moment?
15) Which parameters are contributing to the aerodynamic yawing moment?
16) What is the difference between C_m and c_m?
17) Write the most important flight parameters in the longitudinal plane.
18) What is the relation between UAV CG and NP for a longitudinally statically stable aircraft?
19) Define static stability.
20) Define dynamic stability.
21) Define static longitudinal stability.
22) What is the requirement for static longitudinal stability?
23) Define static lateral stability.
24) What is the requirement for static lateral stability?
25) Define static directional stability.
26) What is the requirement for static directional stability?
27) Define heading stability.
28) Define roll stability.
29) What is the most important component/parameter in providing longitudinal stability?
30) What is the most important component/parameter in providing lateral stability?
31) What is the most important component/parameter in providing directional stability?
32) Define bank angle.
33) Define heading angle.
34) Define sideslip angle.
35) Define wing dihedral angle.
36) What is the aircraft dihedral effect?
37) Explain how a longitudinal trim condition is restored in a longitudinally stable fixed-wing aircraft, if a longitudinal disturbance is applied. Draw a figure.
38) Explain how a directional trim condition is restored in a directionally stable fixed-wing aircraft, if a directional disturbance is applied. Draw a figure.
39) Explain how a lateral trim condition is restored in a laterally stable fixed-wing aircraft, if a lateral disturbance is applied. Draw a figure.
40) What is the typical value for C_{l_β} of a laterally statically stable aircraft?
41) What is the typical value for C_{n_β} of a directionally statically stable aircraft?

Questions 107

42) What is the typical value for C_{m_α} of a longitudinally statically stable aircraft?
43) When and which company developed the first autopilot?
44) Write two aircraft response modes in analyzing longitudinal dynamic stability.
45) Write three aircraft response modes in analyzing lateral/directional dynamic stability.
46) Write the directional static stability derivative.
47) Write the lateral static stability derivative.
48) How is the induced angle of attack calculated in a UAV with a bank angle?
49) In a conventional fixed-wing UAV, what control surface is deflected (and to which direction) to increase the angle of attack?
50) In a conventional fixed-wing UAV, what control surface is deflected to create a rolling moment in order to increase the bank angle?
51) In a conventional fixed-wing UAV, what control surface is deflected to change the UAV nose to the right?
52) What are the three most important dynamic non-dimensional stability derivatives?
53) What is provided by a yaw damper?
54) Briefly describe the Dutch roll.
55) What are four longitudinal/lateral/directional control tools in a fixed-wing aircraft?
56) List three aerodynamic forces in a UAV.
57) Briefly describe characteristics and function of the elevator.
58) Briefly describe characteristics and function of the aileron.
59) Briefly describe characteristics and function of the ruder.
60) What is longitudinal control?
61) What is lateral control?
62) What is directional control?
63) What are primarily tools in longitudinal control?
64) Briefly describe how there is a strong coupling between lateral and directional state variables.
65) For a fixed-wing conventional UAV, briefly describe how all three control surfaces are employed in a turn to make it coordinated.
66) Draw an aircraft to indicate three axes and three moments (include names) with their positive directions.
67) Compare longitudinal control with lateral control; provide main differences.
68) What three UAV flight angles are controlled in the longitudinal control?
69) When ailerons are deflected, briefly describe what happens to the wing lift distribution.
70) What is ruddervator? Explain how it can function as an elevator and a rudder.
71) Write three UAVs that are equipped with a ruddervator.
72) Write two UAVs that are equipped with a rudder, elevator, and aileron.
73) What is adverse yaw? Explain.

6

Propulsion

6.1 Overview

There are mainly four forces acting on an air vehicle in a cruising flight: (1) weight, (2) lift, (3) drag, and (4) thrust (see Figure 3.1). The propulsion system is a vital component to the operation of a UAV; its primary function is to generate a propulsive force or thrust. Two aspects of propulsion are addressed in this chapter. The first is the aerodynamics of generating thrust or what is called "powered lift," which is lift that is directly generated by the propeller or fan and is very similar to upward thrust, but uses somewhat different terminology for historical reasons.

The second aspect of propulsion is the source of the power used to produce the thrust or lift, which is the engine or motor that moves a propeller or rotor or fan or generates a high-speed jet of exhaust gasses.

In this section, both the traditional internal combustion and turbine engines are addressed, as well as the electric propulsion that is almost unique to the UAV world and is becoming more common in both the mini-/micro-UAV and high-altitude, very long endurance segments of that world. Piston/turbine engines are air breathing and mix the oxygen in the atmosphere with fuel – which is already stored in the tank – to support combustion.

6.2 Propulsion Systems Classification

Unmanned aircraft are utilizing different types of propulsion. The core element of a propulsion system is the engine/motor. The term motor is frequently utilized for electric propulsion systems.

There are a number of engine types available in the market for flight operations. They include: (1) electric (battery), (2) electric (solar-powered), (3) piston or reciprocating, (4) turbojet, (5) turbofan, (6) turbo-propeller (or just turboprop), (7) turboshaft, and (8) rocket engines. In general, reciprocating engines have the lowest propulsion efficiency (about 20%–30%) and electric engines have the highest propulsion efficiency (about 60%–80%). Figure 6.1 depicts the classification of air vehicle engines.

In terms of the propeller, engines can be classified into two groups: (1) jet engines and (2) prop-driven engine. A jet engine (e.g., turbojet, turbofan) directly generates thrust via a nozzle. Prop-driven engines (e.g., electric, solar-powered, piston-prop, and turboprop) are generating power and then employ a propeller to convert the engine power to a thrust force.

Introduction to UAV Systems, Fifth Edition. Paul Gerin Fahlstrom, Thomas James Gleason, and Mohammad H. Sadraey.
© 2022 John Wiley & Sons, Inc. Published 2022 by John Wiley & Sons, Inc.
Companion website: www.wiley.com/go/fahlstrom/uavsystems5e

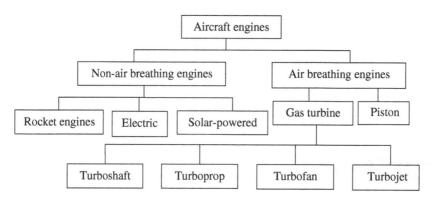

Figure 6.1 Classification of air vehicle engines

In general, most small RC planes and micro- to mini-UAVs and multi-copters employ electric motors, while some small UAVs utilize piston engines. The high-altitude long endurance air vehicles often use a turbofan engine.

6.3 Thrust Generation

The primary function of the propulsion system is to generate a propulsive force or thrust. In this section, the techniques to generate thrust via propellers, and the concept and governing law behind that, will be presented. Fundamentally, an aircraft engine generates thrust based on Newton's third law; there is a reaction for every action. The air/gas is pushed backward; thus, a forward force (i.e., thrust) is created.

Three primary methods used to push the air/gas backward are: (1) propeller, (2) fan, and (3) nozzle. Propellers are mainly employed in piston, turboprop, and electric engines; fans and nozzles are mainly utilized in turbofan engines. Nozzles are used in turbojet and rocket engines.

In Chapter 3, the generation of lift using regular airfoils and wings was introduced. From an aerodynamics point of view, propellers can be thought of as high aspect ratio rotating wings. Propellers generate an aerodynamic force called thrust, just as wings generate an aerodynamic force called lift. There are many ways of describing how this force is actually generated. One explanation is that lower air pressure on the curved surface resulting from an increase in velocity over that part of the surface, as predicated by Bernoulli (an energy equation in low speed), pulls the propeller. While this description is essentially correct, the fact remains that the fundamental principle for the generation of thrust and lift is the reaction to the change in momentum of the mass of air pulled through the propeller disk or wing planform. One must have a momentum generator to produce lift whether it is a wing or an actuator disk (rotor, fan), jet, or a propeller. Based on Newton's second law, in a linear motion, the sum of the forces is equal to the rate of change of linear momentum (i.e., $m.v$). Thus, the force F of a momentum generator is

$$F = T = \frac{dm}{dt}(v_{in} - v_{out}) \tag{6.1}$$

where dm/dt is the mass flowing through the momentum generator per unit time (i.e., mass flowrate), given by

6.4 Powered Lift

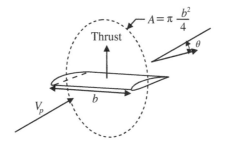

Figure 6.2 Momentum generator

$$\frac{dm}{dt} = \rho v A \qquad (6.2)$$

In Equation (6.1), the subscript "in" indicates the entry conditions and "out" is used for exit conditions. Moreover, in Equation (6.2), parameter A represents the disk area, ρ is air density, and v is velocity at the same point in the flow at which the area and density are measured.

A wing is a vertical momentum generator and produces vertical lift in the manner shown in Figure 6.2. Similarly, a propeller can be thought of as a horizontal lift (thrust) generator. The rate at which an air mass flows past a wing or propeller can be calculated as

$$\frac{dm}{dt} = \rho v \frac{\pi b^2}{4} \qquad (6.3)$$

where $\pi b^2/4$ is the capture area of the air flowing past the wing and v is the velocity.

The change in linear momentum due to the deflection of the air downward is simply $(dm/dt)V \sin\theta$, where θ is the angle by which the air mass is deflected. The force generated by this momentum change in the vertical direction is lift, as shown below:

$$\text{Lift} = L = \frac{dm}{dt} v \sin\theta \qquad (6.4)$$

The power to generate this lift, called induced power, is equal to the rate of change of energy in the downward direction:

$$P = \frac{1}{2} \frac{dm}{dt} v^2 \sin^2\theta \qquad (6.5)$$

Substituting for dm/dt and $\sin\theta$ from Equation (6.4), the induced power in terms of lift becomes

$$P = 2 \frac{L^2}{\rho \pi v b^2} \qquad (6.6)$$

One can see from Equation (6.6) that the power required to produce a given amount of lift is inversely proportional to the square of the wingspan or propeller diameter (b).

6.4 Powered Lift

Lift can also be generated by an actuator disk consisting of a helicopter rotor, quadcopter propellers, or a ducted fan, as shown in Figure 6.3.

For an un-ducted rotor, ambient air is sucked into the disk defined by the spinning fan and passes through it with a velocity v_d and continues to accelerate to a final exit velocity v_e. It is well known and easily proved that

$$v_d = \frac{v_e}{2} \tag{6.7}$$

and the mass flowrate is

$$\frac{dm}{dt} = \rho A v_d = \frac{1}{2}\rho A v_e \tag{6.8}$$

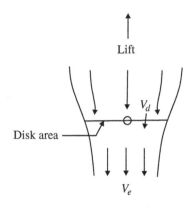

Figure 6.3 Actuator disk

where A is the area of the disk. So the lift of the disk is

$$L = \frac{dm}{dt}v_e = \frac{\rho A v_e^2}{2} \tag{6.9}$$

The induced power is

$$P = \frac{1}{2}\frac{dm}{dt}v_e^2 \tag{6.10}$$

and by substituting for dm/dt from Equation (6.8) and for v_e^3 from Equation (6.8), we find that

$$P = \frac{L^{3/2}}{\sqrt{2\rho A}} \tag{6.11}$$

A slight rearrangement leads to

$$\frac{L}{P} = \sqrt{\frac{2\rho}{L/A}} \tag{6.12}$$

which tells us that the lift per unit power (also called power loading) is inversely proportional to the square root of the disk loading (L/A) and directly proportional to the square root of the density. Plotting power loading (L/P) against disk loading for helicopters, tilt rotor/wing, and fans (Figure 6.4) shows the relative efficiency of each. While the units of power in Equation (6.12) are hp or ft·lb/s, we plot the results of the calculation in the commonly used units of horsepower.

Combining the expressions for lift and power in terms of exit velocity in Equations (6.9) and (6.10), we find that

$$\frac{L}{P} = \frac{2}{V_e} \tag{6.13}$$

This indicates that lift per unit power (i.e., power loading) is inversely proportional to the exit velocity. From this, it is clear that the most efficient powered lift is generated by using a large mass of air at a low velocity.

6.4 Powered Lift

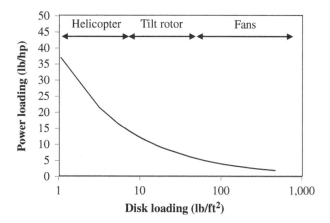

Figure 6.4 Disk loading versus power loading

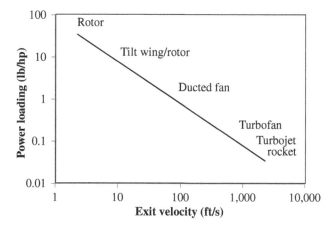

Figure 6.5 Lift-to-power ratio versus exit velocity

Figure 6.5 shows that momentum generators are classified as a function of the lift-to-power ratio or exit velocity. Rotors and propellers, with large disk areas and large amounts of air flowing rather slowly, are the most efficient for hovering. Fans are at a disadvantage compared to rotors, and turbojets at a disadvantage compared to fans, all the way down to rockets, which have the highest exit velocities.

Use of fixed wings with a propeller or turbojet engine producing thrust is a more efficient way to achieve forward horizontal flight than the use of rotary wings that produce only lift. The power used to produce thrust in a fixed-wing configuration directly produces forward motion and indirectly produces lift due to the forward motion of the fixed wings. On the other hand, the power applied to a rotor moves a blade to the rear at the same time that it moves another blade forward, so it does not directly produce any forward motion of the aircraft. The rotor produces only lift and wastes the fraction of the power that is used to overcome the parasitic drag on the rotor and that does not contribute to the lift.

In order to produce forward motion, the plane of rotation of the rotor (and propeller) must be tilted to convert some of the net aerodynamic force on the rotors into thrust, and the total aerodynamic force has to be increased (by increasing the power input to the rotors) to maintain the lift required to support the weight of the aircraft. This situation becomes more pronounced at higher forward speeds, which require more thrust.

This is partially overcome with a tilt rotor/wing that transitions from powered lift using rotary wings into the more efficient wing lift for most of its mission. If the mission requires a lot of time hovering, then this advantage is lost.

A vertically oriented ducted fan can be equipped with wings to aid in the horizontal flight mode. In this case, the fan exit flow is deflected via vanes to provide a horizontal propulsive force. There is a loss, of course, in turning the flow that can range from 10% to 40% depending on the turning angle.

A similar approach that has been used on some fast helicopters is to have short wings that can contribute to lift when the helicopter is in forward flight, reducing the need to increase the total aerodynamic force on the rotors to compensate for using some of the lift as thrust. For this to work, the wings must be set or adjustable to the correct angle to achieve a desired angle of attack when the helicopter is tilted nose downward while in forward flight.

VTOL UAVs (e.g., vertically oriented ducted fan and rotary wing air vehicles) have other disadvantages. They are difficult to control in a hover and they are more mechanically complex, both of which will add to the cost. In addition, engine failure is probably a more serious problem with a VTOL than a fixed-wing air vehicle that can glide or parachute to safety. To minimize this possibility, rotary-wing designers opt for the more reliable, and also more costly, gas turbine power plant. Having said all of this, there are many missions where the VTOL UAV is superior to fixed-wing UAVs. Without the need for launch and recovery equipment, they can attain battlefield mobility that is difficult if not impossible to realize with a fixed-wing vehicle, especially the larger-sized vehicles.

In addition to battlefield mobility, there is the question of transportability or strategic mobility. It is probably more important to the Marine Corps that they can transport extra ammunition or perhaps a tank or two than to have to transport two 5-ton trucks carrying a launcher and a recovery net.

There is also the problem of accomplishing landings in small areas. Even with flaps, which add weight, cost, and complexity, the typical fixed-wing UAV cannot attain a steep enough glide angle to land in very small fields surrounded by trees or other obstacles and be caught by a net. Small ships cannot afford the luxury of a net recovery system (see Chapter 18); there simply is not space available. The larger ships cannot tolerate nets if they interfere with helicopter operations. A VTOL UAV offers a great deal of flexibility for combined UAV and helicopter operations. All of these advantages and disadvantages must be carefully weighed when deciding whether the mission is worth the cost of a VTOL vehicle.

6.5 Sources of Power

There are four primary types of engines used to propel UAVs. They are reciprocating internal combustion engines, rotary engines, gas turbine engines, and electric engines. The application of an electric motor appeared in the past decade and is playing an increasing role on the UAV scene (particularly in small RC planes and quadcopters). Internal combustion and gas turbine engines generate power by burning gasoline, a gasoline/oil mixture, jet fuel (kerosene), or diesel fuel. The electric motors use batteries, solar cells, or fuel cells.

Internal combustion engines have cycles composed of a series of processes that can be plotted as functions of pressure and volume, which give a reasonable indication of the power-generating

6.5 Sources of Power

capability and efficiency of each type of engine. Due to the power generation mechanism, internal combustion engines are also referred to as reciprocating engines and piston engines.

Reciprocating and rotary engines are connected to propellers that provide the thrust to move the air vehicle. Gas turbine engines can either generate direct jet propulsion or be geared to a propeller or rotor. Large fixed-wing and rotary-wing UAVs often use gas turbine power plants (i.e., jet engines) because of their inherent reliability, while small UAVs employ electric engines, since the batteries weight and life are acceptable.

6.5.1 Four-Cycle Engine

The four-cycle internal combustion engine is probably the best understood of all the engines because of its widespread use in the automobile. The cycle is fairly easy to understand and, as the name implies, is made up of four processes. A fixed volume of air/fuel mixture is either injected or sucked into the cylinder cavity during the intake or *induction process* when the piston moves downward from top center. The piston then moves upward and compresses the mixture during the *compression process*. Just before the piston reaches top center, the spark ignites the compressed mixture and additional pressure is generated in the *combustion process*, pushing the piston back to bottom center.

The linear motion is converted into torque via the crankshaft. The piston again moves upward pushing the burnt residue out of the cylinder during the *exhaust process*. Exhaust and intake ports open and close at the appropriate time to allow for the ingress of the fuel/air mixture and the egress of the burnt residue. The volume and pressure during one cycle are shown in Figure 6.6 in what is called an indicator or V–P diagram. The bounded area within the diagram is an indication of the power generated during each cycle. Note the high peak pressure during combustion.

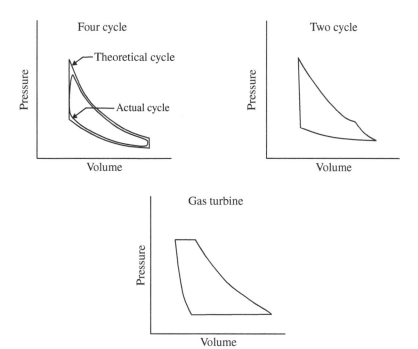

Figure 6.6 Engine cycles

Four-cycle engines may be liquid-cooled or air-cooled and require a fair degree of mechanical complexity because of the valves and the mechanisms to control and move them. The reciprocating motion of the piston also causes considerable vibration, but four-cycle engines are considered to be efficient and reliable. The aerial reconnaissance UAV General Atomics MQ-1 Predator (Figure 11.1) with a wingspan of 14.8 m and takeoff mass of 1,020 kg is employing a Rotax 914F four-cycle piston engine with a maximum power of 115 hp.

6.5.2 Two-Cycle Engine

The two-cycle engine is commonly used with lawn mowers, chain saws, and model airplanes. Although familiar in the household, they are not as well understood as the four-cycle engine and their extensive use in UAVs has led to considerable frustration. The two-cycle engine uses some of the same processes that are executed in the four-cycle engine. When the piston moves toward top center, the fuel/air mixture is sucked into the crankcase (not the cylinder as in the four-cycle case) while, simultaneously, on the opposite side of the piston the burnt residues from the previous cycle are being pushed out of the exhaust port (see Figure 6.7). When the piston has advanced toward top center far enough, both the inlet and exhaust ports are covered and further motion allows compression of the fresh fuel/air mixture.

Again, as in the four-cycle case, just before top center, the spark ignites the mixture, forcing the piston down because of the huge increase in pressure. Power is generated. The down-going piston pushes the fresh fuel/air mixture that was previously sucked into the crankcase up into the combustion chamber through a transfer port in the side of the cylinder opposite the intake and exhaust ports. At this time, the exhaust port is uncovered and the onrushing fresh mixture helps push the burnt residues out of the cylinder. Note that at this time both burnt residues and the fresh mixture are allowed to commingle.

All of the ports must be precisely located so that the opening and closing are timed so the fresh fuel/air mixture pushes out the residues and still have the chamber closed during the combustion process. The two-cycle engine does not need moving valves and their associated mechanisms to accomplish this, and is therefore much simpler than the four-cycle engine. Because the crankcase contains the fuel/air mixture during part of the cycle, it must be sealed.

One of the greatest deficiencies of the two-cycle engine is caused by the commingling of the burnt residues and the fresh fuel/air mixture. There will always be a little adulteration of the latter,

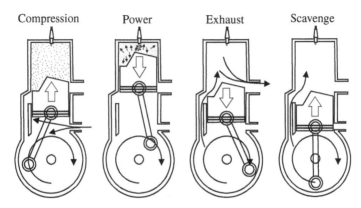

Figure 6.7 Two-cycle process

6.5 Sources of Power

which one can easily envision as leading to increased fuel consumption (one cannot burn fuel that is already burnt) and rough running as will be shown later. The V–P diagram of a two-cycle engine is shown in Figure 6.6.

The friction losses in a reciprocating internal combustion engine are (1) mechanical friction and (2) loss due to the flow of gasses through the intake and exhaust ports (the latter being called pumping loss). Pumping losses in two-cycle engines usually exceed those experienced in four-cycle engines, even at the same piston speed and mean effective pressure (power), for the following reasons, all of which contribute to a less efficient engine:

- The two-stroke engine handles a larger amount of air because some of the air is lost through the exhaust ports during scavenging (when the fresh mixture is pushing out the burned residues).
- Air is drawn into the crankcase in addition to the cylinder.
- There is greater loss during exhaust.

Perhaps the most serious deficiency in a two-cycle engine is its poor performance at low load. Load is defined as the ratio of the actual engine output to the maximum output. When UAVs are in the process of recovery, they are usually operating at low load (i.e., low RPM). In a four-cycle engine, only the compression space is filled with residual gasses at the beginning of the intake stroke. Even with small amounts of fuel–air mixture (associated with low loads), flammable mixtures are maintained. With the two-cycle engine, the entire combustion chamber is filled with residual gasses when intake starts and a lot of fresh mixture must be inducted to ensure burning.

A point is reached at low loads where the fresh charge is so small when mixed with the residual gasses that it will not support combustion. Since the crankshaft is still turning, enough fresh mixture is eventually inducted to cause a firing, but it becomes sporadic. The engine tends to sputter and pop irregularly, and sometimes to quit, at low throttle settings. Fuel injection can help both the problem of higher fuel consumption and low load performance.

The torque produced by a reciprocating engine is normally expressed in terms of average torque occurring throughout the cycle but, in fact, it fluctuates rather dramatically. It reaches a peak during the power stroke but varies between negative and low positive values during intake when the piston is drawing in a fresh charge, compressing it and expelling exhaust gases. The variations in torque during one cycle are shown in Figure 6.8 for one-, two-, and four-cylinder engines. All of these engines have equal mean torque but, as one can see, the greater the number of cylinders the lower the peak torques and, consequently, vibration.

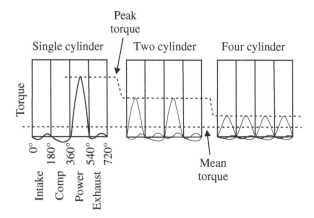

Figure 6.8 Torque variation

6.5.3 Rotary Engine

Vibration is the deadly enemy of electronics and sensitive electro-optic payload systems and is much of the reason for the lack of system reliability of UAV systems. If the reciprocating motion and the cyclical processes of the engine could somehow be alleviated, vibrations would be directly reduced. The rotary engine is a major step in that direction.

The principle of operation of the rotary engine is based on the rotation of a three-sided geometrical shape within a two-lobe geometrical stator. The rotor revolves within this stator such that its three apices make continuous contact with the stator. The stator is an epitrochoid curve based on the path of a point on the radius of a circle that rolls on the outside of a fixed circle. Each face of the rotor completes a four-cycle process identical to the four-cycle engine: intake, compression, combustion, and exhaust.

The cycle takes place during one rotation of the rotor, so one can consider a single bank rotary engine as a three-cylinder engine. As we shall see, there is no reciprocating motion, so vibration can be very low. The end of the rotor is usually provided with an internal gear, concentric with its center, that rotates around a smaller fixed pinion gear mounted on the side cover of the casing.

Referring to Figure 6.9, the cycle of operation will be discussed. The rotor revolves clockwise and, as previously mentioned, each rotation results in three complete Otto cycles. In diagram I, the exhaust and inlet ports are shown to be open, just ending one cycle of exhaust and starting input of a fresh fuel–air mixture to the segment adjacent to side C–A, while the mixture previously drawn into the segment adjacent to side A–B is starting to be compressed. This process continues in diagram II. In diagram III, the mixture adjacent to side A–B has reached its maximum

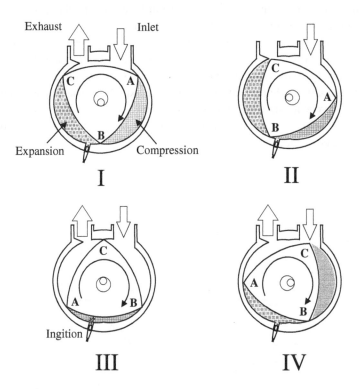

Figure 6.9 Rotary engine

6.5 Sources of Power

compression and is ignited by a spark. At the same time, the segment adjacent to side B–C, which has been expanding and driving the rotor, is opened to the exhaust port.

The burning mixture in the segment adjacent to side A–B now expands and drives the rotor, while the fresh mixture that has been drawn into the segment adjacent to side C–A begins to be compressed, and the combustion products in the segment adjacent to side B–C are expelled through the exhaust port and replaced by a fresh fuel–air mixture entering through the inlet port. Thus, in one revolution, three, four-cycle Otto cycles have been completed, one in each segment.

The satisfactory sealing of the rotor is necessary to ensure reliable operation of the rotary engine. Both side seals and apex seals are required. The side seals are somewhat akin to piston ring seals and are not much of a problem. Apex seals consist of sliding vanes pushed outward against the chamber wall by centrifugal force, sometimes helped by springs to keep them from fluttering. Rotary engines provide nearly vibration-free power for UAVs and, with the exception of the seals, which are becoming less of a problem with the development of new designs and materials, are very reliable.

6.5.4 Gas Turbine Engines

The most reliable of all the engines are gas turbines. They also generate the least amount of vibration because of their steady burning cycle characteristics and pure rotary motion.

The gas turbine engines can generate direct thrust or be geared to turn a rotor or propeller. In either case, the process cycles are essentially the same. The core of a gas turbine engine contains five main elements: (1) inlet, (2) compressor, (3) burner, (4) turbine, and (5) nozzle. Referring to Figure 6.10, air enters an inlet and is compressed by the compressor section of the engine. Compression is obtained either by flinging the air to the circumference of the compressor (centrifugal flow compressor) or grabbing masses of air with small blades and accelerating them rearwards to other blades (axial flow compressor). Centrifugal compressors are cheaper but take up a greater frontal area than axial flow compressors, whose higher cost is associated with the need for all the little blades. Current modern gas turbine engines employ only axial flow compressors.

After the air is compressed, it enters the combustion chamber, or "burner can," where it is mixed with fuel and burned. The resulting hot gas rushes out of the combustion chamber with the energy provided by the burned fuel and impinges on a turbine wheel that is connected to the compressor and turns it. The energy needed to drive the compressor is, of course, not available for propulsion or thrust. Thrust is obtained either by expanding the hot gasses out of a nozzle or by driving a gear train, driven by the turbine, which turns a propeller or rotor.

The gas turbine has even less vibration than a rotary engine, is very efficient at high altitudes, and burns fuel available on the battlefield or on ships without modification. High-speed deep penetrators use gas turbines because of their compactness and thrust-producing capabilities. VTOL vehicles use

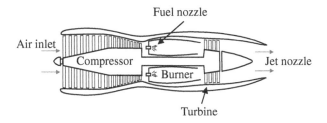

Figure 6.10 Gas turbine engine schematic

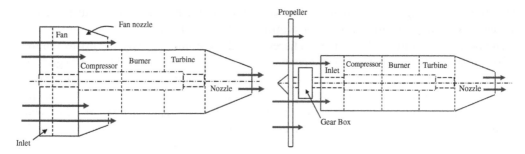

Figure 6.11 Turbofan (left) and turboprop (right) engines

them for these reasons and for inherent reliability. Their major disadvantages are high fuel consumption and limitations on their ability to be miniaturized because of aerodynamic scale effects.

A gas turbine engine with only a core (i.e., without any fan or propeller) mainly generates the thrust via its nozzle. This is referred to as the turbojet engine. When a series of fans in a duct (see Figure 6.11) is connected to a gas turbine engine, a considerable amount of thrust is generated via the ducted fan. This configuration is referred to as the turbofan engine. However, when a propeller (with multiple blades) is connected (see Figure 6.10) to a gas turbine engine, a considerable amount of thrust is generated via the propeller. This configuration is referred to as the turboprop engine. Thus, there are three types of gas turbine engines for UAVs: (1) turbojet, (2) turbofan, and (3) turboprop.

The flow through a turbofan engine is split into two paths. One airflow passes through the fan and flows externally over the core; this air is processed only by the fan. The second airflow passes through the core itself. The ratio of the air mass flow passing through the fan to the air mass flow passing through the core is the bypass ratio.

The HALE Northrop Grumman RQ-4 Global Hawk UAV with a wingspan of 40 m and a takeoff mass of 14,628 kg is equipped with one Rolls-Royce F137-RR-100 turbofan engine. The engine thrust is 34 kN. The aerial reconnaissance General Atomics MQ-9 Reaper (sometimes called Predator B) UAV with a wingspan of 20 m and takeoff mass of 4,760 kg employs a Honeywell TPE331-10 turboprop engine which generates a power of 671 kW. The experimental Boeing X-48 UAV with a wingspan of 6.4 m and a takeoff mass of 227 kg is equipped with three JetCat P200 turbojet engines, each generating 230 N of thrust.

For more details on the characteristics of air breathing engines, the reader is recommended to refer to references such as [18].

6.5.5 Electric Motors

An electric motor is an electro-mechanical machine that converts electrical energy into mechanical energy. In another words, it consumes electric energy to produce mechanical shaft power. In the case of a UAV, a propeller is utilized to convert this mechanical shaft power into an output power, which is thrust times airspeed. Electric motors are much more efficient than reciprocating (piston) engines and gas turbine engines.

Electric motors are found in industrial fans, blowers, pumps, toys, machine tools, household appliances, power tools, disk drives, electric watches, and even cellphones. With the advent of long-endurance, high-altitude loitering UAVs and micro-UAVs, electric motors have become a source of propulsion that can be attractive for a number of reasons. They may have an electric

6.5 Sources of Power

motor that turns a propeller or rotor or may use electric motors to mimic the flight of birds or insects using flapping wings. Except for the emerging class of small AVs that fly using some form of flapping wings, electric motors are mainly used to turn propellers or ducted fans.

Electrically-powered airplanes or model airplanes are not new. Some were said to have flown as early as 1909, although that has been disputed and it has been claimed that the first one flown was in 1957.

The energy supplied to the motor can come from a number of sources. It often comes from batteries but can also come from solar cells and/or fuel cells. The source of power for the small hand-launched AeroVironment RQ-11 Raven with a wingspan of 1.37 m and takeoff mass of 1.9 kg is one Aveox 27/26/7-AV battery-powered electric motor. Each motor generates 400 W of electric power. The small quadcopter DJI Phantom 4 with a diagonal of 350 mm and a takeoff mass of 1.24 kg is equipped with four battery-powered electric engines. The battery pack provides 5,200 mAh of electric energy.

In the early 2000s, NASA and AeroVironment's developed a total of four UAVs as part of an evolutionary series of solar- and fuel-cell-system-powered UAVs. They include Pathfinder, Pathfinder-plus, Helios HP01 (Figure 6.12), and Helios HP03. They were all equipped with a number of solar-powered electric engines with a total power of 2 hp.

The range and endurance characteristics of an electrically-powered aircraft are subject to the aerodynamics of the vehicle in a similar way to airplanes powered by other sources of energy.

Electric motors may be classified by considerations such as power source type, internal construction, number of phases, and type of current. In terms of electric current, there are two types: (1) direct current (DC) motors and (2) alternating current (AC) motors. In DC motors, the current and voltage remain constant as a function of time, while in AC motors, the current and voltage have sinusoidal variations. DC motors are the most popular type for low voltage and/or high starting torque applications.

Moreover, there are two types of electric DC motors commonly used for UAVs. The first type is a "canned" motor. This is a standard DC motor with brushes. The second type is a brushless motor. Brushless motors are much more efficient and lighter than canned motors. Since they have no brushes, there is less friction and there are virtually no parts to wear out, apart from the bearings.

Figure 6.12 NASA and AeroVironment's Helios with solar-powered engines (Source: NASA / wikimedia commons / Public domain)

In an electric motor, the output shaft power is defined as the angular speed times the torque takes to turn the shaft. This power is provided via electric energy. The electric input power to an electric engine is determined by multiplying the voltage (V) and current (I) in the engine:

$$P = VI \qquad (6.14)$$

The unit of power is the volt-ampere (VA) or watt. The input electric power is converted to output mechanical shaft power, which is torque times angular velocity of the shaft. The torque (J) produced by an electric motor is proportional to the current (I) passing through its coils:

$$J = K_t \left(I - I_n \right) \qquad (6.15)$$

where I_n is the no load current, I is the current that produces the torque J, and K_t is the torque constant of the motor, which is a measure of its efficiency. The torque constant is usually provided by the motor manufacturer.

As described earlier in this chapter, the efficiency of a propeller, rotor, or fan is proportional to the area of its disk, which is proportional to the square of its diameter, and the most efficient way to produce thrust or lift with any of them is to have a large diameter and relatively slow rotation. With reciprocating internal combustion engines, it generally is possible to match the revolutions per minute (RPM) of the engine to the desired RPM of a propeller, particularly when using a variable-pitch propeller. For gas turbine engines, the factors that affect the efficiency of the engine itself lead to a need to run the engine at a high RPM and gear down to the desired RPM for the propeller of the rotor.

With electric motors, it is possible to produce the same torque at all RPM, but the size and weight of the motor can be reduced by running the motor at high RPM and gearing it down as needed to produce the desired propeller or rotor torque and RPM.

Most electric motors are designed to run at 50% to 100% of rated load (i.e., output power). The maximum efficiency is usually near 75% of rated load. This implies that, on average, an electric engine wastes about 25% of the electric energy for generating mechanical energy. Power, voltage, thrust, and angular speed are all important considerations when choosing an electric motor.

For more details on the characteristics of electric motors, the reader is recommended to study references such as [19].

6.6 Sources of Electric Energy

We have not felt it necessary to discuss the fuels used for internal combustion engines, which are generally well understood from everyday experience with ground vehicles, but electric motors create a situation in which there are a number of options for how to provide the electrical current that is the "fuel" that makes the motor run. In this section, a brief comparison between various batteries, solar cells, and fuel cells are presented.

6.6.1 Batteries

Batteries can generate a respectable amount of electric power (electric energy per unit time). The limit of their total energy-storage capacity has the same effect on the endurance of an air vehicle as the size of the fuel load has on that of an aircraft using an internal combustion engine. Batteries having a higher energy-storage density per unit weight are the subject of intense research. Battery packs for UAVs are usually rechargeable. The key characteristics of a battery are as follows:

- Capacity: The electrical charge effectively stored in a battery and available for transfer during discharge. Expressed in ampere-hours (Ah) or milliampere-hours (mAh).
- Energy density: Capacity/weight or Ah/weight.
- Power density: Maximum power/weight in watts/weight.
- Charging/discharging rate (C rate): The maximum rate at which the battery can be charged or discharged, expressed in terms of its total storage capacity in Ah or mAh. A rate of 1C means transfer of all of the stored energy in 1 h; 0.1C means 10% transfer in 1 h, or full transfer in 10 h.

6.6.1.1 Nickel–Cadmium Battery

The nickel–cadmium (NiCd) battery uses nickel hydroxide as the positive electrode (anode) and cadmium/cadmium hydroxide as the negative electrode (cathode). Potassium hydroxide is used as the electrolyte. Among rechargeable batteries, NiCd is a popular choice but contains toxic metals. NiCd batteries have generally been used where long life and a high discharge rate is important. Energy density of NiCd batteries is about 150 watts/kg.

6.6.1.2 Nickel–Metal Hydride Battery

The nickel–metal hydride (NiMH) battery uses a hydrogen-absorbing alloy for the negative electrode (cathode) instead of cadmium. As in NiCd cells, the positive electrode (anode) is nickel hydroxide.

The NiMH battery has a high-energy density and uses environmentally friendly metals. The NiMH battery offers up to 40% higher energy density compared to the NiCd. The NiMH has been replacing the NiCd in recent years. This is due both to environmental concerns about the disposal of used batteries and the desirability of the higher energy density. Energy density of NiMH batteries are about 250–1,000 watts/kg.

6.6.1.3 Lithium-Ion Battery

The lithium-ion (Li-ion) battery is a fast-growing battery technology because it offers high-energy density and low weight. Although slightly lower in energy density than lithium metal, the energy density of the Li-ion is typically higher than that of the standard NiCd battery. Li-ion batteries are environmentally friendly for disposal.

Li-ion batteries typically use a graphite (carbon) anode and a cathode made of $LiCoO_2$ or $LiMn_2O_4$. $LiFePO_4$ is also used. The electrolyte is a lithium salt in an organic solvent. These materials are all relatively environmentally friendly. Li-ion is the presently used technology for most electric and hybrid ground vehicles and its maturity and cost are likely to be driven by the large commercial demand. Energy density of Li-ion batteries are about 300–900 kJ/kg.

6.6.1.4 Lithium-Polymer Battery

The lithium-polymer (Li-poly) battery uses $LiCoO_2$ or $LiMn_2O_4$ for the cathode and carbon or lithium for the anode. The Li-poly battery is different from other batteries because of the type of electrolyte used. The polymer electrolyte replaces the traditional porous separator, which is soaked with a liquid electrolyte.

The dry polymer design offers simplifications with respect to fabrication, ruggedness, safety, and thin-profile geometry. The major reason for switching to the Li-ion polymer is the form factor. It

Table 6.1 Energy density of some rechargeable batteries

No.	Battery Type	Energy Density
1	Nickel–cadmium	150 W/kg
2	Nickel metal hydride	250–1,000 W/kg
3	Lithium-ion	300–900 kJ/kg
4	Lithium polymer	360–950 kJ/kg

allows great freedom to choose the shape of the battery, including wafer-thin geometries. Energy density of Li-Poly batteries ae about 360–950 kJ/kg. Table 6.1 demonstrates energy density of various rechargeable batteries.

6.6.2 Solar Cells

The basic principle of a solar cell is that a photon from the sun (or any other light source) is absorbed by an atom in the valence band of semiconductor material and an electron is excited into the conduction band of the material. In order for this to happen, the photon must have enough energy to allow the electron to jump through an "energy gap" that separates the conduction band from the valence band and is due to quantum mechanical effects that create "forbidden" energy states in a crystalline material.

The most common type of solar cell is a silicon positive-intrinsic-negative (PIN) diode. This is created by doping small amounts of selected impurities into a silicon crystal so that it has somewhat higher energy bands than pure silicon (positive doped or "P" material) and then adding additional doping to the surface of the crystal that lowers the energy bands near the surface relative to the undoped material (negative doped or N material). In the region where the doping is intermediate, the energy bands pass through the levels of pure silicon, and the crystal is neither positive nor negative, but "intrinsic" or "I" material. The junction region is shown in Figure 6.13.

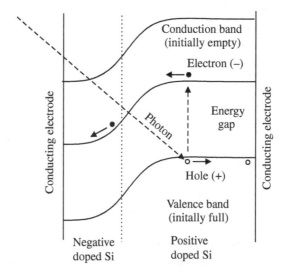

Figure 6.13 PIN junction

6.6 Sources of Electric Energy 125

If a photon is absorbed by an atom in the valence band and an electron from that atom is excited into the conduction band, then the atom becomes a positively charged ion and that positive charge is called a "hole" because it is embedded in a neutrally charged, tightly bound "sea" of neutral atoms. Both the hole and the electron are now free to move through the crystal. The way that the hole moves can be visualized as one in which an electron from the next atom jumps over into the vacancy in the ionized atom's electron shells so that the hole moves one atom over and then this can repeat until the hole moves to the surface of the silicon.

The doping of the junction creates a potential difference in the crystal that causes the electron to move toward the surface and contact on the "N" side and the hole to move to the surface and contact on the P side, so that if these two contacts are connected through a load, a current will flow through that load.

If the photon is not energetic enough to excite an electron into the conduction band, then it may still be absorbed but its energy will be converted into motion of the atoms in the crystal, which heats up the crystal. If the photon is more energetic than required to excite an electron into the conduction band, then it can excite an electron and the remaining energy can go into heating the crystal. The result is that there is a minimum energy for a photon that can be converted into an excited electron, which corresponds to the longest wavelength that can be converted into a current, remembering that a longer wavelength equates to lower photon energy.

Conversion to electrical energy becomes less efficient at shorter wavelengths as more and more of the energy of the photon goes into heating the crystal. What happens for high-energy photons in silicon is that below about 350–400 nm, in the short part of the ultraviolet region, many materials become opaque and the photons are absorbed before they have an opportunity to excite an electron. At the long wavelength end, the cutoff for silicon is at about 1,100 nm, but a sharp roll-off begins at about 1,000 nm.

The total solar "insolation," or energy per unit area, reaching the top of the atmosphere is about 1,400 W/m^2 when measured on a surface that is perpendicular to the sun's rays. Because of atmospheric absorption, this is reduced to about 1,000 W/m^2 at the surface of the Earth at sea level at midday on a clear day. This energy is spread over all wavelengths and it turns out that most of the energy that is absorbed by the atmosphere is at wavelengths that are outside the 400–1,000 nm range in which a silicon solar cell can use them.

This means that there is not a great difference in the maximum energy incident on a solar cell at high altitude and at sea level on a clear day. Of course, if there are clouds or an overcast of haze, the high-altitude cell will still see the full insolation and a cell below the clouds may see very little. Because of the fact that the effective wavelength range of a silicon cell is well matched to the transmission of the atmosphere, it turns out that the round number of 1 kW/m^2 is a useful rule of thumb for the maximum insolation on a solar-cell panel, regardless of altitude, as long as the cell is not below any overcast.

The efficiency of a solar cell is stated in amperes per watt (A/W) for illumination at normal incidence by distribution of light wavelengths that match those of the sun at sea level on a clear day. The efficiency of the cell is dependent on the level of insolation and the 1 kW/m^2 level is used as the standard condition. For many reasons, the efficiency is less than 1. Some of these already have been mentioned and are related to the fact that many of the incident photons have more energy than is required to excite an electron and the excess energy is converted to heat.

In addition, some light is reflected at the surface of the cell and some passes through the junction without exciting an electron. There are losses due to internal resistance within the cell and current leakage when a hole and an electron recombine before they reach the collecting electrodes. There are ways to increase efficiency by stacking junctions to "catch" the photons that pass through the first junction without exciting an electron and by using multiple materials and band gaps to expand the wavelength region in which the cell operates. Some of the techniques are not very applicable to UAVs, such as using concentrating optics, typically curved mirrors, to increase the level of illumination, which turns out to increase efficiency.

At today's state of the art, the useful efficiency of solar cells lies in the range from about 0.20 to about 0.43 when all of the various approaches to increasing efficiency are taken into account. Research and development is intense in this area and the upper limit is likely to increase somewhat. However, there are basic quantum efficiency limits on the process that is the basis for the operation of all of the cells and those limits are well below 1. Efficiency also is not the only factor in a tradeoff between solar cells, particularly for use on a UAV. Some of the less efficient cells are also lighter and more easily configured to be placed on the upper surface of airfoils than some of the more efficient cells and cost may also be an issue, depending on the type of UAV being considered.

6.6.3 Fuel Cells

Fuel cells allow the direct conversion of energy stored in a fuel into electricity without the intermediate stages of burning the fuel to produce heat energy and converting the heat energy into mechanical energy that turns a crankshaft and then using the turning shaft to drive a generator to produce an electric potential and current. Eliminating all of those intermediate steps results in a much simpler system that involves no moving parts (other than fuel valves and peripheral things of that sort) and can be implemented in various sizes from quite small to very large.

It is easiest to visualize this process for a fuel cell using hydrogen gas as its fuel. Instead of combining the hydrogen with oxygen from the air, the fuel cell uses a catalyst to facilitate the ionization of the hydrogen at the anode, creating positively charged hydrogen ions and free electrons. It then uses an electrolyte to pass the hydrogen ions to a cathode that is in contact with oxygen gas. This gives the anode a positive charge and creates a voltage potential between the cathode and anode that drives the free electrons through an external circuit. When the electrons get to the cathode, they combine with the oxygen atoms to form water molecules (Figure 6.14).

All of this works because the binding energy of the water molecules is less than the combined binding energy of the hydrogen and oxygen molecules, so that the final state of the fuel plus oxygen has a lower energy than the initial state. This is exactly the same reason that hydrogen and oxygen will burn in an exothermic reaction if mixed and ignited, but it avoids all the messy things associated with the burning process.

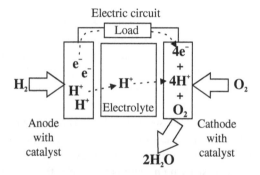

Figure 6.14 Fuel cell

The choice of the electrolyte is very important. Some of the electrolytes that work well in fuel cells need to operate at temperatures as high as 1,000 °C. This clearly requires significant packaging to insulate the cell from its surroundings. For use in a UAS, the more attractive electrolytes are solid organic polymers and solutions of potassium hydroxide in a "matrix." In this context, one could think of the matrix as a layer of some absorbing material that can be permeated by the liquid electrolyte and avoids the issues related to an unrestrained liquid electrolyte.

A fuel cell is not a battery and cannot directly be "recharged." However, if it uses hydrogen as a fuel, the resulting water can be saved and electrolyzed to turn it back into oxygen and hydrogen gas. This makes a fuel cell an attractive way to store and recover energy on an electrically propelled UAV that uses solar cells to provide power during the day but must store energy to remain aloft at night. If the solar-cell subsystem is sized to provide enough energy both to propel the AV and to electrolyze water at a rate high enough to store enough energy for the next night, then the process can go on indefinitely with all of the energy for 24-h operations coming from the sunlight during the day.

The tradeoff between batteries and fuel cells is likely to be partly one of cost at present, with fuel cells probably more expensive than batteries. If cost is not the primary driver for the selection, then the tradeoff is driven by the weight and volume of the required batteries and the total subsystem weight of the fuel cell itself, the fuel storage, the water storage, and the electrolysis system. Water storage is required since there is no external source for water.

The electrolysis system can operate with high efficiency at relatively low voltages (9–12 V) and need not be very large or heavy. Another area that must be considered is the maintenance and/or replacement of the batteries or fuel cell.

Batteries have a limited number of charge/discharge cycles before their energy-storage capability begins to significantly decrease. In addition, most rechargeable batteries need to be fully discharged and recharged periodically to avoid a loss in energy-storage capability. NiCd batteries were notorious for their "memory," which meant that if they were repeatedly only partially discharged before being recharged, they eventually would not deliver any more power once they had been discharged to the level to which they had become accommodated. The newer battery types are not as susceptible to this problem, but manufacturers still recommend full cycling on a periodic basis. A UAS that uses batteries in a long-duration mode may need to make some provision for a full discharge after some number of partial cycles.

Fuel cells can experience "poisoning" by carbon monoxide or carbon dioxide in the atmosphere. There are some approaches to dealing with this, but the simplest approach is to start with very pure water and recycle it without contamination from outside of the cell. Operating time between failures has been an issue for some applications but is also the object of ongoing improvements.

The significant work being done on both batteries and fuel cells in support of ground vehicles is driving the state of the art in both areas at a rate that makes it essential that the tradeoff between batteries and fuel cells be done for each system using the latest state of the art or even the predicted state of the art at the time that the system will go into production. The latter approach is risky, of course, but may be justified when dealing with rapidly evolving technology and if done in a conservative manner with full attention to the risks.

6.7 Power and Thrust

A jet engine (e.g., turbojet, turbofan) directly generates thrust. A prop-drive engine (e.g., electric, solar-powered, piston-prop, and turboprop) employ a propeller to convert the engine power to thrust. In the following sections, the relations between power and thrust, features of propellers, and variations of power and thrust with altitude are explained.

6.7.1 Relation Between Power and Thrust

The product of a piston engine or an electric motor is a shaft torque at a specific angular velocity that can be stated in terms of power (P). The output power of the engine is an input power (Figure 6.15) to the propeller (P_{in}) and is converted into the thrust via the propeller.

In practice, a propeller cannot convert all the shaft input power into propulsive thrust (i.e., aerodynamic force), since part of the input power is wasted due to the propeller drag.

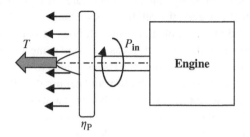

Figure 6.15 Propeller efficiency and engine thrust

A propeller's efficiency (η_P) is defined as the ratio of output thrust power to input torque power and is determined by

$$\eta_P = \frac{P_{out}}{P_{in}} \tag{6.16}$$

Mechanical power is defined as force – that produces the motion – times the speed of the motion. In a fixed-wing aircraft, the propeller output power in a steady-state motion is equal to the propeller thrust times the aircraft speed (TV). Thus,

$$\eta_P = \frac{TV}{P_{in}} \tag{6.17}$$

Thus, in a cruising flight of a fixed-wing UAV, the generated thrust (T) is related to engine power (P_{in} or simply P) as

$$T = \frac{P\eta_p}{V} \tag{6.18}$$

where η_p denotes propeller efficiency and V is the level-flight airspeed.

6.7.2 Propeller

The primary difference between jet engines and prop-driven engines is the propeller. A propeller is a lifting surface – with an airfoil cross-section and at least two blades – which creates an aerodynamic force. This force in aerodynamics is referred to as lift, but in propulsion is called thrust. From an aerodynamics point of view, a propeller is very similar to a wing, as both have airfoil sections. However, one is fixed, but the other one is rotating. Due to this rotation, every section of a propeller experiences a different airspeed. At the root, the airspeed is zero, while at the tip, it is the highest.

The airspeed at each section (V_p) is the vector sum (see Figure 6.16) of two velocities: (1) normal velocity (V_n) to the prop and (2) tangential velocity (V_t):

$$V_p = \sqrt{V_n^2 + V_t^2} \tag{6.19}$$

6.7 Power and Thrust

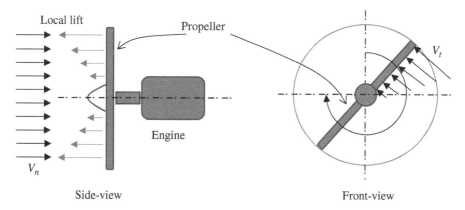

Figure 6.16 Propeller normal and tangential velocities

In a fixed-wing UAV, the normal velocity is also the UAV forward airspeed. A local lift is generated at each station of the propeller. As we move from the root to the tip, the local lift is increased, with the highest value at the tip. The sum of the local lifts is equivalent to the generated thrust.

A propeller needs to be twisted about its radial axis (i.e., in the plane of rotation yz) to provide a constant angle of attack for all stations, with a typical twist angle of about 40–50 degrees. Since the tangential velocity is the highest at the tip of a prop, it has the maximum twist angle at the tip. The lift generated at around the center of a rotating prop is about zero, since its tangential velocity is about zero. To streamline the flow around the center of the prop, a spinner is placed to cover that region.

The propeller efficiency – for a fixed pitch prop – is a function of airspeed (see Figure 6.17) and is the highest at only one airspeed. Typical values for propeller efficiency range from 50% to about 80%. It is recommended that the UAV airspeed for cruise is selected be at the value such that the prop is at its maximum efficiency.

Propellers need to be balanced (both static and dynamic propeller balancing) to nullify or minimize engine vibration and noise from props. All propellers usually come from the factory statically balanced. Dynamic balancing is the process of measuring the vibration of the entire propeller and engine assembly while it is spinning. In this process, a number of accelerometers are attached to

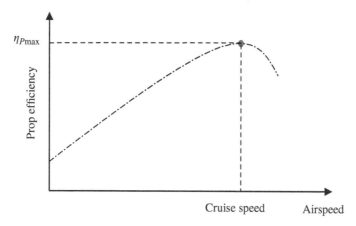

Figure 6.17 Typical variations for fixed-pitch prop efficiencies

the engine, and then the engine is started and brought to full power. An optical tachometer is used to measure the angular velocity of the engine and the propeller. The vibration magnitude of the propeller and engine is measured in meters per second.

The goal is to identify the angle at which the imbalance exists, so a small amount of mass is added to correct the imbalance. The test is repeated until there is no vibration, or the level of vibration is found acceptable within a range (e.g., below 2 mm per second). One of the aviation pioneers in manufacturing propellers is the "Hartzell Propeller," which manufactures composite and aluminum propellers for all classes of aircraft including UAVs.

6.7.3 Variations of Power and Thrust with Altitude

Engine power (P) and thrust (T) are functions of a number of parameters including air density. They have a strong variation with air density, as both power and thrust decrease as air density decreases. Moreover, the air density in the atmosphere is a function of altitude – as we fly higher, the air density decreases. Thus, engine power and thrust decrease with altitude. The variations of T and P with altitude (indeed, air density) is approximated [20] by

$$T_{\text{alt}} = T_o \left(\frac{\rho}{\rho_o}\right)^k \tag{6.20}$$

$$P_{\text{alt}} = P_o \left(\frac{\rho}{\rho_o}\right)^k \tag{6.21}$$

where T_{alt} and T_o denote engine thrust at altitude and sea level respectively, P_{alt} and P_o denote engine power at altitude and sea level respectively, and ρ and ρ_o are air density at altitude and sea level respectively. To find air density at various altitudes, see Reference [12]. The value of variable k is between 0.8 and 1.2; you may assume 1 for initial analysis. The exact value is a function of engine design and its generation. Modern engines are much more efficient than older engines, even at altitudes.

Equations (6.20) and (6.21) are empirical relations and hold for a large number of air breathing jet and prop-driven engines. From Equations (6.20) and (6.21), one can conclude that engine power (P) and thrust (T) are decreasing with altitude. Figure 6.18 illustrates the variations of air

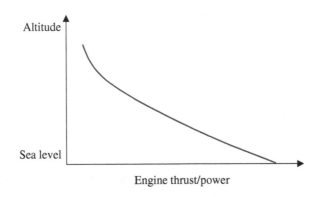

Figure 6.18 Typical variations for engine thrust and power versus altitude

Questions 131

breathing engine power and thrust with altitude. Please note that the power of an electric engine is often not varying with altitude. However, its thrust is decreasing with altitude, which is very similar to other types of engines.

Example 6.1

A small UAV with a prop-driven engine is cruising with a velocity of 74 m/s. The engine propeller is generating 1,100 N of thrust, while the propeller has an efficiency of 81%. How much shaft power is generated by the engine?

Solution:

$$T = \frac{P\eta_p}{V} \Rightarrow P = \frac{TV}{\eta_p} = \frac{1,100 \times 74}{0.81} = 100,494 \text{ W} = 100.5 \text{ kW} \tag{6.18}$$

Questions

1) What do AC, DC, RPM, Ah, mAh, nm, and VTOL stand for?
2) List four main forces acting on an air vehicle in a cruising flight.
3) What is the primary function of the propulsion system of a UAV?
4) What is powered lift?
5) List at least names of six engine types for UAVs. Then provide the name of one UAV for each case.
6) What are three primary methods to push the air/gas backward in a UAV, so that a propulsive force is generated?
7) In terms of propeller, engines can be classified into two groups. What are they? Then, briefly compare them.
8) Write Newton's second law.
9) Write Newton's third law.
10) Propellers generate an aerodynamic force called thrust just as wings generate an aerodynamic force called lift. Explain the similarity.
11) What is the disk loading?
12) What is the power loading?
13) What elements are the most efficient for hovering?
14) Write VTOL UAV's advantages.
15) Write VTOL UAV's disadvantages.
16) Write primary types of engines used to propel UAVs.
17) Write four processes of a cycle in an internal combustion engine. Describe each in brief.
18) Write two main types of cooling for four-cycle engines.
19) Write engine power of the General Atomics MQ-1 Predator.
20) Briefly describe the process of generating power in two-cycle engines.
21) Does the two-cycle engine need moving valves and their associated mechanisms? Why?
22) Write one of the greatest deficiencies of the two-cycle engines.
23) Write two types of friction losses in a reciprocating internal combustion engine.
24) Why are two-cycle engines less efficient engines than four-cycle engines? Provide three reasons.
25) Define load in UAV engines.
26) Briefly compare torque variation in single-cylinder, two-cylinder, and four-cylinder engines.

27) What does vibration do in electronics and sensitive electro-optic payload systems?
28) Write one advantage of the rotary engine compared with piston engines.
29) Briefly describe the process of generating power by a rotary engine.
30) Briefly describe the process of generating the thrust by a piston engine in an air vehicle.
31) Identify the source of power in the following UAVs:
   a) Northrop Grumman RQ-4 Global Hawk
   b) General Atomics MQ-1 Predator
   c) AeroVironment RQ-11 Raven
   d) General Atomics MQ-9 Reaper
   e) DJI Phantom 4
   f) Boeing X-48
   g) AeroVironment's Helios

32) Briefly describe the process of generating the shaft rotation by an electric motor.
33) Briefly describe the differences between turbojet, turbofan, and turboprop engines.
34) Write the main elements of the core of a gas turbine engine.
35) Are current modern gas turbine engines using axial flow compressors or centrifugal compressors?
36) Briefly describe how thrust is obtained in a gas turbine engine.
37) Rate the following three engines in terms of vibration: (a) piston engine, (b) rotary engine, and (c) gas turbine engine.
38) Write major disadvantages of gas turbine engines.
39) What are three types of a gas turbine engine?
40) Define the bypass ratio for turbofan engines.
41) Write the engine thrust of the Northrop Grumman RQ-4 Global Hawk.
42) Write the engine power of the General Atomics MQ-9 Reaper.
43) Write the amount of electric energy for the battery pack of quadcopter DJI Phantom 4 (in mAh).
44) Write two main types of electric motors.
45) Write two main types of electric DC motors.
46) What is the relationship between input power, voltage, and current in an electric engine?
47) List the key characteristics of a battery.
48) Write the characteristics of nickel–cadmium batteries.
49) Write the characteristics of nickel–metal hydride batteries.
50) Write the characteristics of lithium-ion batteries.
51) Write the characteristics of lithium-polymer batteries.
52) What is the typical energy density for nickel–cadmium batteries?
53) What are major techniques to provide energy to electric motors?
54) Explain how the engine vibration problem is solved in a UAV.
55) Briefly describe the basic principle of a solar cell.
56) What is the most common type of solar cell? How is it manufactured?
57) Briefly describe how electrical energy is generated in a solar cell.
58) How much is the total solar "insolation," or energy per unit area, reaching the top of the atmosphere (in W/m^2)?
59) How much is the total solar "insolation," or energy per unit area, at the surface of the Earth at sea level at midday on a clear day (in W/m^2)?
60) What is a useful rule of thumb for the maximum insolation on a solar-cell panel regardless of altitude as long as the cell is not below any overcast?

Questions 133

61) How is the efficiency of a solar cell stated?
62) Write two reasons for the losses in a solar cell.
63) Write one way to increase efficiency of a solar cell.
64) Write the useful range of efficiency for a solar cell.
65) Do fuel cells allow the direct conversion of energy stored in a fuel into electricity without the intermediate stages of burning the fuel?
66) Discuss the process for conversion of energy stored in a fuel into electricity in fuel cells.
67) What is the role of a catalyst in fuel cells?
68) Name four main types of rechargeable batteries which are employed in UAVs.
69) Write two examples for electrolytes used in fuel cells.
70) Explain how a solar-cell subsystem can be employed indefinitely with all of the energy for 24-h operations coming from the sunlight during the day.
71) What were NiCd batteries notorious for? What is the solution to avoid it?
72) Write one approach to deal with poisoning of fuel cells.
73) Compare jet engines with prop-driven engines in terms of the primary output.
74) Define the propeller's efficiency.
75) Briefly describe features of a propeller as compared with a wing.
76) Briefly describe the relation between air velocity at each propeller section and normal and tangential velocities.
77) What is the typical propulsive efficiency range for: (1) reciprocating engines and (2) electric engines?
78) Write typical values for propeller efficiency.
79) Why does a propeller need to be twisted about its radial axis (i.e., in the plane of rotation)? What is a typical value for the twist angle?
80) Why do propellers need to be balanced?
81) Briefly describe the dynamic balancing process for a propeller.
82) Write the name of one of the aviation pioneers in manufacturing propellers.
83) Why do engine power and thrust decrease with altitude? Discuss.
84) A small UAV with a prop-driven engine is cruising with a velocity of 60 m/s. The engine and propeller are generating 1,000 N, while the propeller has an efficiency of 73%. How much power is generated by the engine?
85) A MALE UAV with a prop-driven engine is cruising with a velocity of 460 km/h. The engine is generating 600 kW of power, while the propeller has an efficiency of 68%. How much thrust is generated by the engine?

7

Air Vehicle Structures

7.1 Overview

An essential part of an air vehicle is the structures. We shall consider, in this chapter, the basic concepts and fundamentals of structures. Fundamental relationships in structural analysis are developed in materials science, the mechanics of materials, and aircraft structures. In this chapter, primary functions of the aero-structures, main structural components, primary structural members, loads on structure, structural materials, construction techniques, and basic structural calculations are reviewed.

The primary functions of the air vehicle structure are: (1) to keep the aerodynamic shape of the UAV, (2) to carry the loads, and (3) to protect equipment, payload, and systems from the environmental conditions encountered in flight. In air vehicles, thin shell structures are often employed where the outer surface of the shell is supported by longitudinal stiffening members and transverse frames to resist shear, compressive forces, and bending and torsional moments without failure.

The thin shells (e.g., pressure vessels) that rely entirely on the skins for their capacity to carry loads are referred to as *monocoque*. The most common form of UAV structure is *semi-monocoque* (single shell), which implies that all elements can resist loads, while the skin is also stressed/reinforced. One reason to employ a semi-monocoque structure is to have a lighter structure, which leads to a lower cost. The low-cost structure is considered simultaneously with the objective to have the lowest structural weight.

The structural design and durability of the airframe has not been a significant problem for unmanned air vehicles (UAVs). Although UAVs may have failure and mishaps, they are seldom due to structural failure. Airplane structural design and construction has well-established criteria and techniques based on years of experience. Nevertheless, as one strives for lighter and cheaper materials and simpler fabrication techniques, it is useful to understand the basic structural design principles being used in UAVs. A structural designer needs to determine flight loads, calculate stresses, and design structural elements such as to allow the UAV components to perform their aerodynamic functions efficiently.

7.2 Structural Members

A conventional fixed-wing UAV is generally built up from the basic components such as fuselage/body, wing, horizontal tail, vertical tail, engine, and landing gear. Aircraft structures frequently have arrangements of thin, load-carrying skins, frames, spars, plates, and stiffeners. Many structural

---

Introduction to UAV Systems, Fifth Edition. Paul Gerin Fahlstrom, Thomas James Gleason, and Mohammad H. Sadraey.
© 2022 John Wiley & Sons, Inc. Published 2022 by John Wiley & Sons, Inc.
Companion website: www.wiley.com/go/fahlstrom/uavsystems5e

components of an aircraft consist mainly of thin plates stiffened by arrangements of ribs and stringers. A structural component such as a wing or fuselage is comprised of a number of elastic members. The landing gear can be loosely assumed to be part of the aircraft structure. In this section, primary structural elements of fuselage, wing, and tail are introduced, while the skin is treated independently.

7.2.1 Skin

One of the main elements of all primary structural components such as the wing, tail, and fuselage is the skin. The structural skin is a thin plate that is defined as a sheet of material whose thickness is small compared with its other two dimensions. The skin is expected to be capable of resisting bending moment as well as surface forces.

Air pressure forces on the skin generate various loads on the skin, including shear force, tensile/compressive forces, and bending moment. The *skin* is efficient for resisting shear and tensile loads; however, it may buckle under comparatively low compressive loads. The primary function of the *wing skin* is to form an impermeable surface for supporting the *pressure forces* from which the wing lift is produced.

To help the skin to carry various aerodynamic loads, it is often reinforced by supporting plates (e.g., ribs), stringers, and frames. Thus, the skin is transmitting the aerodynamic forces to the ribs and stringers through plate and membrane action. Another reason to attaching stringers to the skin is to divide the skin into small panels. This technique not only improves the skin buckling stress, but also avoids the weight penalty due to increasing the skin thickness.

The combined function of *stringers and skin* is to resist axial forces and bending moments, and to make the combination more stable. Moreover, resistance to shear and torsional loads can be achieved by using a combination of skin and spar webs. The wing skin is often riveted to the spars and longitudinal stiffeners and stringers. A similar method is employed in attaching skin to tail structural members. In the case of fuselage, the skin is riveted to the flanges and longitudinal stiffeners.

The fuselage/wing/tails skin can be made from a variety of materials, ranging from impregnated fabric to plywood, aluminum, or composites. Under the skin and attached to the structural components are the many components that support airframe function.

7.2.2 Fuselage Structural Members

A basic fuselage is frequently a single cell thin-walled tube, and its structural members are essentially comprised of skin, transverse frames, longerons, and stringers (see Figure 7.1b). Fuselage usually consists of frame assemblies, bulkheads, and formers. The skin is reinforced by longitudinal members called longerons. The transverse parallel frames, which extend across the fuselage, are known as bulkheads.

A much more common structural element is the ring frame, which is two-dimensional and supports loads applied in its own plane. Frames form the basic shape of semimonocoque fuselages. The ring frames – which are reinforced by stringers and longerons – are primary members used to keep the fuselage diameter against all internal and external forces as well as bending and torsional moments.

The basic loads on frames are: (1) reacting shear forces from the fuselage skin, (2) distributed forces from floor beams, (3) shear forces (lift, drag) from wing spar attachments, (4) weight of components such as the engine and wing, (5) bending moment from the wing/tail attachment, and (6) internal pressure from the pressurized compartment.

7.2 Structural Members

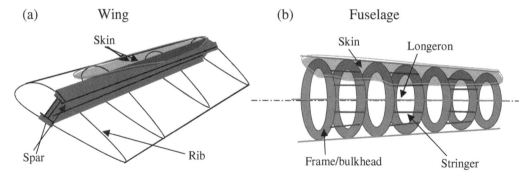

Figure 7.1 Primary structural members of a wing/fuselage for a fixed-wing air vehicle

The horizontal tail lift is generating bending moment on the fuselage, while the vertical tail lift is generating torsion. In the fuselage of some HALE UAVs (e.g., Global Hawk), there should be a pressurized compartment for sensitive equipment (e.g., gyros). This requirement necessitates more supporting members and requires a higher thickness for the fuselage skin.

7.2.3 Wing and Tail Structural Members

Wing, horizontal and vertical tails are usually referred to as the lifting surfaces and share a similar structural configuration. Often wings/tails are of full cantilever design and are made of two halves, left and right segments, which are connected together at the fuselage. The structural members in a wing/tail are spar, rib, stiffeners, stringers, and skin (see Figure 7.1a) in a relatively simple arrangement.

After skin, the spar is the most important structural element of a wing/tail. The spar is a long beam with a high second moment of area cross-section such as I, O, or Z. Spars are referred to as the principal structural members of the wing and tail. In general, wing construction is based on one of three fundamental designs: (1) monospar, (2) two spars (a primary and a secondary), and (3) multispar. They correspond to the longerons of the fuselage. Spars (Figure 7.1a) run parallel to the lateral axis of the aircraft, from the fuselage toward the tip of the wing, and are usually attached to the fuselage as a beam, or a truss.

A wing *spar* is the principal structural member of the wing/tail and is along the wingspan from wing root (at the fuselage) to wing tip. The main function is to withstand bending and torsion caused by the forces acting on the wing (e.g., lift). The wing spar also carries the weight of the wings while on the ground. Its cross-section provides a high second moment of area to carry the wing bending moment.

Generally, a wing structure is designed with a two-spar configuration. One spar (main) is usually located at the maximum thickness and the other (secondary) about two-thirds of the distance toward the wing's trailing edge (in front of the flap/control surface). The main spar is responsible for protecting the maximum thickness of the wing at various stations. The braced beams are still found in some light air vehicles in the form of braced wing structures.

The spars of the wing usually comprise an upper and a lower flange connected by thin, stiffened webs. The role of the spar web is to develop shear stress to resist shear force and torsional moment. Moreover, the spar web – along with the skin, the spar flanges, or caps – performs a secondary but significant function in stabilizing the combination of structures. The spar web also exerts a stabilizing influence on the skin in a similar manner to the stringers.

After the spar and skin, the third primary member of a wing/tail is the *rib*. The main function of the rib is to maintain the shape of the wing/tail cross-section for all combinations of loads, which is governed by aerodynamic requirements. Ribs are attached to the main spar (Figure 7.1a) and are repeated at frequent intervals. The rib geometries are exactly the same as the shape of airfoil sections at each station. Holes are often cut in the ribs at positions of low stress to reduce weight and to accommodate wires, fuel pipes, and electro-mechanical systems.

The wing skin is often reinforced by stringers and plates. A weakness of the *stringer* is the *buckling* due to the high load. They mainly rely on rib attachments to prevent buckling.

7.2.4 Other Structural Members

Other than skin, spar, frames, stiffeners, and stringers, there are a number of structural members that play various roles in the air vehicle structure. A few examples are: (1) engine nacelle, (2) undercarriage/landing gear bay door, (3) cutouts, (4) hard points, (5) pylon, (6) cowling, and (7) inspection doors/panels. In this section, forms and functions of such members are briefly presented.

Nacelles (i.e., pods) are streamlined enclosures used primarily to house the engine and its components. Engine mounts are also expected in the nacelle. These are the assemblies to which the engine is fastened. *Cowlings* are the detachable panels covering those areas into which access must be gained regularly, such as the engine and its accessories.

Engine *pylons* (for podded engines) are structural members used to attach a podded engine to the wing or fuselage. In a supersonic air vehicle, or a subsonic air vehicle with stealth capabilities, the engine inlet is designed with a long intake, and blended with fuselage.

Aircraft structures often need *cutouts* to allow: (1) hydraulic or electrical lines to pass through, (2) easy access to the integrated engines, (3) a camera to rotate easily, (4) satellite communication signals to pass through cutout, and (5) sensors to operate. Other reasons for cutouts are to facilitate fuel pipes/tubes to pass through sections and to reduce weight. Figure 7.2 illustrates an inspection door and cutout in the fuselage of RQ-4 Global Hawk.

Cutouts have a shared role with skin in resisting the distributed aerodynamic (pressure and friction) loads. A cutout will cause stress concentration and distribute concentrated loads around discontinuities into the structure. Examples of discontinuities in the wing/fuselage surface are: (1) retracting/extending landing gear bay, (2) inspection panels, (3) payload/sensor holes, (4) fuel tank door and cap, and (5) bomb and gun bays.

Inspection doors and doors covering landing gear and weapon bays are not capable of resisting flight loads, so a provision must be made for transferring the loads from skin and frames around

Figure 7.2 Inspection door and cutout in Fuselage of RQ-4 Global Hawk (Source: U.S. Air Force / wikimedia commons / Public domain).

the cutout. To allow for cutout effectiveness, either strong flanges/bulkheads are inserted or spar flange areas are increased. These provisions will consequently increase the structural weight and vehicle cost. To support panels against buckling loads, some panels are stiffened.

To mount stores, fuel tanks, and weapons under the wing, *hardpoints* are designed to allow carrying, dropping, or launching. They must be attached at both sides of the wing at intervals. At hardpoints, the structure must be sufficiently reinforced to carry the weight of the store as well as the shear force due to the drag they generate.

7.3 Basic Flight Loads

In the design of an airframe, several factors such as aerodynamic loads (due to pressure distribution and friction), weight loads (e.g., fuel and engine), weight distribution, gust load, load factor, ultimate load, propulsive force, landing loads (e.g., shock and brake), and aero-elasticity effects must be considered.

Primary loads on an air vehicle include basic aerodynamic loads (during level flight): (1) lift, (2) Drag, (3) maneuvering load – dynamic loads – during turn and maneuver, (4) engine thrust, (5) *inertia* load, (6) gust load – when a gust hits the UAV, (7) weight, (8) bird strike, and (9) landing load. Figure 7.3 shows aerodynamic forces and moments on an aircraft during flight.

Wing, fuselage, and tails are each subjected to direct/shear force and bending/torsional moments, and must be designed to resist against critical combinations of these loads. These forces/moments are creating various stresses such as normal, shear, bending, and torsional stresses.

Over the external flight loads, *fuselage* may have a pressurized compartment, and thereby will have internal load and should withstand *hoop stress*. Some UAVs designed for high-altitude flight must withstand internal pressure, since some sensitive equipment need atmospheric pressure to perform normally. Engine thrust and weight of engines will generate force and moment on the fuselage and wing, the value depending on their relative positions.

Two basic forces on wing sections are lift and drag. These two forces will consequently generate a banding moment in two planes (*yz* and *xy*) at the wing roots. Moreover, a torsional moment will be applied to the wing due to aileron deflection during rolling. *Wings* may also carry store/weapons

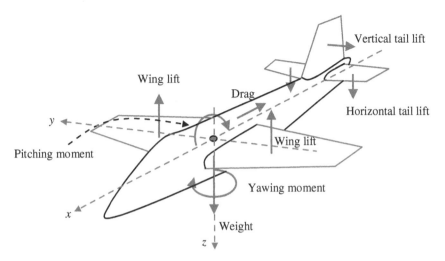

Figure 7.3 Aerodynamic forces and moments on an aircraft during flight

and/or extra fuel tanks, with resulting additional aerodynamic and body forces contributing to the existing bending, shear, and torsion.

The majority of air vehicles have tricycle landing gear, with a nosewheel in the vertical plane of symmetry and their main wheel located under left and right parts of the wings. The air vehicle is encountered with ground loads including concentrated shock forces during landing, through the landing gear.

The engine thrust in a single engine UAV – always along the fuselage centerline – acts in the plane of symmetry. However, in a multiengine UAV, the thrust location (which is along the rotor/shaft) is a function of the engine's location. The engine may be located on the wings or beside the fuselage. However, in the case of engine failure, active engine thrust will cause severe wing/fuselage bending moments.

Lift and drag are surface forces; lift is derived from the air pressure × area, while friction drag is primarily calculated from surface (shear) stress × area. In contrast, weight is a body force, derived from gravitational and inertia effects.

In order to select a structural material and determine its dimensions, first it is necessary to determine the forces that cause the structure to bend, shear, and twist. These forces are created by launch forces, aerodynamic forces due to pressure, inertia, maneuver, and the propulsion system, and their magnitude is determined by balancing the individual components using force diagrams.

Consider, for example, a wing as viewed from the front and a simplified distribution of lift along the span, as shown in Figure 7.4. One may also consider the weight of the wing itself and any concentrated loads such as the landing gear or engines. These forces and weights cause the wing to bend, shear, and twist. The bending moment around any point along the span is obtained by calculating the product of the force (conveniently broken down into small increments) and their distance from the point in question along the span, as shown in Figure 7.4. In this example, we will not consider external stores or other concentrated loads.

The bending moment is calculated around an axis. In Figure 7.4, we take that axis to be through the center of the fuselage and directed out of the plane of the paper. The moment then is given by

$$M = \sum_i F_i d_i \quad (7.1)$$

The bending moment must be resisted by the wing structure (usually a spar). If the spar is visualized as a simple beam, one can see from Figure 7.5 that the bottom surface of the beam tends to be stretched and the top surface compressed for the loading condition shown. If the ability of the material at the bottom of the spar to resist stretching, called tensile strength, is not exceeded, then the wing or beam will not fail provided the upper surface does not buckle. The tensile strength of various materials is found in engineering handbooks. The possibility of the upper part of the member buckling adds considerable complexity and is beyond the scope of this analysis.

Calculation of the pressure distribution over the various aerodynamic surfaces of an aircraft's structure (e.g., the wing) is presented in Chapter 3.

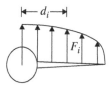

Figure 7.4 Wing lift distribution

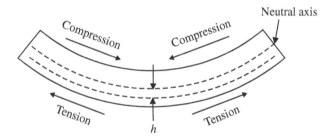

Figure 7.5 Bending stress

It is common practice to refer to the elements of a structural member as "fibers" regardless of whether or not the material actually has a fibrous structure, so the layers along the top and bottom of the spar are the top and bottom fibers even if the beam is a metal forging.

We can see from Figure 7.5 that the strain is greatest at the top and bottom fibers, so the stress must be greatest there as well. This is the reason that the so-called "I-beam" is so universal in load-bearing structures. It concentrates the material at the two outer limits of the beam, putting the most material in the place that experiences the greatest stress.

The stress on the fibers of the beam is proportional to their distance from the "neutral axis" that is halfway from the top to the bottom of the beam and is a fiber along which there is no stretching or compression. That distance is labeled h in the figure.

In addition to the compressive and tensile stress due to bending, there are stresses due to shear. The shear forces are calculated by simply summing the forces at each increment (without regard to distance) up to the point in question, or F_v equals the sum of the individual F_i. This force is resisted by the cross-sectional area of the spar or beam.

There can also be a twisting or torque if the forces are not aligned with the center-line of the beam. All of these forces must be considered in order to determine whether the spar can resist all of the loads imposed upon it.

The primary aerodynamic force on the wing is the lift, which is distributed along the wingspan. Let us assume, for the sake of a simple example, that a wing is uniformly loaded, that is, the lift distribution is a rectangle, and that there is no twisting (see Figure 7.6).

In this simple case, the bending moment around the x-axis at the fuselage centerline is given by

$$M = \sum_i F_i d_i = \int_0^R F(r) \cdot r \, dr \qquad (7.2)$$

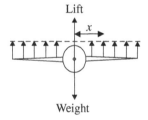

Figure 7.6 Uniformly loaded wing

For a simple case in which the wing loading is uniform as a function of r, the integral gives $M = \frac{1}{2}FR^2$, where F is the sum of the forces, which will be $W/2$ for the case shown in Figure 7.6, and R is the half-wingspan ($b/2$).

The shear force, on the other hand, is simply a linear function of the distance along the wing. This implies that the shear force is increased from the tip to the root, since the wing is modeled as a single support beam (at the root).

If we assume that the half-wingspan of the air vehicle is 5 m and that the vehicle has a mass of 200 kg, resulting in a weight of 1,960 N, we can plot both the bending moment and the shear force as a function of where we measure them along the wing, starting at the center of the wing spar, assumed to be at the center of the fuselage. In this calculation, the axis around which the bending moment is calculated or at which the shear force is calculated is a distance x from the center of the fuselage, as illustrated in Figure 7.6. The resulting curves are shown in Figure 7.7.

If the allowable stresses from the handbook are greater than those calculated, the beam will not fail. One can see that if a constant depth spar does not fail at the root, it will not fail anywhere along the span, because both the bending moment and the shear decrease as one moves toward the tip, as shown in Figure 7.7. One could, in fact, taper the spar to save weight, which is often done in practice.

We can also see why it may be advantageous from a structural standpoint to taper the planform of the wing, so that more of the lift is produced near the wing root than near the tip, which reduces the bending stress at the wing root. Real wings rarely are uniformly loaded and there are other issues, such as the need to support the upward load on the wing while the UAV is on the ground from landing gear mounted on the wing, or to hang other things under the wing, such as a bomb, missile, or sensor pod. All of these concentrated loads must be added to the bending moment and shear diagram.

The structural analysis of the wing, tail, and fuselage of the air vehicle is very complex and is beyond the scope of this book. The concepts and fundamentals of structural analysis are provided in this chapter. For detailed calculations, the interested reader may refer to references such as [21].

In practice, the curvature and shape of the structural elements are such that computer analysis is usually required for final calculations. However, it is relatively easy to determine whether or not the wings will stay on the fuselage.

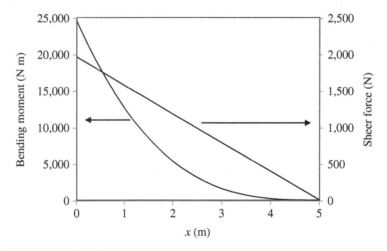

Figure 7.7 Shear force and bending moment diagram

7.4 Dynamic Loads

We have tacitly assumed in the discussion thus far that the air vehicle is in straight-and-level flight and not experiencing wind gusts. It must be recognized that turns, pull-ups, and gusts influence the loads on the structure by upsetting or modifying the balance of forces and must be accounted for. Maneuvering always involves acceleration and acceleration adds or magnifies forces. The acceleration is measured in multiples of the acceleration due to gravity (g) and a 3-g pull-up will magnify the vertical forces by a factor of three. If the spar was designed to carry only the loads in straight and level flight it will not only fail during a 3-g pull-up but will also fail in a 3-g turn. Figure 7.8 shows the forces in straight-and-level flight as well as the forces in a turn.

Notice that the weight is always directed down, but the lift is always perpendicular to the wing (when you look at the front view). Therefore, to turn without losing altitude, the vertical component of the lift must always equal the weight and consequently the total lift must be increased to make up for the bank angle. The larger the bank angle, the greater the required total lift and, therefore, the force on the wing.

$$W = L\cos\phi \tag{7.3}$$

where ϕ is the bank angle. The relationship is simply:

$$\frac{L}{W} = \frac{1}{\cos(\phi)} = n \tag{7.4}$$

where n is called the "load factor" and is equal to 1 when $L = W$ (as in a cruising flight). The n has no unit, but is often expressed in terms of g's (or gravity). The *load factor* in a turn for a 30-degree bank is $n = 1.15$, and the structure is subjected to a 1.15-g load perpendicular to the wing; all the loads on the span must be multiplied by 1.15.

The operating flight strength of an air vehicle can be presented in the form of a V–g or V–n diagram, also called the maneuver flight envelope, where V is the airspeed.

The operating flight load limits on a UAV are usually presented in the form of a V–n diagram. Structural designers will construct this diagram with the cooperation of the flight dynamics engineers. The diagram will determine the structural failure areas and areas of structural damage/failure. The UAV should not be flown out of the flight envelope, since it is not safe for the structures.

The diagram has airspeed (V) on the horizontal axis and structural load, n, in units of g, on the vertical axis. The diagram or envelope is applicable to a particular altitude and air-vehicle weight. The load factor is defined as lift-over-weight. In level, steady flight the load factor is 1, since the load equals the weight under those conditions.

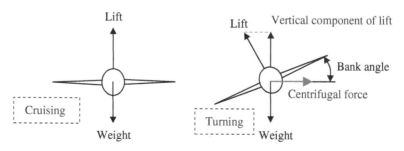

Figure 7.8 Forces during roll

Two of the lines in its construction are related to aerodynamics and are called stall lines. They show the load at a maximum rate of climb, just before stalling. The aircraft cannot fly at any larger rate of climb, so cannot experience any load larger than that shown along the stall lines, which is a function of the maximum lift coefficient and velocity squared. The load lines take the form of a parabolic curve with positive and negative branches that meet at zero airspeed and zero load. The two branches, lines O–A and O–B in Figure 7.9, represent regular flight and inverted flight respectively.

The horizontal lines starting at A and B are the limiting loads for positive and negative forces respectively. In other words, any increase in speed at point A that was accompanied by an increase in attack angle to remain on the stall line would overstress the aircraft and risk structural failure. Airspeed may be increased, but it is necessary to lower the nose to hold the rate of climb constant. The vertical line labeled "V_D or V dive" is a limiting velocity for a vertical dive, which stresses the aircraft along its axis.

The load levels associated with the maximum positive and negative maneuver load levels and maximum vertical dive speed are based on the strength of the air vehicle and are somewhat arbitrarily assigned. The US Federal Aviation Administration provides a method to calculate maneuver loads for various kinds of aircraft. For acrobatic aircraft, they specify $n = 6$ for normal flight and $n = 3$ for inverted flight. Maximum vertical dive velocity is specified as 1.5 times cruise velocity.

Gusty air creates additional loads on the airframe and must be accounted for. Gusts cause an abrupt change in the angle of attack (for a vertical gust) or the true airspeed (for a horizontal gust), or both in the general case, because the aircraft cannot instantaneously change its velocity to match the suddenly changed velocity of the surrounding air mass. The change in airspeed and/or angle of attack leads to a change in lift, which changes the load on the wing, as illustrated in Figure 7.10. Gust loads are directly proportional to airspeed, which is why aircraft pilots reduce their airspeed when they encounter severe turbulence. For a UAV, precautions must be taken with the design of the autopilot to ensure that it will take the same kind of precautions in order to prevent overstressing of the vehicle.

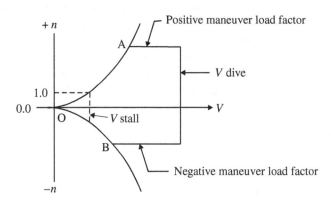

Figure 7.9 Maneuver load diagram

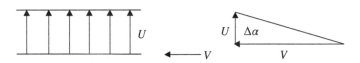

Figure 7.10 Gust diagram

7.5 Structural Materials

7.5.1 Overview

Airframe structural components are constructed from a wide variety of engineering materials. The main groups of materials employed in aircraft fabrication have been wood, aluminum, and fiber-reinforced composites. Early RC planes are constructed primarily of wood and plastic. The most common materials – aerospace aluminum and steel tubing – have followed. Many newly designed UAVs are built from advanced lightweight composite materials, such as epoxy/glass, honeycomb, and carbon fiber. Structure of most current small RC planes – made by hobbyists – are fabricated using balsa wood and plastic cover (e.g., Monokote).

In supersonic and hypersonic UAVs, the flight loads as well as the skin temperature will significantly increase. The material candidates for handling such tough flight conditions are often titanium alloys, steel, and ceramic-based composite materials. They are mainly incorporated in some UAV component assemblies such as fuselage nose and wing leading edge.

Several factors influence the selection of the structural materials for an air vehicle, but the probable top three are: (1) strength, (2) lightness, and (3) cost. Other properties having varying significance are resistance to corrosion, fatigue, environmental issues, ease of fabrication, and consistency of supply. Engine mounts are usually constructed from chrome/molybdenum steel tubing in light UAV and forged chrome/nickel/ molybdenum assemblies in larger UAVs.

UAVs are constructed of many different materials, but the current trend seems to be toward composites. Composite construction offers several advantages that account for its almost universal use in the fabrication of UAVs. The primary benefit is the unusually high strength to weight ratio. In addition, a molded composite construction allows for simple, strong structures that can be built without requiring expensive equipment and highly skilled assemblers. Being aerodynamically smooth, compound curvature panels increase strength and can be easily fabricated with composites as compared to other types of materials. Fiber composites are highly durable, often require no maintenance, and can be used in harsh chemical environments.

7.5.2 Aluminum

Aluminum alloys are the most widely used materials for aircraft structures. To improve mechanical properties, aluminum is alloyed with less than 10% of one or more of the following metals: copper, magnesium, manganese, silicon, lithium, iron, nickel, and zinc. For instance, aluminum 2124 alloy has 4.4% copper, 1.5% magnesium, and 0.6% manganese.

Aluminum alloys feature strong corrosion resistance characteristics, high strength, light weight, and high fracture toughness. The 2000 series alloys (e.g., the aluminum–lithium–copper alloy 2095) are good candidates for aircraft structures. They have a proof stress of 510 N/mm^2, tensile strength of 585 N/mm^2, and an elongation of 8%. Titanium alloys possess high tensile strength and a good fatigue-strength/tensile-strength ratio, and can retain considerable strength at high temperatures (up to 500 $^\circ$C).

In large UAVs, even when the airframe is constructed of composite materials, it is metalized to prevent radar waves from penetrating the skin and reflecting off the electronic boxes inside. In the Northrop Grumman RQ-4 Global Hawk, the fuselage is made of aluminum, a semi-monocoque construction with a V-tail, while the wings are made of composite materials. However, its bulbous nose is made of composite materials, since it houses communications (a large, steerable satellite antenna) and sensor payloads.

7.6 Composite Materials

Composite materials are formed from two dissimilar parts: (1) matrix and (2) fiber. The fiber consists of *laminates/filler/fiber* to provide stiffness and strength, and the *matrix* is used to hold the elements together. Therefore, each part contributes differently to the final properties of the structural member. Three most popular materials for matrix are: (1) polymers, (2) elastomers (e.g., epoxy), and (3) foam. A few popular materials for fiber are: (1) glass, (2) carbon (graphite), (3) ceramics, and (4) aramid. In some aircraft, metals such as aluminum are employed as matrix (e.g., graphite–aluminum and boron–aluminum). The constituent elements in a composite retain their identities (they do not dissolve or merge completely into each other) while acting in concert to provide a number of benefits.

In general, composite materials are expensive to produce. However, composite construction does offer these advantages in the fabrication of UAVs: (1) light weight, (2) a high strength-to-weight ratio, (3) corrosion resistance, (4) aerodynamically complex shapes and compound curvature that can easily be fabricated, (5) molded composite construction that allows for simple strong structures that can be built without requiring expensive equipment and highly skilled assemblers, (6) stiffness and strength of a composite material that can be adjusted (e.g., directional strength), (7) complex internal geometry that can be created to maximize the function it can be used for, (8) high electric strength (insulator), (9) radar transparent, (10) non-magnetic, (11) low maintenance, (12) long-term durability, (13) dimensional stability, (14) customized surface finish, and (15) rapid installation.

It is estimated that replacing an aluminum alloy structure by carbon-fiber-reinforced plastics would result in about 20–30% saving in total structural weight. The strength of carbon-fiber-reinforced plastics is three times that of aluminum alloy.

7.6.1 Sandwich Construction

There are various techniques in constructing an aircraft structure using composite materials. The list of techniques is presented in the next section (7.7). This section is devoted to the sandwich construction. Sandwich structures have smooth external surfaces and are capable of developing high normal stresses, while requiring small numbers of supporting rings or frames.

An effective form of fabrication of aircraft structure is the sandwich construction, which comprises a light honeycomb or corrugated core sandwiched between two outer skins of the stress-bearing sheet. Thus, this construction has three main segments with different functions: (1) core, (2) skins, and (3) bond materials. A sandwich panel, illustrated in Figure 7.11, has two outer

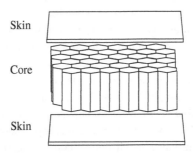

Figure 7.11 Sandwich panel

7.6 Composite Materials 147

surfaces or working skins separated by a lightweight core. The primary function of the core is to stabilize the outer skins, while bond materials are utilized to bond/glue the skins to the core.

When loads are applied to any beam, such as a wing spar, most of the stress occurs at the outer fibers or surfaces. Taking advantage of this fact by using sandwich techniques is the reason for the effectiveness of a composite construction.

The skin can, of course, be metal (e.g., aluminum), but composite laminates such as fiberglass, kevlar, and graphite fibers are used extensively because they can be "draped" around oddly shaped cores and hardened in place.

Core materials can be polystyrene, polyurethane, polyvinyl chloride, aluminum honeycomb, or balsa wood. Various kinds of resins are used to bond the skin to the core and transfer stresses throughout the skin. They include epoxy, polyester, and vinyl ester. Honeycomb structured wing panels are often used in composite wings.

7.6.2 Skin or Reinforcing Materials

The strength of a composite structure is almost entirely dependent on the amount, type, and application of the skin or reinforcing material. The skin fabrics come in two primary configurations or patterns: unidirectional (UD) and bidirectional (BD). A unidirectional fabric has almost all of its fibers running in one direction so the tensile strength would be greatest in that direction.

Bidirectional fabrics have some fibers woven at angles relative to others and therefore have strength in multiple directions. Of course, the UD fibers can be combined at various angles to also provide greater strength in all directions. In addition, multiple layers of material or fabric sheets can be applied to give greater strength where needed and lesser weight where less strength is needed. The composite skins are usually made of the materials as listed in Table 7.1.

In is recommended to manufacture appropriate areas of the UAV (e.g., radomes, in which radar scanners and SatCom radar dish antenna are housed) from radar-translucent materials, such as honeycomb, Kevlar, or glass composites. In either a down-looking radar or an up-looking satellite communications antenna, or a forward-looking seeker, the radome should allow radio waves to pass through skin without any distortion.

7.6.3 Resin Materials

The resin is used to bond or "glue" the skin to the core material and transfer the stresses throughout the skin. Resins irreversibly harden when cured and provide high strength and chemical resistance to the structure. Table 7.2 presents three main resin types. Resin is also the most popular element as the matrix for fiber-reinforced composites.

Table 7.1 Reinforcing (skin) materials for sandwich construction

No.	Materials	Features
1	E Glass	Standard fiberglass, the workhorse of composites.
2	S Glass	Fiberglass similar in appearance to E but 30% stronger.
3	Kevlar	An aramid organic chemical material, very strong but also difficult to work with.
4	Graphite	Long-parallel chains of carbon atoms, very strong and expensive.

Table 7.2 Resin materials

No.	Resin	Features
1	Polyester	A common resin that is used to make everything from boats to bathtubs.
2	Vinyl ester	A resin that is a polyester–epoxy hybrid.
3	Epoxy	A thermosetting resin used extensively with home-built aircraft, RC model UAVs.

Table 7.3 Foam materials

No.	Foam	Features
1	Polystyrene	A white-colored foam that is easy to cut with a hot wire to produce airfoil shapes. It is easily dissolved by fuel and other solvents.
2	Polyurethane	A low-density foam that is easily carved but cannot be cut with a hot wire. Used for carving detailed shapes.
3	Urethane polyester	Foam used in surfboards that has good resistance to solvents.

7.6.4 Core Materials

Core materials used in UAV construction are usually foams, but balsa wood is sometimes used. Table 7.3 presents three main foam materials for the core.

7.7 Construction Techniques

The entire metal airframe and its components may be fastened/joined by rivets, bolts, screws, and other fasteners. Welding, adhesives, and special bonding techniques are also employed. In general, one strives for lighter and cheaper materials and simpler fabrication techniques. In the past decade, a new technique referred to as 3d (three-dimensional) printing has been developed and utilized.

There are two types of composite manufacturing processes: (1) open molding and (2) closed molding. With open molding, the laminate is exposed to the atmosphere during the fabrication process. In closed molding, the composite is processed in a two-sided mold set, or within a vacuum bag. Three basic techniques for open molding are: (1) hand layup, (2) chopped laminate process, and (3) filament winding.

In closed molding, a number of methods are employed: (1) compression molding, (2) pultrusion, (3) reinforced reaction injection molding, (4) resin transfer molding, (5) vacuum bag molding, (6) centrifugal casting, and (7) continuous lamination, where, due to the volume of production, an appropriate technique is selected. For instance, for a low-volume production, open molding and vacuum bag molding are preferred.

The most common construction practice is to cut the foam core to the desired shape, either with a hot wire or with a saw, if the material is not amenable to hot wire cutting. The foam is then sealed to prevent too much absorption of the resin and a supply of resin is mixed. The resin is spread over the surface and a pre-cut piece of reinforcing material (skin) is laid over the wet resin at the proper orientation. The liquid resin will seep through the skin material and the excess is removed.

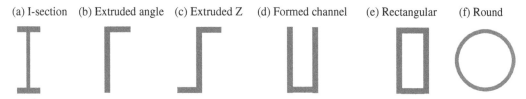

Figure 7.12 Common cross-sections of spars

Layers of material are added in the specified direction and numbers to obtain the final laminate having the desired strength.

Another method is to work with a mold or cavity, draping the skin fabric and resin inside the mold so as to form hollow structures. Molded panels and substructures can be bonded to make a completed structure. Care must be taken to ensure that the proper attachments are chosen for structures carrying concentrated loads.

Thin-walled columns (as in a wing spar or fuselage longerons) in the shape of longitudinal stiffeners are frequently fabricated by forming from a flat plate or directly by the extrusion process. Figure 7.12 illustrates six common cross-sections of thin-walled columns. These columns are mainly used in spars of wings and longerons of fuselages.

7.8 Basic Structural Calculations

In most structural problems, the primary objective is usually to find the distribution of stress in any member produced by applied loads. Moreover, other than deflections and displacements, one needs to determine if the structures can withstand the flight loads without failure by calculating the safety factor. All structural members are elastic, but the solution of problems in elasticity often presents difficulties. In developing an engineering approach to analyze the structures, various constraints, such as allowable wingtip deflection, variations in atmospheric temperature, and safety factor, must be considered.

An air vehicle structure usually consists of many components generally arranged in an irregular manner. In theory, these components possess a number of degrees of freedom and redundancies, which make it very hard to analyze. In order to analyze aircraft structure, an idealized mathematical model should be developed to determine stresses and/or displacements by using equations of equilibrium. The structural weight is significantly important, so an accurate knowledge of component loads and stresses is essential.

In aircraft structural design, there is an approach called fail-safe, using a *damage tolerant* structure. Based on this approach, the fail-safe structure is designed to have a minimum life during which no catastrophic damage occurs. The fail-safe approach relies on the fact that the failure of a structural member does not necessarily lead to the collapse of the complete structure. The remaining un-failed members should be able to withstand the load shed by the failed member until the failure of that member is discovered.

Structural engineers employ powerful methods of structural analysis in conjunction with the force–displacement and stress–strain relationships using boundary conditions. The following is a list of relevant engineering tools and theories in structural analysis: the theory of elasticity, equations of equilibrium, energy methods, principle of virtual work of a rigid body, principle of superposition, a dummy or fictitious load method, strain–displacement and stress–strain relationships, indeterminate structures, and strain energy.

The behavior of each linear element can be analyzed by basic structural analysis techniques, and then the behavior of the complete structure is determined by superposition. For instance, distributed loads in supporting beams are replaced by a series of statically equivalent point loads at a selected number of nodes. Moreover, wing/fuselage skins are idealized into a number of *finite elements*.

Basically, there are two approaches in formulating a structural problem: (1) the *stiffness* (or *displacement*) method, where displacements of the elements are treated as the unknowns and are then used to determine internal forces and (2) the *flexibility* (or *force*) method, where internal forces are treated as being initially unknown.

An effective engineering technique for a complex combination of elements in structural analysis is referred to as the finite element method (FEM). The essence of this method involves dividing the structure into a number of small pieces called finite elements. The structural governing equations with appropriate boundary conditions are applied at the nodes. The simultaneous solutions of all equations yield stress and deflection at each node and element.

There are various commercial software packages in structural analysis; a few examples are ANSYS, Nastran, Abaqus, Matlab, and Autodesk. In this section, a number of fundamental structural calculation parameters and techniques are briefly presented.

7.8.1 Normal and Shear Stress

Each wing section (left or right) is usually attached to the fuselage at the root, so it can be modeled as an end-loaded cantilever. The structural members are designed to carry the flight loads or to withstand stress without failure. In designing the structure, every square inch of wing and fuselage must be considered in relation to the physical characteristics of the material of which it is made. Every part of the structure should be planned to carry the load that is applied on it.

The stress analysis is the basic calculation to determine the safety factor. There are five major stresses [21] to which structural members are subjected: (1) tension, (2) compression, (3) torsion, (4) shear, and (5) bending. A single member of the structure is often subjected to a combination of stresses.

For tension and compression, the normal or direct or axial stress is defined as

$$\sigma = \frac{N}{A} \tag{7.5}$$

where N denotes the normal force and A denotes the cross-sectional area of the beam (e.g., wing spar). The direction of the force N in Figure 7.13 is such that it produces tensile stress on the faces of the beam. In a direct shear, the shear stress is defined as

$$\tau = \frac{S}{A} \tag{7.6}$$

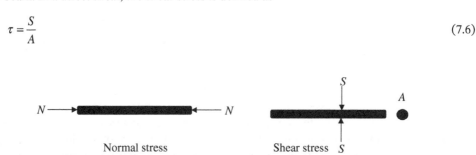

Figure 7.13 Beam under normal and shear stresses

7.8 Basic Structural Calculations

where S denotes the shear force, and A denotes the cross-sectional area of the beam (here, a spar). A uniform shear stress through the cross-section of a beam may be assumed.

Recall that stress is defined as the applied force over a unit area and strain is the unitless resulting deformation. We can see that both depend not only on the bending moment but also on the cross-sectional shape and, more importantly, the depth (thickness from top to bottom in the figure) of the beam.

The external and internal forces cause linear and angular deflections/displacements in a deformable body. These displacements are generally defined in terms of strain (ε), which relates to changes in length. The proportionality constant between stress and strain is Young's modulus (E), which is a characteristic of the material:

$$E = \frac{\text{stress}}{\text{strain}} \tag{7.7}$$

Typical values of E for a number of popular materials are: (1) aluminum, 71.7 GPa, (2) glass, 46.2 GPa, and (3) carbon steel, 207 GPa. The symbol GPa stands for giga Pascal, where 1 Pascal is 1 N/m².

Some simple structural elements are comprised of members capable of resisting axial loads only. However, wing structures consist of beam assemblies (i.e., spar/rib/skin) in which the individual members resist shear force and bending moments, in addition to axial loads.

The bending moment, M, acting on the member section in Figure 7.14 produces a distribution of direct stress, σ, through the depth of the member cross-section. The bending moment at the fixed end (e.g., wing root) is

$$M = F \cdot L \tag{7.8}$$

where F denotes the applied force and L denotes the length (i.e., the distance between the force to the wing root). It is customary to assume that the bending moment is positive when it produces compression at the upper surface of the beam and tension at the lower surface. The maximum bending stress is at the wing root:

$$\sigma_{max} = \frac{Mc}{I} \tag{7.9}$$

where M denotes the bending moment, c, magnitude of the greatest y, and I, the second moment of area about the z-axis. For a beam of rectangular cross-section ($b \times h$):

$$I = \frac{1}{12} bh^3 \tag{7.10}$$

For a beam with a circular cross-section,

$$I = \frac{\pi D^3}{64} \tag{7.11}$$

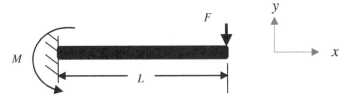

Figure 7.14 Bending moment at the wing root

Figure 7.15 Global Hawk wing tip deflection (Source: Tom Miller / NASA)

Global Hawk's long wing (wingspan of 39.9 m) droops under 15,000 pounds of fuel, which accounts for 60% of the aircraft's weight. However, the wingtip is deflecting up (see Figure 7.15) due to the lift force. This behavior is frequently true for most fixed-wing UAVs.

When a torsion is applied on a solid circular bar, the maximum shear stress:

$$\tau_{max} = \frac{Tr}{J} \qquad (7.12)$$

where T denotes torque, r is radius of the bar, and J is polar second moment of area of the beam cross-section (torsion constant). For a beam of round cross-section:

$$J = \frac{\pi r^4}{2} \qquad (7.13)$$

In practice, a structural member frequently carries a distributed load (w) and, simultaneously at a particular point, a concentrated load (N), a bending moment (M), and a torque (T). Hence, the impact of all these loads must be considered in calculating the maximum stress.

7.8.2 Deflection

When a force (e.g., lift) is applied to a structural element (e.g., a beam), it will deflect. The generated deletion should be limited to avoid the long-term structural failure and to avoid any interference in the payload operations. The force is either along the longest dimension of a beam or perpendicular to it. In both cases, one should determine the deflection and make sure it is within the permissible values.

For a concentrated load (e.g., engine weight on the wing, fuel tank inside the wing), the maximum beam (e.g., spar) deflection (δ) is

$$\delta = \frac{Fl^3}{3EI} \qquad (7.14)$$

where F denotes the applied force (here, mainly, lift, drag, and weight), l is length (here, the wing semi-span), I is the second moment of the area, and E is Young's modulus of elasticity.

For a uniformly distributed load (e.g., the lift on the wing and tails), the maximum beam deflection (δ) is

$$\delta = \frac{wl^4}{8EI} \qquad (7.15)$$

7.8 Basic Structural Calculations

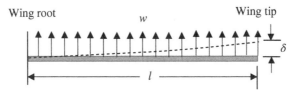

Figure 7.16 Deflection of a cantilever beam under a uniform distributed load

where w is the load per unit length. For a wing, this is equivalent to the lift divided by the wingspan (i.e., $w = L/b$). Indeed, the lift distribution is elliptic (see Figure 7.4). To utilize this equation, you need to find the equivalent uniform distribution (see Figure 7.16). For a half-wing section (left or right half), the maximum deflection will be at the wingtip.

For aluminum, $E = 71.7$ GPa, glass, $E = 46.2$ GPa, and carbon steel, $E = 207$ GPa. The unit GPa is giga Pascal, where 1 Pascal = 1 N/m^2.

Example 7.1

A HALE UAV left wing with a semi-span of 20 m has generated a lift of 30,000 N. Assume the load is mainly carried by its main spar (made of aluminum) with a second moment of area of 0.001 m^4. Determine the deflection at the wing tip. Assume that the lift is a uniform distributed load along the span. Ignore the wing weight.

Solution:
The load per unit length is

$$w = \frac{L}{l} = \frac{L}{b/2} = \frac{30,000}{20} = 1,500 \text{ N/m}$$

The deflection at the wing tip is

$$\delta = \frac{wl^4}{8EI} = \frac{1,500 \times 20^4}{8 \times 71.7 \times 10^9 \times 0.001} = 0.418 \text{ m} = 41.8 \text{ cm} \tag{7.15}$$

7.8.3 Buckling Load

If an increasing normal compressive force is applied to a slender column, there is a value of the force at which the column will suddenly bow. This failure is referred to as buckling.

A large proportion of an air vehicle's structure is modeled as a slender column under a compressive force. For instance, the wing spar comprises an upper and a lower flange connected by thin, stiffened webs. Other examples are thin webs stiffened by slender longerons or stringers, thin-walled columns, and stiffened plates.

These elements are susceptible to failure by buckling at a buckling stress, which is frequently below the yield stress of the material. This implies that such a thin web or long column will buckle before it yields. For the case of the web, it will buckle under shear stress at a fraction of its ultimate load. Therefore, the most critical mode of failure for slender columns, thin plates, and even stiffened panels is buckling rather than yielding.

154   *7 Air Vehicle Structures*

One solution for a thin plate to avoid buckling failure under a relatively small compressive load is to be stiffened. The prediction of buckling loads of slender columns and thin plates/panels is very important in aircraft structural design.

7.8.4 Factor of Safety

The safety factor for a single fixed (i.e., static) normal load is

$$SF = \frac{S_y}{\sigma} \tag{7.16}$$

The safety factor for a single fixed (i.e., static) shear load is

$$SF = \frac{S_{ys}}{\tau} \tag{7.17}$$

where SF stands for safety factor (factor of safety), S_y, yield strength, S_{ys}, shear yield strength, σ, normal stress, and τ, shear stress. When $SF > 1$, the structure is safe, and will not fail. For aerospace aluminum, $S_y = 345$ MPa and for steel, $S_y = 505$ MPa.

To determine the safety factor for a combination of various loads, one needs to use a failure theory such as the maximum shear stress theory (von Mises) and the distortion energy theory. As the safety factor increases (greater than 1), the structure is safer, but it will be heavier, and will incur a higher cost. For a wing structure, a recommended safety factor is about 1.2 to 1.5. For some more critical elements (e.g., engine mount), the safety factor could be well above 2.

Example 7.2

Consider the wing of a small UAV made of aerospace aluminum with one spar. The wing spar has a square cross-sectional area (10 cm × 10 cm) and a half wingspan of 6 m. The half wing has created 5,000 N of lift in a cruising flight. Can this wing withstand such a load? Assume that the lift is a distributed load along the span; you may assume it is located at the wingtip. Ignore the wing weight.

Solution:
The maximum bending moment is at the wing root:

$$M = F.L = 5,000 \times 6 = 30,000 \, \text{J} \tag{7.8}$$

The second moment of area of this beam about the z axis is

$$I = \frac{1}{12}bh^3 = \frac{1}{12} \times 0.1 \times 0.1^3 = 8.33 \times 10^{-6} \, \text{m}^4 = 833.3 \, \text{cm}^4 \tag{7.10}$$

The value for c, the magnitude of the greatest y, for this square beam is the height (h) divided by 2, which is $10/2 = 5$ cm or 0.05 m.

$$\sigma_{\max} = \frac{Mc}{I} = \frac{30,000 \times 0.05}{8.33 \times 10^{-6}} = 180 \times 10^6 \, \frac{\text{N}}{\text{m}^2} = 180 \, \text{MPa} \tag{7.9}$$

The yield strength for aerospace aluminum is 345 MPa. Thus,

$$SF = \frac{S_y}{\sigma} = \frac{345}{180} = 1.92 \tag{7.16}$$

Since the safety factor is greater than 1, yes, this wing is able to withstand such a load.

7.8.5 Structural Fatigue

Aircraft structural members are frequently subjected to repetitive loads over a long period of time. For example, a horizontal/vertical tail suffers variations in loading (positive and negative tail lift), possibly hundreds of times in a single flight, due to wind gust. After a number of years, this structural member may fail at a stress level much below the yield stress for a nonrepetitive load. This phenomenon is referred to as structural *fatigue*.

The fatigue analysis is another important task of an aircraft structural engineer to make sure the structure will not fail due to fatigue. A repeated load could be one of the following: (1) alternating, (2) fluctuating, or (3) completely reversed. In either case, the fatigue strength (σ_{alt}) of the material for each member is determined to withstand a specified number of cycles.

Bending, torsion, and axial stresses may be present in alternating loads. The safety factor (n) for repeated (i.e., dynamic) loads using distortion energy fatigue failure theory is determined by the following relationship:

$$n = \frac{S_y}{\sigma'_{max}} \tag{7.18}$$

where σ'_{max} is the von Mises maximum stress for midrange and alternating loads and S_y denotes the yield strength. Further discussion on fatigue analysis is beyond the scope of this text, but the interested reader may refer to references such as [22, 23].

Questions

1) What do FEM, SF, and GPa stand for?
2) What are primary functions of the structure of an air vehicle?
3) What is the most common form of UAV structure?
4) List the basic components of a conventional fixed-wing UAV.
5) Define the structural skin.
6) What is the skin efficient in resisting?
7) What is the primary function of the wing skin?
8) What is the combined function of stringers and skin?
9) What is the typical attachment form between the wing skin and a spar?
10) What are the main structural elements of a typical fuselage?
11) Briefly describe the relation between the main structural elements of a typical fuselage.
12) List the basic loads on fuselage frames.
13) What load is generated by the horizontal tail lift on a fuselage?
14) What load is generated by the vertical tail lift on a fuselage?
15) What wing, horizontal and vertical tails are usually referred to?
16) What are the main structural elements of a typical wing?
17) Briefly describe the wing spar and its main function.
18) Briefly describe the wing rib and its main function.
19) Briefly describe the relation between the main structural elements of a typical wing.
20) A wing has two spars. Where are the best locations for each spar?
21) Why are holes often cut in the wing ribs?
22) Briefly describe the nacelle and its main function.
23) Briefly describe the engine pylon and its main function.

24) Briefly describe the inspection door and its main function.
25) Briefly describe the hardpoint and its main function.
26) List at least three reasons for cutouts in aircraft structures.
27) List primary loads on an aircraft.
28) What are two basic forces on wing sections?
29) List primary loads on a wing.
30) Briefly describe how the bending moment on a wing is calculated.
31) Briefly describe why the strain is greatest at the top and bottom fibers of a wing.
32) Briefly describe how shear forces on a wing are calculated.
33) Briefly describe why both the bending moment and the shear decrease as one moves toward the tip of a wing.
34) What is a 3-g pull-up or a 3-g turn in terms of acceleration?
35) What are the directions of lift and weight?
36) Define load factor.
37) What is the load factor in a 30-degree bank?
38) What is the maneuver flight envelope? Discuss.
39) Discuss features of two of the lines in a $V-n$ diagram that are related to aerodynamics.
40) What is the maximum load factor for acrobatic aircraft for normal flight?
41) Briefly describe the gust loads.
42) List typical materials for aircraft structures.
43) What is the structure of most current small RC planes – made by hobbyists – made of?
44) What are the top three factors that influence the selection of the structural materials for an air vehicle?
45) What advantages do composite construction offer in the fabrication of UAVs?
46) What are the most widely used materials for aircraft structures?
47) Briefly describe how to improve the mechanical properties of aluminum.
48) Write the tensile strength for aluminum alloys.
49) In the Northrop Grumman RQ-4 Global Hawk; what are the fuselage and the wing made of?
50) Composite materials are formed from two dissimilar parts. What are they? Then write the function of each part.
51) Write the three most popular materials for a matrix of composite materials.
52) Write three popular materials for the fiber of composite materials.
53) Write five examples of discontinuities in the wing/fuselage surface which cause stress concentration.
54) Briefly describe the format and features of sandwich structures.
55) Write three main segments for sandwich structures. Then write the function of each segment.
56) The skin fabrics come in two primary configurations or patterns. What are they?
57) Write popular reinforcing materials for a sandwich construction.
58) What should be the feature of the radome with respect to radio waves?
59) What is the primary function of resin in composite materials?
60) Write popular resin materials.
61) Write popular foam materials.
62) There are two types of composite manufacturing processes. What are they?
63) Write three basic techniques for open molding.
64) Write six methods employed in closed molding.
65) Write five common cross-sections of thin-walled columns.
66) Briefly describe the fail-safe approach in aircraft structural design.

Questions 157

67) Write five engineering tools and theories in structural analysis.
68) Briefly describe the superposition technique in simplifying distributed loads for supporting beams.
69) Basically, there are two approaches in formulating a structural problem. What are they? Describe each approach briefly.
70) Briefly describe the finite element method.
71) Write three commercial software packages in structural analysis.
72) Write five major stresses.
73) Briefly compare normal stress and shear stress.
74) Briefly compare stress and strain.
75) Define Young's modulus.
76) What is Young's modulus for: (a) aluminum and (b) carbon steel?
77) Briefly explain how the maximum bending stress is determined at the wing root.
78) What is the impact of fuel weight and lift at the wing structure for a Global Hawk?
79) Briefly describe buckling and buckling stress.
80) Define the safety factor for a static load.
81) Define the safety factor for a dynamic load.
82) What is the yield strength for aerospace aluminum?
83) Write the names of two failure theories employed in the calculation of the safety factor.
84) What is the recommended safety factor for a wing structure?
85) Define the phenomenon of structural fatigue.
86) Write three forms of a repeated load.
87) A MALE UAV left wing with a semi-span of 12 m has generated a lift of 20,000 N. Assume the load is mainly carried by its main spar (made of aluminum) with a second moment of area of 0.0009 m^4. Determine the deflection at the wing tip. Assume that the lift is a uniform distributed load along the span. Ignore the wing weight.
88) A HALE UAV left wing with a semi-span of 24 m has generated a lift of 34,000 N. Assume the load is mainly carried by its main spar (made of aluminum) with a second moment of area of 0.0015 m^4. Determine the deflection at the wing tip. Assume that the lift is a uniform distributed load along the span. Ignore the wing weight.
89) Consider the wing of a small UAV made of aerospace aluminum with one spar. The wing spar has a square cross-sectional area (8 cm × 8 cm) and the half wingspan is 5 m. The half wing has created 4,000 N of lift in a cruising flight. Can this wing withstand such a load? Assume that the lift is a distributed load along the span; you may assume it is located at the wingtip. Ignore the wing weight.
90) Consider the wing of a small UAV made of aerospace aluminum with one spar. The wing spar has a rectangular cross-sectional area (width of 12 cm × height of 15 cm) and a half wingspan of 8 m. The half wing has created 18,000 N of lift in a cruising flight. Can this wing withstand such a load? Assume that the lift is a distributed load along the span; you may assume it is located at the wingtip. Ignore the wing weight.

Part III

Mission Planning and Control

Mission planning and control are critical elements in the successful completion of any task by a UAS. Chapter 8 addresses the subsystems, configuration, and architecture of the mission control station, the interfaces within the control station and with the source of tasking and the users of information generated by the UAS, and the functions that are performed in the control station or the tasking organization.

Chapter 9 discusses the operational features of how the AV and the payloads are controlled in terms of the degree of automation or "autonomy" that is possible and/or desirable. The options range from complete remote control through complete autonomy. As might be expected, the most common levels of operational control are not at either of these extremes, although both are possible and may be desirable in specific situations. In mini and very small UAVs and RC planes, the UAV is primarily controlled by a ground pilot remotely. However, in long-range and long-endurance UAVs, parts of the flight operations are left to the autopilot, which, via an automatic flight control system and a navigation system, controls and navigates the air vehicle. The elements and sensors of the autopilot, as well as flight control classifications, are also discussed in Chapter 8.

Introduction to UAV Systems, Fifth Edition. Paul Gerin Fahlstrom, Thomas James Gleason, and Mohammad H. Sadraey.
© 2022 John Wiley & Sons, Inc. Published 2022 by John Wiley & Sons, Inc.
Companion website: www.wiley.com/go/fahlstrom/uavsystems5e

8
Mission Planning and Control Station

8.1 Introduction

The mission planning and control station, or MPCS, is the "nerve center" of the entire UAV system. It remotely controls the launch, flight, and recovery of the air vehicle (AV); receives and processes data from the internal sensors of the flight systems and the external sensors of the payload; remotely controls the operation of the payload (often in real time); and provides the interfaces between the UAV system and the outside world.

In this chapter, MPCS subsystems, architecture, physical configurations, and interfaces are addressed. Moreover, elements of the local area network, essential features of a standard local area network architecture, standards that are applied to UAV systems equipment including open system interconnection architecture, various layers of open system interconnection standards, and preflight mission planning using various planning aides will be discussed. Furthermore, roles of the human pilot in a UAS via MPCs are described.

8.2 MPCS Subsystems

The planning function can be performed at some location separate from the control function. The MPCS is sometimes called the ground control station or GCS. However, some capability for changing plans in real time to adapt to ongoing events during the mission is essential, and we will assume that at least a simple planning capability is available at the control site and use both terminologies as appropriate.

To accomplish its system functions, the MPCS incorporates the following subsystems:

a) Readouts and controls of AV status.
b) Displays and controls of payload data.
c) Map displays for mission planning and for monitoring the location and flight path of the AV.
d) The ground terminal of a data link that transmits commands to the AV and payload and receives status information and payload data from the AV.
e) One or more computers that, at a minimum, provide an interface between the operator(s) and the AV and control the data link and data flow between the AV and the MPCS. They may also

---

Introduction to UAV Systems, Fifth Edition. Paul Gerin Fahlstrom, Thomas James Gleason, and Mohammad H. Sadraey.
© 2022 John Wiley & Sons, Inc. Published 2022 by John Wiley & Sons, Inc.
Companion website: www.wiley.com/go/fahlstrom/uavsystems5e

162   *8 Mission Planning and Control Station*

perform the navigation function for the system, and some of the "outer loop" (less time sensitive) calculations associated with the autopilot and payload control functions.

f) Communication links to other organizations for command and control and for dissemination of information gathered by the UAV.

Experience has indicated that even for the simplest system, it is highly desirable to provide the operators with a "user-friendly" interface that integrates some of the basic flight and navigation functions and provides as much automation as possible in the control and navigation functions.

However, many operational requirements have evolved into a form that requires that the system be operated by personnel who need not have the degree of skill and training implied by either of those classes of "pilots."

The discussion in this chapter primarily addresses the configuration for an MPCS that automates the piloting of the AV to the extent that the operator needs only to make inputs telling the AV where to go, at what altitude, and perhaps at what speed, while the computers in the MPCS and the autopilot on the AV take care of the details of actually flying the desired path. Functions of an autopilot, various sensors and actuators in an autopilot, and automatic flight control system are discussed in Chapter 9.

There is greater leeway in the level of automation for operating the payload. In the simplest systems, an imaging payload such as a TV camera may be under almost complete manual control. The lowest level of "automation" would be to provide some inertial stabilization for the line of sight of the camera. Higher levels of automation include automatic tracking of objects on the ground to stabilize the line of sight or automatic pointing at a position on the ground specified by the operator as a set of grid coordinates.

At the highest level of automation, short of autonomy, the payload may automatically execute a search pattern over a specified area on the ground. At this level, the navigation, flight, and payload automation may be tied together in such a way that the AV flies a prespecified standard flight path that is coordinated with the automated payload pointing in such a way as to efficiently and completely search a specified area. Autonomous operation, in which the real-time supervision and participation of a human operator is replaced by artificial intelligence (AI) in software on the computers in the AV and control station is also possible. These levels of control are further discussed in Chapter 9.

Except for the rare case of a free flight system, the MPCS incorporates a data-communication link with the AV to control its flight. The flight may be controlled at a rather long distance or only within the line of sight. In the latter case, the AV may continue beyond the line of sight by following a preplanned flight path and preprogrammed commands to its mission area. If the mission area is within communication range (which is usually the line of sight for UHF systems), commands can be supplied to the AV to control the flight path and activate and control various sensor packages. If the UAV is to provide information, such as video imagery in the case of a reconnaissance air vehicle, the MPCS contains the means to receive the down-coming signal and display the information collected by the payload, such as a TV picture.

Command signals to the AV and sensors use the uplink of the data-link system and status, and sensor signals from the AV use the downlink. The MPCS therefore includes the antenna and transmitter to send uplink signals, and the antenna and receiver to capture downlink signals, along with any control functions that are required to operate the data link.

The data-link transmitter and receiver may have a second function related to AV navigation, particularly if the data link operates in a line-of-sight mode. It may measure the azimuth and range to the AV to determine the position of the AV relative to the ground station. This information may be used either as the sole source of position data for navigation or as supplementary data to correct drifts in an onboard AV navigation system. The almost universal use of the global positioning system (GPS) navigation has largely replaced both the inertial navigation system and the use of the data link for navigation, but in some systems intended for use where GPS might be jammed, these capabilities might be retained.

The MPCS must display two types of information to the operators: (1) flight data and (2) payloads and sensors data. Control of the AV itself requires a display of basic status information such as position, altitude, heading, airspeed, and fuel remaining. This may be displayed much as it would be in the cockpit of a manned aircraft, using anything from analog gauges to digital text and graphics displays, but new systems are likely to use digital display screens for all information presented to the operators, even if some of it is presented as images of analog "gauges" or displays. This is consistent with the movement to "glass cockpits" in most manned systems.

The reason for this trend is that the digital displays can be reconfigured in real time to show whatever is needed and provide great flexibility in adapting a control station to different payloads and missions or to different AVs. On the human interface side of this choice, operators are likely to be very comfortable with a graphical user interface and navigation through various "windows" using a mouse and keyboard.

The second type of information to be displayed consists of the data gathered by the onboard sensors of the payload. These displays can have many and varied features, depending on the nature of the sensors and the manner in which the information is to be used. For images from TV or thermal cameras, the display is a digital video screen. The frames can be held stationary (freeze frame) and the picture can be enhanced to provide greater clarity. Other types of data can be displayed as appropriate. For instance, a radar sensor might use either a pseudo-image or a traditional "blip" radar display.

A meteorological sensor might have its information displayed as text or by images of analog gauges. An electronic warfare sensor might use a spectrum analyzer display of signal power versus frequency and/or speakers, headset, or digital text displays for intercepted communications signals. It is generally desirable to add alphanumeric data to the sensor display, such as the time of day, AV position and altitude, and payload pointing angles.

It is desirable to provide recording and playback capability for all sensor data, to allow the operators to review the data in a more leisurely manner than is possible in the real-time displays. This also allows the data to be edited so that selected segments of the data can be transmitted from the MPCS to other locations where it can be used directly or further analyzed.

Control inputs from the operators for both the AV and the sensor payload may be accomplished by any of a large variety of input devices (such as joysticks, knobs, switches, mice, or keyboards). Feedback is provided by the status and sensor displays. If joysticks are used, some tactile feedback can be provided by the design of the joystick. Airborne visual sensors can be slewed, fields of view can be changed, and the sensors themselves can be turned on and off.

The position of the AV over the ground must be known in order to carry out the planned flight path and to provide orientation for the use of the sensors. Furthermore, one common use for a UAV is to find some target of interest and then determine its location in terms of a map grid. The UAV sensor typically provides the location of the target relative to the AV. This information must be combined with knowledge of the location of the AV in order to determine the target location on the map grid.

In the simplest system, the MPCS might display the grid coordinates of the AV as a numerical readout, allowing the operators to plot its location on a paper map and to determine target locations relative to that position by manually plotting the azimuth and range of the target from the AV position. Most UAV systems automate at least part of this function by automatically plotting the position of the AV on either a paper or digital video display and automatically calculating the location of the target, which may be displayed on the same plot and/or provided as numerical text on a video display.

Finally, since the information obtained from the AV, and/or its status, is important to someone outside of the MPCS, the equipment necessary to communicate with whoever provides tasking and commands to the UAV operators and with the users of the data is an essential part of the MPCS.

From its name, it is evident that pre-mission planning, that is, determination of optimum flight routes, target and search areas, fuel management, and threat avoidance, is a function carried out in the MPCS. Also included in modern MPCS systems are a feature for self-test and fault isolation as well as a means for training operators without requiring actual flight of the AV (built-in simulators).

A block diagram of an MPCS is shown in Figure 8.1. Most of the elements of the MPCS will be connected by a high-bandwidth bus. The unconnected block for communications with the rest of the organization to which the UAS belongs represents voice and other links to upper levels of the command structure and to any other elements that provide support to the UAS in the form of supplies or services. It may also include voice communications with users of the information produced by the UAS. All of this may be included in a network connection of some sort, either the same network as is used to distribute the video and other high-bandwidth data to users or a separate network that may be lower in bandwidth but may have a broader domain.

Power to run the system is provided by various sources, ranging from a standard power network for fixed locations though generators and down to batteries for the smallest and most portable control stations. In general, there are two groups of functions for an MPCS: (1) planning and (2) operation. In summary, the functions of an MPCS can be grouped as listed in Table 8.1.

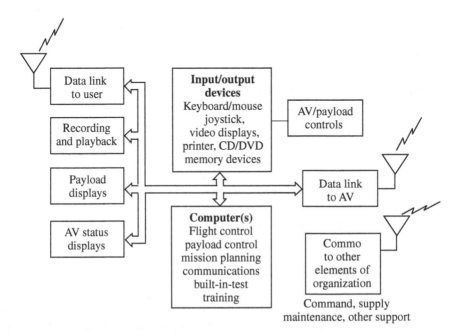

Figure 8.1 MPCS block diagram

8.3 MPCS Physical Configuration

Table 8.1 Functions of an MPCS

Groups	Functions
Planning	1) Process tasking messages
	2) Study mission area maps
	3) Designate flight routes (waypoints, speeds, altitudes)
	4) Provide operator with plan
Operation	5) Load mission plan information
	6) Launch UAV
	7) Monitor UAV position
	8) Control UAV
	9) Control and monitor mission payload
	10) Recommend changes to flight plan
	11) Provide information to the commander
	12) Save sensor information when required
	13) Recover UAV
	14) Reproduce hard copy or digital tapes or disks of sensor data

The functions in the planning group should be implemented prior to the launch of the UAV.

8.3 MPCS Physical Configuration

In general, the control station does not necessarily need to be located on the ground; it also may be located in a submarine, ship, ground vehicle, or another aircraft. A ground control station physical configuration (i.e., type) range from a small handheld remote controller to a mobile truck or to comprehensive fixed central command stations.

All of the equipment of the MPCS is housed in one or more containers that almost always must be portable enough to displace and set up a new base of operations rapidly. Ground control stations can be in the following configurations: (1) handheld radio controller, (2) portable ground control station (GCS), (3) close-range UAS GCS, (4) medium-range UAS GCS, (5) long-range UAS GCS, and (6) central station.

Most small homebuilt UAVs, quadcopters, and RC model airplanes are controlled via a small handheld radio controller, which weighs about 1–2 pounds. This type of GCS (see Figure 8.2) has often two joysticks/levers and a number of buttons with an antenna and uses a radio signal to send a command to the air vehicle. The pilot should have a visual line of sight and the control is in real time. In order to change the flight path, the pilot will push/pull or move to the left/right stick/lever to create a moment/force on the UAV, which causes a change in the UAV attitude. There is a stick for activating engine(s) in order to increase/decrease airspeed. A typical range for such a GCS is about a few hundred feet.

A handheld controller may only have a transmitter to just send the pilot commands. This is the lowest level of remote control. As a commercial example, the Spektrum DSMX 6-channel 2.4 GHz is a dual stick radio transmitter. The next level for this type of controller is to have a transmitter to send the pilot commands, plus a receiver for receiving the payload data (e.g., an image or video

Figure 8.2 A radio-controlled model aircraft with its controller

taken by a UAV camera). As a commercial example, the Radiolink AT10II 2.4 GHz with 12 channels and dual stick and the Flysky FS-i6 2.4 GHz with 6 channels and dual stick are equipped with both a radio transmitter and receiver. Some handheld controllers may have an LCD display to illustrate the control mode and setup options to enable the pilot to program the controller.

In a rudimentary form, the MPCS could consist of something not much more sophisticated than a radio-controlled model aircraft control set, a video display for payload imagery, paper maps for mission planning and navigation, and a tactical radio to communicate with the world outside the UAV system. This might be adequate for a UAV that flies within short-range line of sight (LOS) and can be controlled much like a model airplane.

Traditional and high-end RC planes rely on dedicated handheld controllers to operate, but modern technology made cellphones (an extremely capable smart device) as the second option. A number of manufacturers are planning to use cellphones for their new quadcopters. There are modern small quadcopters in the market that can be directly controlled and monitored via cellphones. For instance, the following quadcopters can be controlled by a cellphone: (1) DJI Mavic 2 Pro, (2) Yunee Mantis G, (3) Parrot Anafi, (4) Ryze Tello, (5) Hubsan H501s, and (6) DJI Mini 2. These quadcopters require nothing more than an app and a WiFi or Bluetooth connection.

The payload data – that a receiver is receiving – can be stored or displayed in real time. For this goal, either a computer with display (e.g., laptop) or a cellphone with an appropriate app is required. When a pilot is able to view the camera output in real time, he/she can control the UAV through this display. This type of control – where there is a first-person view (FPV) or a remote-person view (RPV) – is referred to as video piloting. As a commercial example, the Holy Stone HS175 quadcopter is equipped with a 2K HD camera that has the capability for FPV via a cellphone.

The third level in line is to upgrade the receiver to add a feature to receive the UAV live flight data (e.g., location and airspeed). The flight data can just be shown as values for flight parameters or to depict the UAV location on a background map, with other data on a side window. Now there are generally two groups of data for displaying in a GCS: (1) payload (e.g., camera) output and (2) UAV flight data. A larger display (e.g., laptop) is necessary to demonstrate both information. An on-screen display (OSD) readout can be used to show the navigational data with analog FPV video feed. Figure 8.3 illustrates a Worthington Sharpe portable GCS that includes a large display.

A portable MPCS can be placed in a suitcase or briefcase/backpack size containers. However, more sophisticated mobile MPCS use one or two shelters mounted on trucks that can range from

8.3 MPCS Physical Configuration

Figure 8.3 Worthington Sharpe portable GCS (Source: Samworthington / Wikimedia Commons / CC BY-SA 4.0)

light utility trucks or tactical vehicles of the HMMWV (High Mobility Multipurpose Wheeled Vehicle, Humvee) class up to large trucks in the 5-ton and up class. The shelter must provide working space for the operators and environmental control for both people and equipment. The AAI RQ-7 Shadow 200 GCS is mounted on a light four-wheel truck. Two operators inside the GCS control the flight of Shadow 200 and handle/monitor its sensors and payloads. Another example is the RQ-2 Pioneer whose ground control station could be housed in a shelter on an HMMWV or a truck.

The General Atomics MQ-1 Predator GCS is housed in a commercially available trailer or a lightweight truck. Figure 8.4 shows the operator's workstation for a Predator UAV with positions for the

Figure 8.4 MQ-1 Predator Operator's workstation (Source: General Atomics Aeronautical Systems Inc.)

pilot and payload operator and multiple digital displays showing maps, AV and payload status information, sensor imagery, and anything else needed to allow the operators to control the functions of the AV. This particular workstation is located in a truck and is designed for fixed installations. Similar workstations for mobile control stations would share displays between the pilot and payload operator and take other steps to reduce the total space required, but would still have to offer all the functionality as this complete system.

The size of the MPCS shelter is driven by the number of personnel and the amount of equipment that must be housed. As electronics and computers have become smaller and smaller, the number of personnel and desired displays has become the primary driver. It is usually desirable to have an individual AV operator and a payload/weapons operator seated side by side. There is often a mission commander who supervises and directs the air vehicle and payload operators and acts as an overall coordinator.

The mission commander usually also operates the interfaces between the UAV and the command-and-control system. It is convenient if the mission commander is located so that he/she can see both the AV status and sensor displays. This can be accomplished either with a separate workstation that can call up both sets of displays or by locating the mission commander so that he can look over the shoulders of the two operators and use their displays.

When observing something interesting, the payload operator can freeze the frame, or slew the sensor in the proximity of the interesting observation to see if additional information is available. An intelligence officer or other user must have access to this information in order for it to be useful. This can be accomplished either by locating the user within the ground station or by making provisions for remote displays.

Some users of data may want the ability to make direct, real-time inputs to sensor or AV control. This usually is not a good idea. In most cases, control of the mission by persons outside of the control station should be limited to providing tasking carried out by the dedicated operators within the station. That is, if a commander wants to look again at a particular scene, it is better to require the information to be requested from the mission commander rather than to give the commander a duplicate joystick to slew the payload in real time.

Only the crew within the control station has the full situational awareness and training to know how best to carry out the tasking without placing the AV in jeopardy or disrupting the flight plan. Often, the best way to provide a second look will be to play back a recording of the first look – hence the importance of recording the scene and providing editing and routing capabilities for selected data.

The manner in which all of the equipment is connected and placed in the shelter is called the equipment configuration, so as not to confuse it with the computer architecture or software configuration. Figure 8.5 shows a typical equipment configuration (i.e., ground station setup) for a MALE UAV.

Many of the functions and equipment described, such as the mission monitor, map display, AV status readouts, control input devices (joystick, track ball, potentiometer), and keyboard, can be combined into one or more common consoles or workstations. All of the electronic interfaces to communicate with the other workstations (if any), the data link, a central computer (if one is used), and communications equipment are contained within the workstation.

An unmanned aircraft pilot may be referred to as: (1) remote controller, (2) remote pilot, (3) operator, or (4) pilot-in-command.

Table 8.2 provides ground control station configurations with their primary characteristics.

The small Holy Stone HS720E quadcopter is remotely controlled via a handheld controller, while the pilot can watch the camera output through a cellphone. The pilot of an AeroVironment RQ-11 Raven is often controlling the vehicle via a *portable* MPCS in an open field.

8.4 MPCS Interfaces

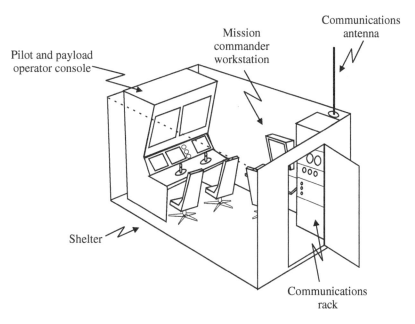

Figure 8.5 Ground station setup for a MALE UAV

Table 8.2 Functions of an MPCS

No.	Configurations/Types	Weight/Size/Feature	Typical Range
1	Handheld radio controller	0.2–1 pounds, requires line-of-sight	Few hundred feet
2	Portable ground control station	Back-packed, comparable to the size of a laptop, in suitcase, requires line-of-sight	Few thousand feet
3	Close-range UAS GCS	In a van, requires line-of-sight	10–20 miles
4	Medium-range UAS GCS	In a light four-wheel truck, could be beyond line-of-sight	100–1,000 miles
5	Long-range UAS GCS	In a large truck, could be beyond line-of-sight	Hundreds to thousands of miles
6	Central station	In a building/hangar or a number of trucks	Unlimited

Some UAVs such as Northrop Grumman RQ-4 Global Hawk and Boeing Insitu ScanEagle can be remotely controlled by a pilot in an office inside a building thousands of miles away. To provide such very long-distance remote control, satellite communication is employed.

8.4 MPCS Interfaces

The MPCS must interface with other parts of the UAV system and with the outside world. Some of these interfaces have already been discussed in some detail. The required interfaces can be summarized as follows:

- *The AV*: The "logical" interface from the MPCS to the AV is a bridge or gateway from the MPCS LAN to the AV LAN, via the data link. The physical interface may have several stages: (1) from the MPCS LAN to a data-link interface within the MPCS shelter; (2) from the shelter-mounted part of the data link to the modem, radio frequency (RF), and antenna parts of the data link at a remote site; (3) from the data-link transmitter via the RF transmission to a data-link RF and modem section in the AV (the air data terminal); and finally (4) from the modem in the AV to the air vehicle LAN. In some systems, the link from the ground transmitter to the air vehicle may itself involve several stages from the ground to a satellite or airborne relay and from there to other satellites or airborne relays and finally to the AV.
- *The launcher (catapult or rail)*: This interface can be as simple as a voice link (wire or radio) from the MPCS shelter to the launcher. In some systems, there will be a data interface from the MPCS LAN to the launcher, and perhaps to the AV, while it is still on the launcher, either via the launcher or directly to the AV. This interface allows the MPCS to confirm that the AV is ready to launch, command the AV to execute its launch program, and command the launch itself. When the AV takes off from a runway or aircraft carrier deck, this link is likely to be a simple voice link to the ground or deck crew supporting the AV.
- *The recovery system*: This interface can vary from a voice link to the recovery system up to more elaborate data links. In the simplest case, the AV automatically flies into some type of net and the only communication between the MPCS and the recovery system is to confirm that the net is ready and that any beacons on the net are operating. Another possibility is a manual landing involving a pilot who can see the AV and flies it in the manner of a radio-controlled model aircraft, in which case there will be a remote AV control console that is used by an operator to fly the AV and must be linked to the AV either by its own short-range data link or through the MPCS.
- *The outside world*: The MPCS must have the communications interfaces to operate within whatever communications nets are used for tasking and reporting. If the UAV is being used for fire control, this may include dedicated fire-control networks such as the Army's tactical fire-control network. In addition, if the MPCS is responsible for remote distribution of high-bandwidth data (such as a live or recorded video), it requires special data links to the receivers of the remote users. In a simple case, this might consist of coaxial or fiber-optic cables to a nearby tactical headquarters or intelligence center. If long distances are involved, high-bandwidth RF data links may be used, with their own special requirements for antennas and RF systems.

All of these interfaces are important, but the two interfaces that reach outside the immediate vicinity of the MPCS (the interface to the AV via the data link and the interfaces to the outside world) are the most important and critical. These two interfaces are the ones that will be least under the control of the MPCS designer, and are most likely to involve significant external constraints on data rates and data format.

The interface to the AV via the data link is the subject of Part Five of this book. The interface to the outside world is equally important, but is beyond the scope of this book, and is not further discussed.

8.5 MPCS Architecture

8.5.1 Fundamentals

The word "architecture," when applied to the MPCS, is generally used to describe the data flow and interfaces within the MPCS. Every MPCS has an architecture in this sense. However, the importance and visibility of this architecture is closely linked to the importance ascribed to three basic concepts in UAV system design:

8.5 MPCS Architecture

1) "Openness" describes the concept of being able to add new functional blocks to the MPCS without redesigning the existing blocks. For instance, an "open" architecture would allow the processing and display needed for a new AV sensor, as well as the data flow to and from that sensor, to be added to the MPCS simply by plugging a new line-replaceable unit into some type of data bus within the MPCS or even by just adding new software. This process is similar to the addition of a new functional board to a desktop computer.
2) "Interoperability" describes the concept of an MPCS that is capable of controlling any one of several different AVs and/or mission payloads and of interfacing with any of several different communications networks to connect with the outside world.
3) "Commonality" describes the concept of an MPCS that uses some or all of the same hardware and/or software modules as other MPCS.

These three concepts clearly are not independent. In many ways, they are different ways of describing the same goal from different viewpoints. An open architecture facilitates interoperability by accepting new software and hardware to control a different AV or payload and facilitates commonality by the very act of accepting that software or hardware. Interoperability and commonality are easier to achieve in an open than in a closed architecture. However, none of the three concepts automatically include the other two. One could, in principle, have a completely open architecture that had no interoperability or commonality with other UAV or "outside world" systems.

As the nerve center of a UAV system, the MPCS must carry much of the burden for establishing openness, interoperability, and commonality. The MPCS generally is the most expensive single subsystem of the overall UAV system, and is the least exposed and expendable part of the system. Therefore, it makes sense to maximize its utility and to concentrate the investment in interoperability and commonality in the MPCS.

Within a single UAV system, the second most "profitable" target for commonality and interoperability is the AV, where the ability to accept common payloads, data links, navigation systems, and even engines can have a major impact on the cost and utility of both the single system and of an integrated family of UAV systems operated by a single user. Many of the architectural concepts discussed below for the MPCS apply directly to the AV as well.

The data link, despite being treated in this book as a separate subsystem of the UAV, has as its primary function the "bridging" of the gap between the MPCS and AV subsystems. When viewed in this sense, the data link would ideally be a transparent link in the overall data architecture of the system. In fact, practical limitations make the link non-transparent in most systems, whose characteristics must be taken into account in the architecture and design of the rest of the system.

The architectural issues related to how an MPCS addresses openness, interoperability, and commonality requirements are most easily visualized in terms of the concept of a local area network (LAN). Within this concept, the MPCS and AV can be visualized as two LANs that are "bridged" with each other (via the data link), and "gateways" connect the UAV system with other command, control, communication, and intelligence systems of the user organization (the outside world). The MPCS architecture determines the structure that allows functional elements to operate within the MPCS LAN, interfaces to the AV LAN through the data link "bridge," and provides the "gateways" required to interface with other networks in the outside world.

The concepts of LAN, bridging, and gateways are all part of the jargon in common use by the telecommunications community. It is beyond the scope of this book to describe them in detail. However, a general understanding of these concepts provides a background that allows a UAV system designer to visualize how the MPCS performs its function as the system nerve center and forms a basis for understanding the architectural issues raised by any specific set of system requirements.

8.5.2 Local Area Networks

LANs originated in the 1970s when microcomputers began to proliferate in our society. Prior to the microcomputers, offices and companies maintained large mainframe computers connected to dumb terminals (terminals that have no built-in computing capability). The central computer shared time with each terminal but directly handled all external information flow to printers and the users located at the terminals. The introduction of microcomputers allowed computing functions to be distributed among a large number of "smart" terminals and "smart" peripheral devices such as printers, displays, and special-purpose terminals with embedded central processing units (CPUs), memory, and software.

Each of these nodes might be performing a variety of independent functions at its own rate, but might also need to interchange data or to make use of functions available only at another node (e.g., printing). Sharing of data and facilities such as memory was possible if a means was provided to interconnect all the independent processing nodes. This function is performed by the LAN.

An MPCS is in effect a miniature office. Information in the form of AV status, wide-band video signals, communications with other elements of the organization, and other signals are received and processed to provide video imagery, target data, control payloads and AVs, stored, printed, and sent to intelligence centers and operational commanders. Just as in the office, information is shared within the MPCS and sent to other offices (UAV and military systems). LAN concepts are quite appropriate to describe MPCS communication architectures.

LANs have three critical elements: (1) layout and logical structure, (2) communications medium, and (3) network transmission and access. These three elements are discussed in Section 8.6.

8.5.3 Levels of Communication

Communication between devices can consist of the transmission of unformatted data between the two. For instance, text from a computer using word processor brand A might be transmitted to a second computer using word processor brand B on a simple wire circuit. If the two word processors are incompatible, then a common set of characters must be found that both word processors understand. In this case an ASCII set of characters can be utilized, but since this set is limited, some of the information used by a word processor, such as underlining or italics, may be missing. There are three levels of communications between devices: (1) basic level, (2) enhanced level, and (3) open communication system.

The words and sentences in the message would be retained, but some essential information may be lost when formatting and emphasis are deleted. This level of communication is called the basic level. The problems related to unformatted data are serious even for text. They are essentially insurmountable for graphics or specialized command or sensor data.

A second level, called an enhanced level, is communication between devices using a common format that retains all special coding. Many proprietary network architectures exist that operate on the enhanced level with proprietary formats, and thus are not able to communicate with one another. This is something that the UAV community does not want to happen with an MPCS.

The level of communications in which any device can communicate with any other device in a format that retains all information, regardless of manufacturer and their internal formats and protocols, is an open communication system.

Realizing and implementing the critical characteristics necessary for the operation of an open LAN is a major undertaking. If all the devices, software, cabling, and other hardware were manufactured and operated by one entity, it would not be difficult to make them all work together.

8.5 MPCS Architecture　　173

However, even if one company manufactured all UAV system's hardware and software, the problem would remain because the UAV system must operate with other weapon and communications systems that may come from different countries and use different data protocols.

To provide a level of uniformity, it is necessary to design and operate using a set of standards. De facto standards exist in the telecommunications industry today. They are set by the leaders of the industry and everyone else follows. Standards are also set by mutual agreement among governments, manufacturing groups, and professional societies.

Many different standards presently are applied to UAV systems equipment. In the United States, the Unmanned Vehicle Joint Project Office (JPO) and Joint Integration Interface (JII) Group has recommended standardization using the International Organization for Standardization (ISO) Open System Interconnection (OSI) architecture. At a minimum, the OSI model provides the framework from which more detailed standards can be applied. Other standards such as MIL standards and RS-232C standard still apply within the OSI architecture standard. A discussion of the OSI standard illustrates the essential features of a standard LAN architecture. In Section 8.7, layers of the Open System Interconnection (OSI) standard are briefly presented.

8.5.4　Bridges and Gateways

Bridges are connections between LANs that have similar architectures, such as a UAV ground station and its AV. In the UAV case, they are connected via the data link. Unless the data link is designed originally to interface directly to the LAN, it will require a processor at the interface to the LAN that converts the data addressed to the data link or the AV into whatever format is required by the data link and converts downlinked data into the formats required by the LAN in the MPCS. A similar processor will be required at the AV end of the data link.

The data link has two identities within the LAN. It is a "peripheral device" within the LAN that may receive requests from other nodes in the LAN that consist of commands to the data link with regard to antenna pointing, use of anti-jam modes, and so on. It may also provide data to other nodes within the MPCS, such as antenna azimuth and range to the AV. In its other role, it is the bridge to the AV. In this role, it should be relatively transparent to the LANs in the MPCS and on the AV.

If the LAN in the AV has a different architecture than that in the MPCS, then the data link becomes a gateway. The interfaces to the outside world will generally be gateways.

A gateway connects diverse architectures. UAV ground stations may be required to communicate with other communication stations such as the Joint Surveillance and Target Acquisition Radar System (JSTARS). Until the time comes when all systems are designed to the same standard, communication between JSTARS and a typical MPCS is similar to a computer that employs a Windows operating system talking to a LINUX computer or an Apple computer or an Apple cellphone (which uses the iPhone operating system, iOsS). They do not understand each other unless there is an explicit interface that does the necessary translation.

A gateway is a node within the LAN that converts formats and protocols to connect to a different architecture outside of the LAN. Note that the distinction between gateways and bridges within the UAV system may blur. One could consider the data link an outside network and construct gateways to it at both the MPCS and AV ends. These gateways would function in a manner very similar to the interface from the data link to the bridge interface of the LAN when the data-link interface is considered a bridge. The difference is that the interface would now be within the LAN instead of within the data link.

As discussed in the chapter on data links, it is usually desirable to make the details of the data link transparent to the MPCS and AV. This suggests making the data link accept the formats and protocols of the LANs at both ends (act as a bridge). This approach makes it much easier to exchange data links, since the bridge interface in the LAN does not change. If the LAN must provide a gateway interface to each data link, then changing data links format also requires changing the gateway.

Four chapters, Chapter 13, Data-Link Functions and Attributes, Chapter 14, Data-Link Margin, Chapter 15, Data-Link Reduction, and Chapter 16, Data-Link Tradeoffs, are dedicated to various features of data link.

8.6 Elements of a LAN

Local Area Networks were introduced in Section 8.5.1. Here, critical characteristics and elements of LANs are reviewed. LANs have three critical characteristics: (1) layout and logical structure, (2) communications medium, and (3) network transmission and access. These three characteristics are discussed in the following sections.

8.6.1 Layout and Logical Structure (Topology)

A set of workstations, computers, printers, storage devices, control panels, and so on can be connected in parallel on a single cable to which they all have simultaneous access. This is called a "bus" topology. Alternatively, they can be connected sequentially on a single cable that is in the shape of a loop, called a "ring" topology. Finally, a network in which each device is connected directly to a central controller is called "star" topology.

A bus uses a single linear cable to connect all the devices in parallel. Each device is connected by a "tap" or a "drop" and must be able to recognize its own address when information in the form of a packet is broadcast on the bus. Since all devices are attached linearly to the bus, each one must be checked in sequence to find a fault.

Since all devices have simultaneous access to the bus, there must be some protocol to avoid conflicts if more than one device wants to broadcast at the same time. This typically is accomplished by introducing random delays between receipt and transmission of messages from each device to ensure that there are openings for other devices to use the bus. This does not ensure a lack of conflict, so a bus system also has a means of determining that a conflict has occurred and some type of methodology for trying again with a lower probability of conflict. Sometimes this consists of increasing the length of the random delays in transmission. Clearly, when a bus becomes busy, it may become a very slow way to interconnect the devices.

A ring is on a single cable like a bus, but the cable closes on itself to form a ring. The devices are connected to the ring by taps similar to a bus, but the connections are sequential rather than parallel. Each device can communicate directly only with the next device in the ring. Information packets are passed along the ring to a receiver/driver unit, in which the receiver checks the address of the incoming signal and either accepts it or passes it to the driver where it is regenerated and sent to the next device in the ring.

A special packet called a token is sent around the ring and when a device wants to transmit, it waits for the token and attaches its message to the token. The receiving device attaches an acknowledgment to the token and reinserts it into the ring. When the transmitting device receives the token with an acknowledgment, it knows that its message has been received. It removes the message and sends the token to the next device. The token can be "scheduled" to go to some device other than the one physically next around the ring.

8.7 OSI Standard   **175**

The routing of the token can provide some devices with more opportunities to transmit than others. For instance, if device A has a great deal of high-priority data to transmit, the token might be scheduled to return to device A every time it is released by any other device on the ring. This would effectively allocate about half of the total ring capacity to device A. This "token ring" is a simple way of preventing two or more devices from transmitting information at the same time. In other words, the token-passing concept prevents the collision of data or information.

A star system is one in which each device is connected directly to a central controller. The central controller is responsible for connecting the devices and establishing communications. It is a simple and low-cost method of interconnecting devices that are in close proximity such as those in a mission planning and control system.

8.6.2 Communications Medium

The movement of signals within a LAN can be via ordinary wires, twisted pairs, shielded cable, coaxial cable, or fiber-optic cable. The choice of medium affects the bandwidth that can be transmitted and the distance over which data can be transmitted without regeneration. Fiber-optic cable is far superior in bandwidth to any electronic medium and has the additional advantage of being secure against unintentional emissions and immune to electromagnetic interference.

8.6.3 Network Transmission and Access

The way in which devices access the network (receive and transmit information) is of paramount importance. Data must not collide (two devices transmitting at the same time) or it will be destroyed. A device must also be able to determine if it is the intended recipient of the data, so it can either receive it or pass it on.

8.7   OSI Standard

The communications between devices and levels of communication were introduced in Section 8.5.2. In this section, layers of the Open System Interconnection (OSI) standard are discussed. The OSI model or standard has seven layers: (1) physical layer, (2) data-link layer, (3) network layer, (4) transport layer, (5) session layer, (6) presentation layer, and (7) application layer.

8.7.1 Physical Layer

The physical layer is a set of rules concerning hardware. It addresses the kind of cables, level of voltages, timing, and acceptable connectors. Associated with the physical layer are specifications such as RS-232C, which specifies which signal is on which pin.

8.7.2 Data-Link Layer

The first layer (physical) gets the bits into the transmission system, rather like the slot in the mailbox. The second layer (data-link) specifies how to wrap them and address them, so to speak. This second layer adds headers and trailers to packets (or frames) of data and makes sure the headers and trailers are not mistaken for the data. This layer provides a protocol for addressing messages to other nodes on the network and for providing "data about the data" that will be used in error-correction routines and for routing. A MIL standard may be used to spell out the details as to how this will be done. One standard in general use is MIL-STD-1553 "Aircraft Internal Time Division Command/Response Multiplex Data Bus."

8.7.3 Network Layer

The network layer establishes paths between computers for data communications. It sets up flow control, routing, and congestion control.

8.7.4 Transport Layer

The transport layer is concerned with error recognition and recovery.

8.7.5 Session Layer

The session layer manages the network. It recognizes particular devices or users on the network and controls data transfers. This layer determines the mode of communication between any two users such as one-way communication, two-way simultaneous, or two-way alternating.

8.7.6 Presentation Layer

The presentation layer makes sure that the data can be understood among the devices sending and receiving information by imposing a common set of rules for presentation of data between the devices. For example, if a device provides color information to both a color and monochrome monitor, the presentation layer must establish a common syntax between the two so that a particular color could represent highlighting on the monochrome screen.

8.7.7 Application Layer

The application layer acts as the interface between software and the communications process. This layer is the most difficult to standardize because it deals with standards that interface with a particular device and by their very nature are nonstandard. The application layer contains many of the underlying functions that support application-specific software. Examples include file and printer servers. The familiar functions and interface of the operating system (e.g., Microsoft Windows, Apple macOS, Linux, Android, and Apple's iOS) are part of the application layer.

8.8 Mission Planning

The functions are performed in the ground control station or the tasking organization can be divided in three groups: (1) preflight mission planning, (2) navigation and target location, and (3) remote control of the UAV and the payloads. The first function is discussed in this section, while the other two functions (navigation and control) will be presented in Chapter 9.

As with manned aircraft flights, preflight planning is a critical element in successful mission performance. The complexity of the planning function depends on the complexity of the mission. In the simplest case, the mission might be to monitor a road junction or bridge and report traffic

8.8 Mission Planning

passing the monitored point. Planning for this mission would require determination of flight paths to and from the point to be monitored and selection of the area in which the AV will loiter while monitoring the point. This may involve avoidance of air-defense threats on ingress and egress, and almost always will require an interaction with an airspace management element. In a fairly simple environment, it may be no more complicated than preparation of a straightforward flight plan and filing of that flight plan with an appropriate command element. Main operations and the trajectory in a typical flight mission were depicted in Chapter 4 (Figure 4.1).

It may be necessary to select one or more loiter points prior to takeoff in order to avoid airspace conflicts in the vicinity of the target area. In this case, the planning function must take into account the type of sensor to be used, its field of regard and field of view, and its effective range. If the sensor is a video camera, the position of the sun relative to the targets and AV position may be a factor in selection of the loiter point. In rough terrain or heavy vegetation, it may also be important to predict what loiter point will provide a clear line of sight to the target area. It sometimes may be acceptable to fly to the area of the target and then find a good vantage point, but at other times it may be necessary to determine the vantage point before takeoff.

Even in this simple case, it is likely to be valuable to have automated planning aides within the MPCS. These aids may take the form of one or more of the following software capabilities:

- Digital map displays on which flight paths can be overlaid using some form of graphical input device (such as a light pen, touch screen, or mouse).
- Automatic calculation of flight times and fuel consumption for the selected flight path.
- Provision of a library of generic flight segments that can be added to the flight plan and tailored to the specific flight.
- Automatic recording of the flight path in forms suitable for control of the AV during the mission and for filing of the flight plan with the airspace management element.
- Computation of synthetic imagery, based on the digital map data, showing the views from various loiter positions and altitudes to allow selection of an acceptable vantage point for performance of the mission.

Storage of the flight plan for later execution means that once the plan is completed, it is stored within the MPCS in such a way that each phase of the plan can be executed simply by recalling it from memory and commanding that it be carried out. For instance, the mission plan might be broken down into segments such as flight from launch to loiter point, loiter at a given point, move to a second loiter point, and return to recovery point. The operators would then only have to activate each segment in turn in order to carry out the mission as planned.

A flexible software system would allow exits and entries into the preplanned mission at various points with minimum operator replanning. For instance, if an interesting target were seen while on route to the preplanned loiter point, it should be possible to suspend the preplanned flight segment, go into one of several standard orbits, examine the target, and then resume the preplanned flight segment from whatever point the AV has reached when the command to resume is issued.

More complicated missions may include several submissions with alternatives. This type of mission may put a premium on the ability to calculate times and fuel consumption so that all submissions can be accomplished on time and within the total endurance of the AV. To assist in such planning, it is desirable to have a "library" of standard task plans. For instance, there could be a library routine for searching a small area centered on a specified point.

The inputs to the library routine would include the map coordinates of the point, the radius to be searched around the point, the direction from which the area should be viewed (overhead, from the east, from the west, and so on), the clutter level anticipated in the target area, and the class of target being searched for. Based on known sensor performance against the class of target in the specified clutter, the library routine would compute the flight plan required to place the sensor at an optimum range from the target, the sensor search pattern and rate, and the total time to search the area.

The resulting plan would be inserted into the overall flight plan and the fuel consumption and time required for this segment of the mission would be added to the mission summary. The digital scene generator might be used to select the direction from which the area will be searched. As each segment was added to the mission summary, the planner could monitor the total scheduling of the mission and compatibility with times specified in the tasking and with the total mission time available from the AV.

While all of this planning can be performed manually, with the assistance of handbooks or by applying "rules of thumb" used to estimate search times and other key elements of the mission plan, experience with early UAV systems indicates that the effort put into automation of mission planning is likely to have a major payoff in terms of operator acceptance of the system and efficiency of the use of limited AV resources.

The target for a surveillance mission of a UAV could be:

1) Fixed (e.g., house, bridge, or power tower) or moving (e.g., car, individual human, or another aircraft).
2) Small (e.g., one house) or large (e.g., a farm).
3) Single (e.g., one tower) or multiple (e.g., a few towers).
4) No risk associated, or with associated risk (e.g., natural fire, volcano, or enemy fire).
5) Easy to find (a house in a large farm), or challenging to identify the target (e.g., needs a face recognition in a populate residential area).

To finalize a flight plan for such target(s), it sometimes requires a fair amount of calculations to ensure a safe and secure flight operation, and a successful payload result is achieved. To design an optimum flight plan, a number of factors are involved that could be optimized:

a) Energy consumption – this should be minimized.
b) Overall flight duration to be spent – this should be minimized.
c) Overall flight distance needs to be travelled – this should be minimized.
d) Number of probable obstacles – this should be minimized.
e) Efficiency of payload output – this should be maximized.

When there are multiple mission objectives, it is very challenging to optimize all factors with one flight plan. One method to incorporate all relevant mission requirements is to employ a technique known as the multidisciplinary design optimization (MDO). The MDO is a mathematical method that formulates the flight mission problem. Then the mission planner must choose a model to relate the constraints and the objectives to the flight variables. To derive the flight plan, this mathematical problem (often a series of differential equations) is normally solved using appropriate techniques from the field of optimization.

In 2021, the US Navy [3] conducted its first-ever aerial refueling between a manned aircraft and an unmanned tanker, with a Boeing-owned MQ-25 Stingray test vehicle performing its first midair tanking mission with a Navy F/A-18E-F Super Hornet. There are a number of challenges for this mission to plan, including controlling the unmanned aerial vehicle (e.g., height, distance, attitude angle, and speed) from the GCS by an MQ-25 ground operator.

8.9 Pilot-In-Command

In a non-full-autonomous UAV, a human is needed to remotely control the air vehicle and payload at some level. The human is communicating with the UAV via the ground control station. This section is dedicated to briefly describe the roles of the human pilot in a UAS via MPCs. Some organizations that operate UAV systems require a pilot-rated operator (or a qualified radio-controlled model airplane operator), who could, if required, actually fly the AV based on visual estimates of AV position and attitude.

Unmanned aircraft pilots often receive less training than "regular" pilots, but are extensively trained as pilots with an emphasis on piloting unmanned aircraft. Titles can include: (1) remote controller, (2) remote pilot, (3) operator, and (4) pilot in command (PIC). The term "pilot in command" is mainly utilized by Federal Aviation Administration (FAA) in dealing with UAV pilots.

The Department of Transportation's FAA has regulated operational rules [24] for commercial use of small unmanned aircraft systems (sUAS). From FAR Part 107, a small UAV is defined as one whose weight is less than 55 lb (i.e., its mass is less than 25 kg). Under Part 107, an sUAS can be operated only below 400 ft above ground level (i.e., local altitude), 500 feet below clouds, and must have at least 3 statute mile visibility.

Part 107 does not apply to model aircraft that satisfy all of the criteria specified in section 336 of Public Law 112-95. The rule codifies the FAA's enforcement authority in part 101 by prohibiting model aircraft operators from endangering the safety of the National Airspace System (NAS). The following are a number of basic rules from FAR part 107:

1) Visual line-of-sight (VLOS) only; the unmanned aircraft must remain within VLOS of the remote pilot in command and the person manipulating the flight controls of the small UAS. Alternatively, the unmanned aircraft must remain within VLOS of the visual observer.
2) At all times the small unmanned aircraft must remain close enough to the remote pilot in command and the person manipulating the flight controls of the small UAS for those people to be capable of seeing the aircraft with vision unaided by any device other than corrective lenses.
3) Small unmanned aircraft may not operate over any persons not directly participating in the operation, not under a covered structure, and not inside a covered stationary vehicle.
4) Daylight-only operations, or civil twilight (30 minutes before official sunrise to 30 minutes after official sunset, local time) with appropriate anti-collision lighting.
5) Foreign-registered small unmanned aircraft are allowed to operate in the US under part 107, if they satisfy the requirements of Part 375.

For the list of complete rules and regulations, the reader may study Reference [24]. Chapter 8 is dedicated to full coverage of piloting the air vehicle and payload control.

As shown in Table 8.1, there are various functions of an MPCS that are all conducted by PIC. The pilot in an MPCS must be able to access all relative information when needed. To allow a remote pilot to effectively control a UAV, the fundamentals of ergonomics (or human factors) should be observed.

Ergonomics is the science of designing human interaction with machine/equipment and workplaces to fit the user. The field of human factors engineering employs scientific knowledge about human characteristics in specifying the design and use of a human–machine system. The aim is to improve system efficiency by optimizing performance, while providing comfort and safety. The operator work environment (i.e., MPCS) should be comfortable and the interface must be effective; otherwise, there will be errors and operative fatigue and may lead to the loss of the UAV.

Proper ergonomic design is necessary to prevent repetitive strain injuries, which can develop over time and can lead to a long-term disability. For further information, you may refer to Reference [25], which provides an introduction, principles, fundamentals, and useful data for various aspects of human factors. Mode discussion on remote piloting is presented in Chapter 9 (Section 9.3.1).

Questions

1) What do AI, MPCS, LAN, GCS, GPS, LOS, VLOS, MDO, RPV, iOS, and OSI stand for?
2) What do PIC, NAS, FAA, HMMWV, sUAS, FPV, OSD, and CPU stand for?
3) What are the main functions of an MPCS?
4) Describe MPCS in one paragraph.
5) List MPCS subsystems.
6) Briefly describe two types of information that the MPCS must display to the operators.
7) Control inputs from the operators for both the AV and the sensor payload may be accomplished by any of a large variety of input devices. Write at least five devices.
8) Write at least two power sources to run the MPCS.
9) Write physical configuration of GCSs listed in this chapter.
10) What are main elements of a handheld remote control of a small UAV?
11) What is the typical range for a small handheld radio controller?
12) How many sticks does a Radiolink AT10II handheld controller have?
13) Does the Spektrum DSMX 6-channel 2.4 GHz handheld controller have a receiver?
14) Briefly describe the video piloting.
15) Briefly describe the characteristics of a portable MPCS.
16) There are generally two groups of data for displaying in a GCS. What are they?
17) How many operators are there inside the GCS of Shadow 200?
18) Briefly describe the characteristics of the MPCS of Shadow 200.
19) Briefly describe the characteristics of the workstation (i.e., MPCS) of General Atomics MQ-1 Predator.
20) What are three human jobs in a workstation for a MQ-1 Predator UAV?
21) Briefly describe the task of the mission commander in a GCS.
22) Provide a figure to illustrate the ground station setup for a MALE UAV.
23) What are the advantages of a truck for a medium- and long-range UAS GCS?
24) What are the differences between equipment contained in a close-range UAS GCS and a long-range UAS GCS?
25) What is the typical range for a long-range UAS GCS?
26) What is the typical range for a portable ground control station?
27) List four main required interfaces of an MPCS.
28) Briefly describe the MPCS interface with the AV.
29) The physical interface from the MPCS to the AV may have four stages. What are they?
30) The MPCS of a HALE UAV needs to have an interface with the outside world. Write a possible interface form.
31) What does the word "architecture" mean when applied to the MPCS?
32) Which three basic concepts in the UAV system design is closely linked to the MPCS architecture?
33) Write features of an "open" architecture in an MPCS.
34) Write the main feature of an "interoperable" architecture in an MPCS.

Questions **181**

35) Write the main characteristic of an architecture in an MPCS that features commonality.
36) What element plays the role of the nerve center in a UAV system?
37) What element bridges the gap between the MPCS and AV subsystems in a UAV system?
38) Briefly describe the need and the role of a Local Area Network in an MPCS.
39) Write at least two smart peripheral devices for microcomputers.
40) An MPCS is in effect a miniature office. Discuss why.
41) LANs have three critical elements. What are they?
42) Briefly describe what we mean by "communication between devices."
43) Write three levels of communication between devices.
44) Briefly describe the "enhanced level" of communication between devices.
45) Briefly describe the "open communication system" level of communication between devices.
46) What standardization is recommended by the Unmanned Vehicle Joint Project Office (JPO), Joint Integration Interface (JII) Group for MPCS architecture?
47) Briefly describe what we mean by "bridge" in communication.
48) Briefly describe what we mean by "gateway" in communication.
49) The data link has two identities within the LAN. What are they?
50) A computer that employs a Windows operating system is required to talk to an Apple cellphone (which uses an iPhone operating system). What communication element is needed to facilitate this communication?
51) LANs have three critical characteristics. Write down these three.
52) A set of workstations, computers, printers, storage devices, and control panels can be connected in three different topologies. What are they?
53) Write features of a "bus" topology in a LAN.
54) Write features of a "star" topology in a LAN.
55) What is the role of a special packet called a "token" in a LAN?
56) What is the role of a "token ring" in a LAN?
57) List typical communication mediums for the movement of signals within a LAN.
58) Write three advantages of a fiber-optic cable in communications.
59) Write layers of the OSI model or standard.
60) What is addressed by the physical layer of an OSI standard?
61) What is provided by the data-link layer of an OSI standard?
62) What does the MIL-STD-1553 standard cover?
63) What is the transport layer of an OSI standard concerned with?
64) What layer of an OSI standard is concerned with two-way simultaneous communication?
65) Which layer of an OSI standard is the most difficult to standardize? Why?
66) Name at least three UAVs that can be controlled by a cellphone.
67) Name at least three operating systems that may be employed for computers and communication systems.
68) Write at least four automated planning aides within the MPCS.
69) Write at least three missions mentioned in this chapter for mission planning.
70) Consider the following UAV flight mission. Monitor a road junction or bridge and report traffic passing the monitored point. What would a planning for this mission typically require? Discuss.
71) Consider the following UAV flight mission: a complicated mission that may include several sub-missions with alternatives. What would a planning for this mission typically require? Discuss.
72) Write at least five alternative targets for a surveillance mission of a UAV.

182   8 Mission Planning and Control Station

73) To design an optimum flight plan, a number of factors are involved that could be optimized. Write at least five factors.
74) Briefly describe the "multidisciplinary design optimization" technique.
75) Write at least four challenges in designing the mission plan for an aerial refueling by a UAV from a manned fuel tanker.
76) Write titles that are often used for unmanned aircraft pilots.
77) What part of FAR regulates the commercial use of small UAVs?
78) What is the definition of a small UAV by FAR 107?
79) Does Part 107 apply to model aircraft?
80) Write at least four basic rules from FAR part 107.
81) What is ergonomics?
82) Identify the type of MPCS of the following UAVs:
    a) General Atomics MQ-1 Predator
    b) AeroVironment RQ-11 Raven
    c) Holy Stone HS720E quadcopter
    d) Northrop Grumman RQ-4 Global Hawk
    e) AAI RQ-7 Shadow

9

Control of Air Vehicle and Payload

9.1 Overview

Fundamentals of autopilots, its primary elements, and particularly closed-loop control systems were presented in Chapter 5. This chapter discusses how the human operators exercise control over the UAV and its payloads. To organize that discussion, we will make use of the fact that the remote operators, usually assisted by processors/computers located both on the ground and in the AV, must perform the functions of the aircraft commander, pilot, copilot, radar and/or weapons operator, and any other functions that would be performed by humans onboard for a manned system.

While not all of these functions are present in every manned aircraft or every UAV, a pilot is always required, and for all but the most basic UAV missions a separate payload operator is commonly used. There must always be an aircraft commander, but in many manned aircraft that function is combined with that of the pilot.

The fact that the pilot, copilot, and aircraft commander are not in the aircraft and able to look outside through the windows to maintain awareness of the situation and surroundings is important. It alters the roles of these three functions relative to the payload operator function because all become dependent on the payloads for much of the information that they need about the situation external to the AV. For that reason, some unmanned systems combine the aircraft commander function with that of the payload operator.

However, these functions are divided between the "air crew," who are all required in a UAS. There are significant differences in the issues and tradeoffs associated with how each is performed. For the purposes of this discussion, we define the key functions (see Figure 9.1) as follows:

A) *Mission planning*: determining the plan for the mission based on the tasking that comes from the "customer" for whom the UAS is flying the mission.
B) *Piloting the aircraft*: making the inputs to the control surfaces and propulsion system required to take off, fly some specified flight path, and land.
C) *Controlling the payloads*: turning them on and off, pointing them as needed, and performing any real-time interpretation of their outputs that is required to perform the mission of the UAS.
D) *Commanding the aircraft*: carrying out the mission plan, including any changes that must be made in response to events that occur during the mission.

The main feature of these definitions that is different from most manned systems is to separate the "pure" piloting function from any discretionary functions associated with commanding the

Introduction to UAV Systems, Fifth Edition. Paul Gerin Fahlstrom, Thomas James Gleason, and Mohammad H. Sadraey.
© 2022 John Wiley & Sons, Inc. Published 2022 by John Wiley & Sons, Inc.
Companion website: www.wiley.com/go/fahlstrom/uavsystems5e

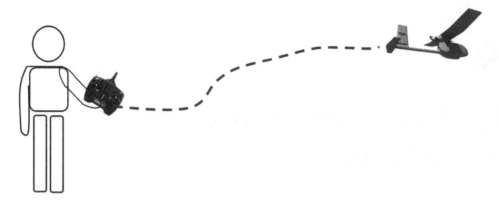

Figure 9.1 Remote piloting of the aircraft

aircraft. The pilot is responsible only for getting the aircraft from one point to the next. This includes dealing with any temporary upsets such as gusts, wind shear, or turbulence and continuing to fly the aircraft successfully, if possible, after loss of power or damage to the airframe, but does not include making decisions about where to go next or what to do next.

In this chapter, we also discuss operational control in considerable detail and describe the various levels of automation that may be applied to each of the functions that are required to complete a successful UAV mission.

9.2 Levels of Control

There are a number of levels of control that require various modes of operator interaction with the AV. They can be classified into six levels:

- *Level 1 – Partial remote control*: the humans do all the things that they would do – to control the flight – if they were onboard the AV, basing their actions on sensor and other flight instrument information that is downlinked to the operator station and implemented by direct control inputs that are uplinked to the AV. Payload data are saved on the memory card. It will be downloaded and used after the UAV is recovered.
- *Level 2 – Full remote control*: the humans do all the things that they would do – to control both the flight and payload – if they were onboard the AV, basing their actions on sensor and other flight instrument information and payload output. Payload data are transmitted in real time to the pilot. It will be used live and saved on GCS. For the case of a small quadcopter, both jobs (i.e., control both the flight and payload) can be performed by one human. However, for the case of a MALE UAV, at least two individuals (one flight pilot and one payload operator) are required.
- *Level 3 – Assisted remote control*: the humans still do all the things that they would do if they were on the AV, based on the same information downlinked to them, but their control inputs are assisted by automated inner control loops that are closed onboard the AV.
- *Level 4 – Exception control*: the computers perform all the real-time control functions based on a detailed flight plan and/or mission plan and monitor what is happening in order to identify any event that constitutes an exception to the plan. If an exception is identified, the computers notify the human operators and ask for directions about how to respond to the exception.

9.3 Remote Piloting the Air Vehicle

Table 9.1 Intellectual levels of autonomous flight

Level	Modes of Control	Automation	Dealing with Obstacles
1	Partial remote control	Zero automation	No obstacle detection
2	Full remote control	No automation	No obstacle detection
3	Assisted remote control	Low automation	No obstacle detection
4	Exception control	High automation	Detecting and alarming obstacles
5	Full flight control	Full automation	Detecting and avoiding obstacles at some flight phases
6	Full flight and payload control	Intelligent full automation	Detecting and avoiding obstacles at all flight phases

- *Level 5 – Full flight automatic control*: the flight is fully automated; the only function of the humans is to prepare a flight mission plan that the UAS performs without human intervention. However, the payload is controlled remotely.
- *Level 6 – Full flight and payload automatic control*: the flight and payload control are fully automated; the only function of the humans is to prepare a flight and payload mission plan that the UAS performs without human intervention.

Table 9.1 shows the intellectual levels of autonomous flight. Each of these levels can be applied to each of the functions individually. It is assumed that mission planning may be performed using software tools that automate many of the details, as discussed in the previous chapter. However, it is inherently a human function and the decision-making part of the planning is not automated.

In dealing with obstacles, the control levels of UAVs in terms of detecting and avoiding obstacles can be compared. In levels 1 through 3, there are no obstacle detection and avoidance in the control structure. In level 4 (exception control), there is a feature of detecting and alarming obstacles, but obstacles are not avoided. Finally, in levels 5 and 6, the control structure includes detecting and avoiding obstacles at all flight phases.

In general, the payloads are only utilized within the target area; they do not need to be activated during takeoff, climb, and cruise. For the case of a camera in an electric UAV, if the camera is on for the entire duration of the flight, the flight endurance will be heavily reduced. However, in the remote control of a UAV, if the payload (e.g., camera) is needed for flight safety or navigation, the payload should be on from takeoff to landing.

In this chapter, we discuss some of the issues and tradeoffs that determine how these levels are applied to the other core functions of pilot, payload operator, and aircraft commander.

9.3 Remote Piloting the Air Vehicle

All of the possible levels of human control listed above can be used in a UAS. At the most basic level, remote piloting the air vehicle requires the real-time control of every flight parameter using GCS sticks via the line of sight (as in RC model planes, quadcopters, and very small UAVs). At the next level, the air vehicle is remotely controlled by the pilot, but with much less intervention (as in Boeing ScanEagle, Northrop Grumman RQ-4 Global Hawk, and General Atomics MQ-9 Reaper). This is possible because there is a relatively well-defined set of situations and events that call for an equally well-defined set of pilot responses.

Most pilots would say that this oversimplifies the role of the pilot and neglects the "art" and nuance that a good pilot applies to his/her control of the aircraft. That certainly is true. However, for the rather routine flying that is involved in most UAV missions today, the software in the autopilot may be adequate to fly the aircraft in a manner that would be hard to distinguish from what would have happened with a live pilot at the controls. Even if an unanticipated situation or a software error were to cause a crash, it is not clear that one could conclude that the autopilot was inferior to a human pilot as we know, unfortunately, that manned aircraft sometimes crash due to pilot error.

In fact, under normal circumstances, an autopilot may be able to fly the aircraft better than the best human pilot. Many state-of-the-art fighter aircraft operate near the boundary of instability, and always have an autopilot assisting the human pilot to maintain stability by making small control adjustments with a bandwidth and sensitivity that a human cannot match. The situation is less clear for some possible future unmanned missions that might require extreme aerobatics. Some of these are discussed in Chapter 11 on Weapon Payloads.

Using our criterion, it appears that present-day autopilot technology is sufficient to provide a fully automated piloting function. That does not mean that all UAVs provide such a capability. Modern autopilots are capable of taking off, flying any desired flight plan, and landing without any human intervention.

9.3.1 Remote Manual Piloting

The lowest level of UAV control is the remotely manual control using control sticks in the GCS. It is possible directly to pilot the AV remotely with little or no autopilot assistance, as was implied by the now-abandoned terminology "remotely piloted vehicle (RPV)." This is particularly applicable to small AVs using technology similar to model airplanes, particularly within line of sight.

A simplified functional block diagram of a closed-loop automatic control system was shown in Chapter 5. As shown in Figure 5.7, the control system of a plant in a closed-loop form is generally made up of four basic elements: (1) plant, (2) controller, (3) actuator (or servo), and (4) measurement device. For the case of the remotely controlled flight of a UAV (Level 1), the controller is the ground pilot, the plant is the AV, the measurement device is the human eye, and the actuator is his/her hand via the GCS joystick. The block diagram of a generic remote flight control system with one negative feedback is shown in Figure 9.2.

The flight control output is measured by a feedback element (measurement device) to produce the primary feedback signal, which is then compared with the reference input. The difference between the reference input and the feedback signal (i.e., error) is the input to the controller. For a remote control, the controller and the comparator are in the pilot brain. Then the controller is generating a signal to the actuator based on a control law. Some of the measurement devices are placed on the UAV, and one of them is the pilot eye. The plant is the UAV in the air and the actuators are the joystick on the handheld control box.

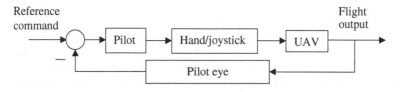

Figure 9.2 Block diagram of a generic closed-loop remote flight control system

9.3 Remote Piloting the Air Vehicle

In reality, there are multiple outputs (e.g., UAV airspeed, altitude, and attitude angles) and so multiple feedbacks (e.g., inner and outer loops). All of the outputs are measured, some of them are demonstrated on the display to the pilot.

When a pilot is able to view the camera output in real time, he/she can control the UAV through a display. This type of control – where there is a first-person view (FPV) or a remote-person view (RPV) – is referred to as video piloting. The air vehicle can be either driven or piloted remotely from a first-person perspective via an onboard camera, fed wirelessly to video FPV goggles or a video monitor.

In Figure 9.2, each arrow represents a signal and its direction. When remotely controlling the flight of a UAV (Level 1), there are various types of signals, as depicted in Figure 9.3. In the process of sending the pilot command from a handheld control box to a UAV, the signals are of various types:

1) S1: mechanical force applied by pilot to deflect the GCS joystick.
2) S2: electric signal generated by the potentiometer when rotating along with the stick – this signal is sent to the GCS microcontroller.
3) S3: digital signal produced by the microcontroller and sent to the GCS transmitter. In modern GCSs, there are analog-to-digital converters (ADC) to convert analog (e.g., electric) signals to digital signals. Moreover, a digital-to-analog converter (DAC) converts digital signals to analog (e.g., electric) signals.
4) S4: electric signal generated by a transmitter and sent to the GCS antenna.
5) S5: electromagnetic (i.e., radio) wave generated by an antenna and communicated to the UAV.
6) S6: light is reflected and an infra-red signal is generated by the UAV. These two signals are available to the GCS.

These six signals are just examples for one autonomy level. In various UASs, there may be slightly different signals for communicating between control system elements.

However, beyond the line of sight, piloting must be based on visual cues from onboard cameras and flight instruments using information from onboard sensors transmitted on the downlink. In this case, the human pilot has to have significant piloting skills, to include a capability to fly the AV based on the instruments alone in case the imaging sensors fail or be rendered useless by fog or clouds. In the early days of military UAVs, this mode was often used for takeoff and/or landing, with the remainder of the flight being performed using one of the more automated modes.

There can be serious issues in directly piloting the aircraft if there are significant delays in the up- and downlinks of the data link, which is certain to be true when the data link uses satellite relays to allow the "pilot" to be on another continent from the AV. These issues relate to responding to turbulence and other rapidly changing conditions. The most straightforward, and perhaps only,

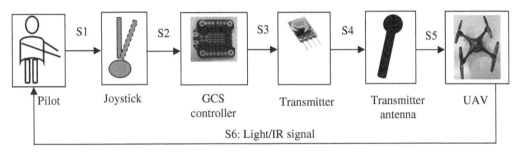

Figure 9.3 Signals in a remotely closed-loop flight control system

solution to that problem is to use an autopilot-assisted control mode when there are significant delays in the remote-control loop.

9.3.2 Autopilot-Assisted Control

At the next higher level of automation, a UAV may retain at least a semblance of the operator piloting the air vehicle in the form of operator commands that are relative to the present attitude and altitude of the AV. In this case, the operator commands a turn right or left and/or climb or descent, including some indication of the rate of turn, ascent, or descent, and the autopilot converts that command into the set of commands to the control surfaces that will accomplish the intent of the operator while maintaining AV stability and avoiding stalls, spins, and excessive maneuvering loads.

This is a much–more-assisted mode than the stability augmentation that, now, is relatively common in state-of-the-art fighter aircraft and has already been mentioned. It is sufficient to allow a non-pilot to "fly" the aircraft, at least under routine flight conditions.

Autopilot-aided manual control can be combined with autonomous navigation from waypoint to waypoint and can be used even for large UAVs operated well outside the direct line of sight. In that case, the "pilot" on the ground is presented with video imagery in at least the forward direction and a set of flight instruments providing airspeed, heading, altitude, and attitude, as well as engine, fuel, and other indications needed to fly the aircraft.

In addition, a display of the ground position and track of the air vehicle is available on some sort of map. This mode provides great flexibility in real-time control of the flight path similar to that provided by direct remote control, but takes care of all the details onboard the aircraft with control loops that have sufficient bandwidth to deal with any transient condition and with the autopilot providing most of the piloting skills.

The assisted mode may be the primary mode for small systems using very simple control consoles and intended for operation largely within the line of sight of the operator. It is simple to implement, flexible in operation, and suitable for controls similar to those of a video game. This allows operation by personnel in the open, possibly wearing gloves.

When used with a small and simple control console, this mode leaves the control of the track over the ground in the hands, and head, of the operator, and may allow the operator to fly the AV into the ground or other obstructions. The assisted mode requires more pilot training and skill than a fully automated mode, and some users have required AV operators using such systems to be pilot qualified. However, other users have trained operators specifically to control the UAVs without requiring them to be able to pilot even a light manned aircraft.

One of the major tradeoff areas with regard to operator qualifications is how much of the landing process is automated. Landing is in many ways the hardest thing that a pilot does, particularly in bad weather, gusting, and/or crosswinds. If the landing is fully automated, whatever mode may be used during the rest of the flight, then the piloting qualifications of the operator can be relaxed.

9.3.3 Complete Automation

Many modern UAV systems use an autopilot to automate the inner control loop of the aircraft, responding to inputs from onboard sensors to maintain aircraft attitude, altitude, airspeed, and ground track in accordance with commands provided either by a human AV operator or contained in a detailed flight plan stored in the AV memory.

9.3 Remote Piloting the Air Vehicle

The human inputs to the autopilot can be stated relative to the Earth as the map coordinates of waypoints, altitudes, and speeds. In a modern system using a GPS navigator, it is not even necessary to require that the operator deal with airspeed and headings, taking into account the direction and speed of the wind through which the AV is flying. Using GPS, the autopilot can implement the necessary variations in airspeed and heading to keep the AV moving at a desired ground speed along a desired track on the ground.

In this case, one might say that the function of piloting the aircraft is completely automated. The lowest level at which a human would be involved in this process is that of the aircraft commander, who would tell the autonomous autopilot where to go next, at what altitude, and at what speed.

This mode of control could be called "fly by mouse" or perhaps "fly by keyboard" as it is basically a digital process in which coordinates, altitudes, speeds, and, perhaps, preplanned maneuvers contained in libraries, such as orbits of various shapes, are strung together on a computer on the ground and the autopilot does the rest.

A pure fly-by-mouse mode may not provide enough real-time flexibility to adapt to a dynamic flight plan. For example, if something were seen using one of the sensors and it was desired to alter the flight path to take another look from a different angle, a pure fly-by-mouse mode of operation would require changing a flight plan. Even with software tools, this might be an awkward way to respond. A more user-friendly approach would be to have an autopilot-assisted mode available and to suspend the flight plan while the pilot or, perhaps, the aircraft commander, took semi-manual control of the AV.

9.3.4 Summary

The fly-by-mouse mode represents the highest level of automation with regard to piloting the AV and can be described as "completely automated flight." To the extent that it is successful in executing the flight plan without incident, despite any turbulence or other unexpected events, it might be said that a "passenger" on the AV would not be able to tell that there was no pilot in the cockpit.

There is a continuum of options for the level of automation of the pilot function, running from no automation to full automation. These options may be applied differently for different phases of the flight so that some of the more difficult stages of the flight, such as takeoff and landing, are fully automated and others are handled with a combination of fly by mouse for preplanned flight segments and autopilot-assisted operations in response to real-time events.

Small and inexpensive autopilots and onboard acceleration sensors, combined with GPS navigation, result in quite affordable implementation of a full fly-by-mouse mode. Therefore, the trade-offs between that mode and the various lower levels of autonomy may well be driven by the nature of the system (AVs operating within the line of sight versus those that operate beyond the line of sight) and the nature of the ground control station.

Small and simple ground control setups may make it easier to use a game-controller mode than to enter the data for a detailed flight plan. Short and highly flexible mission requirements may also tip the balance away from detailed planning and toward direct operator control over maneuvers.

Full automation for the piloting function requires a detailed flight plan. An issue arises if it is necessary to change the flight plan while the mission is in progress. An example of a very significant unexpected event that might occur would be as a loss of power. This would force a major change in the flight plan that might be very hard to plan in advance.

We consider changes to the flight plan in response to events during the mission as being part of the mission control process, which is the function of the aircraft commander rather than a function of the pilot (recognizing that there may not be a separate aircraft commander for some UAVs).

If the autopilot is capable of carrying out whatever altered flight/mission plan is provided either by a human or by a computerized aircraft commander without human aid, then the "flight" might be considered completely automated.

9.4 Autopilot

9.4.1 Fundamental

The method almost universally applied for the control of today's UAVs is use of an automatic, electromechanical computer system in the form of an autopilot. The autopilot is the most important subsystem in a UAV compared with a manned aircraft. The first aircraft autopilot was developed by Sperry Corporation in 1912 with the objective of keeping the wing level.

An autopilot is the integrated software and hardware, and primarily performs four functions: (1) control (i.e., follow commands), (2) navigate, (3) guide the UAV, and (4) stabilize the UAV (if needed). The relation between the control system, the guidance system, and the navigation system is shown in Figure 9.4.

The control system is the most challenging and sensitive element of an autopilot. Automatic control systems employ a feature called feedback or closed-loop operation. The definitions of navigation, functions of the navigation system, navigation equipment, and related processes are presented in Section 9.6.

Guidance is defined as the process for guiding the path of an air vehicle towards a desired target. Guidance is the means by which a UAV steers, or is steered, to a destination and follow a trajectory. A guidance system can be assumed as the 'driver' of a UAV that exercises planning and decision-making functions to achieve an assigned flight mission. The output of a guidance system is often a path to follow based on a commanded mission, such as waypoints.

A simplified functional block diagram of a closed-loop automatic control system is shown in Figure 9.5. In this generic block diagram, there is one negative feedback; however, there are more

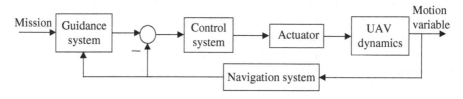

Figure 9.4 Control, guidance, and navigation systems in an autopilot

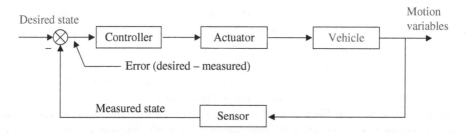

Figure 9.5 Block diagram of the closed-loop control system

than one feedback in an actual UAS. There are mainly four elements in a closed-loop control system: (1) controller, (2) actuator, (3) sensor, and (4) vehicle.

The actual state of the UAV flight path, attitude, altitude, airspeed, etc. is measured (by sensor) and electrically fed back and compared to (subtracted from) the desired state. The difference, or error signal, is amplified and used to position the appropriate control surface, which, in turn, creates a force/moment to cause the air vehicle to return to the desired state, driving the error signal to zero. The reader is advised to study references such as [26] to have sufficient knowledge on terminology, fundamentals, and challenges of control systems.

9.4.2 Autopilot Categories

In general, there are four categories for UAV autopilots: (1) stability augmentation, (2) hold functions, (3) navigation, and (4) combined category. Stability augmentation is definitely required for inherently unstable rotorcraft (including quadcopters) to provide and maintain stability. Hold functions are primarily employed in a long cruising flight and a long turning flight. Modern quadcopters are often equipped with a combined category to allow the UAV to return to its base if the connection/command is lost.

In general, there are three stability augmentation systems: (1) roll damper, (2) yaw damper, and (3) pitch damper. There are two groups of hold functions: (1) longitudinal hold functions and (2) lateral-directional hold functions. In the longitudinal plane, there are primarily three functions: (1) pitch attitude hold, (2) altitude hold, (3) auto-throttle, and (4) airspeed hold. However, in the lateral-directional mode, three hold functions are dominant: (1) bank angle hold, (2) heading angle hold, and (3) turn rate mode.

For the navigation, there are two groups of modes/functions: (1) longitudinal navigation functions and (2) lateral-directional navigation functions. In the longitudinal plane, there are primarily four modes: (1) automatic flare mode, (2) glide slope hold, (3) terrain following, and (4) automatic landing. In the lateral-directional mode, three hold functions are dominant: (1) localizer, (2) VOR hold, and (3) tracking a series of waypoints.

9.4.3 Inner and Outer Loops

In order for an autopilot to simultaneously perform all functions, its structure requires more than one feedback and more than one loop. In such cases, there are often inner loops and outer loops, each for controlling a signal and a variable. If the UAV is already stable, inner loops are dedicated to slower state variables (e.g., airspeed) and outer loops are for faster state variables (e.g., angle of attack).

If stabilization is required, the control and stabilization loops should be accurately inter-related. In general, primary stabilization is accomplished in the inner loop while maintaining the air vehicle in its prescribed attitude, altitude, and velocity state via the outer loop. In addition, there is an outer loop that performs the task of maneuvering and navigating the air vehicle. Another outer loop is also used to capture guidance beams for electronically-assisted or automatic recovery. One example of a block diagram of a feedback control system showing both the inner and outer loops is shown in Figure 9.6.

Since there are three axes in aircraft dynamics, there are three control systems working simultaneously: (1) longitudinal control system, (2) directional control system, and (3) lateral control system. Figure 9.6 demonstrates a general representative example of all three control systems.

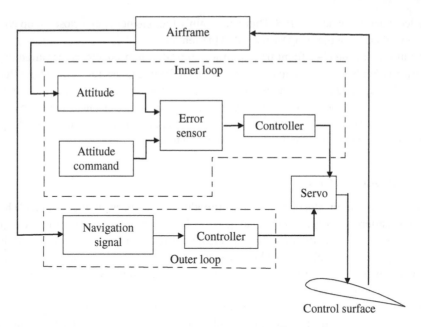

Figure 9.6 Flight control system block diagram for a fixed-wing UAV

Table 9.2 Control elements in three control planes

No.	Control Plane	Control Tools	Sensors	Navigation Signals
1	Longitudinal (xz plane)	Elevator and engine throttle	Vertical gyro, pitch rate gyro, pitot tube, altimeter, GPS	Angle of attack, pitch angle, pitch rate, airspeed, vertical speed, altitude, x and z coordinates
2	Directional (xy plane)	Rudder	Directional gyro, yaw rate gyro, GPS, compass	Heading angle, yaw rate, y coordinate
3	Lateral (yz plane)	Aileron	Vertical gyro, roll rate gyro	Bank angle, turn rate, roll rate

For instance, in the longitudinal control (i.e., xz plane), the control surface is the elevator; the sensors are altimeter and pitot tube; navigation signals are x, z, pitch angle, pitch rate, and airspeed; and the actuator is the related stick. For other control planes, Table 9.2 demonstrates navigation signals (i.e., typical outputs), control surfaces, sensors, and control plane. In all axes, and in general, the servo (i.e., actuator) often could be either a mechanical/electric jack (to generate and apply a force) or an electric motor (to generate and apply a torque).

9.4.4 Overall Modes of Operation

In addition to maintaining the attitude and stabilizing the air vehicle, the automatic flight control system can accept signals from onboard sources or from the ground (or satellite) to control the flight path, navigate, or conduct specific flight maneuvers. Such an operation is accomplished through the outer loop. The provision of these signals is called coupling, and their operation is called "mode of operation." For instance, the "air-speed mode" means that the airspeed of the vehicle is controlled or held constant automatically. This requires a sensor to measure the airspeed (e.g., pitot tube).

Attitude mode means that the air-vehicle attitude in pitch (e.g., pitch angle), roll (e.g., bank angle), and yaw (e.g., heading angle) is automatically maintained using gyros or other devices. Automatic mode implies that the air vehicle is completely controlled automatically and manual mode implies human intervention. In some cases, switching from one mode to another is automatic; thus, after intercepting a glide slope beam, the pitch channel is switched from altitude hold to glide slope track and the air vehicle automatically flies down the glide slope beam.

The outer control loop includes the human operator, if there is one, and implements the operational control of the air vehicle. Operational control also includes the control of any payloads and the overall direction of the mission.

9.4.5 Control Process

Theory, concept, definitions, fundamentals, and governing equations of the control process were presented in Chapter 5. In this section, the application of control in an autopilot is discussed.

As the control surfaces move, they create aerodynamic forces that cause the air vehicle to respond. The sensors sense this response, or air vehicle motion, and when the flight parameters (e.g., attitude, speed, or position) fall within the prescribed limits, their error becomes zero and the actuators in turn cease to move the surfaces.

The error signal is compensated so that the desired position or attitude of the air vehicle is approached slowly and will not overshoot. The system continuously searches for and adjusts to disturbances so that the air vehicle flies smoothly. The navigation system operates in much the same manner, but the sensors are often pitot tube, compasses, inertial platforms, radar, and GPS receivers.

The sensors measure the air vehicle's attitudes (via vertical/directional gyro), angular rates (via rate gyros), airspeed (via the Pitot-static system), heading angle (via the compass), altitude (via the barometer or altimeter), and other functions as desired or necessary. Section 9.5 is dedicated to flight sensors.

The measured attitudes, altitudes, rates, etc. are compared to the desired states and if they deviate beyond a prescribed amount, an error signal is generated, which is used to move a control surface such that the deviation is eliminated. When the comparison function is done by the microprocessor, it is usually entered to a controller to generate the command signal.

9.4.6 Control Axes

Automatic flight control systems (AFCS) are also classified on the basis of the number of axes/ planes they control. All of these systems can also incorporate throttle control to maintain a desired airspeed, as well as to control altitude.

- *Single axis*: A single-axis system usually controls motion about the roll axis only. The control surfaces forming part of this system are the ailerons, and such a system is often called a "wing leveler." The "pilot" in the ground control station can inject commands into the system, enabling him to turn the air vehicle and thereby navigate the vehicle. Sometimes signals from the magnetic compass or a radio beam are used to maintain a magnetic course or heading automatically. This type of operation is part of the outer loop, which will be discussed later.

- *Two axes*: Two-axes control systems usually control the air vehicle about the pitch and roll axes. The control surfaces used are the elevator and the ailerons, although rudders alone are sometimes used as "skid to turn" devices. With pitch control available, the altitude of the air vehicle can be maintained in straight and level flight. Steep turns, which lead to a loss in altitude when using the roll control only (see discussion earlier in Chapter 5 under Section 5.4.2. Pitch Control) can be made without that loss by controlling the pitch attitude.
- *Three axes*: As the name implies, a three-axes system controls the air vehicle about all three axes and incorporates the use of the rudder for yaw control. Some UAVs do not use a three-axes system. This reduces cost without much reduction in capability because yaw control does not contribute significantly to the overall system (only coordination of a turn with the rudder). If missiles and other ordnance are to be used with the UAV, yaw control (a three-axis control system) becomes more attractive.

With the advance and miniaturization of computer hardware packages, current modern industrial UAVs are all equipped with a three-axis autopilot. One- and two-axes autopilots are still utilized by garage-built model UAVs.

Please note, in the RC plane terminology, that the number of axes sometimes means the number of channels. However, sometimes, the number of channels means the number of communication signals. When you see the term "axis" or "channel," pay attention to the community that is employing these terms, to discover what is meant by the term.

9.4.7 Controller

As shown in Figure 9.5, the controller is an element to receive the error signal and to generate the command signal for the actuator. The controller contains the necessary electronics to process the error signal described above, amplify it, and prepare it for the actuators. In addition, the modification and combining of signals from the different axes is accomplished in the controller. The controller also usually contains the electronics for processing commands and housekeeping outputs of the flight control system.

In analog systems, the controller is often an electric circuit, while in digital systems – where analog-to-digital and digital-to-analog elements are required – the error signal is digitally processed in a digital processor (e.g., computer and microcontroller). In digital systems, a computer code (programming) should be developed to process the digital signal.

Some control techniques are applied in the time domain, while others are utilized in the frequency domain. The control engineer should decide about the domain (time or frequency) that best suits the environment or architecture in the control system. In the frequency domain, the mathematical model is derived as a function of "s", while in the time domain ("t"), differential equations or a state space model are employed.

A controller is processing/regulating/compensating the error signal based on a control law that needs to be selected by a control engineer. There are various control laws and techniques to deal with the variety of dynamic systems. For linear systems, control laws such as simple gain, PI, PD, PID, gain scheduling, lead-lag, lag-lead, and rate feedback can be employed. For nonlinear systems, there are a number of nonlinear control laws (e.g., dynamic inversion, quantitative feedback theory (QFT), neural network, and H_∞) that are complex and challenging. To deal with uncertainties, the robust control technique is chosen and to have an optimized output, the optimal control technique (e.g., the linear quadratic regulator, LQR) is recommended.

All of these closed-loop control techniques can be considered under the umbrella of an advanced research topic titled artificial intelligence (AI). In general, AI and its various sub-fields are simulating human intelligence centered around particular goals and the use of particular tools.

The most widely used and popular control law in industry and even in UAVs is PID, which stands for Proportional, Integral, and Derivative. PID control is a simple, but powerful method for controlling a variety of processes, including flight operation. In this controller, three operations are applied to the error signal (i.e., $e(t)$): (1) proportionally (P) amplified, (2) integrated (I), and (3) differentiated (D). Thus, the control signal, $u(t)$, in time domain is

$$u(t) = K_P \left(e(t) \right) + K_I \int e(t) \mathrm{d}t + K_D \frac{\mathrm{d}e(t)}{\mathrm{d}t} \tag{9.1}$$

where K_P, K_I, and K_D are three control gains. The simplest part of PID control is the proportional component, which amplifies the error signal by K_P. The integral component is to nullify the steady-state error in the process. The derivative component is to damp the oscillations in the response, and to address a multiplier based on the rate of change in the error. Various performance deficiencies can be resolved by applying appropriate values of PID gains. Reference [26] provides techniques to determine PID gains, as well as application of a number of popular control laws.

9.4.8 Actuator

Another main element in an automatic flight control system, as seen in Figure 9.5, is the actuator. The actuators are electro-mechanical elements to produce the force/torque necessary to move the control surfaces when commanded as a result of signals coming from the controller. Actuators used in large aircraft are usually hydraulic, but UAVs often use electric motors (often referred to as servomotors or just servos (see Figure 9.6)), thereby obviating the need for hydraulic pumps, regulators, tubing, and fluid, all of which are heavy and often leak.

In all three axes, the linear actuator is often a mechanical/electric jack to generate and apply a force (e.g., to push the elevator up or down). However, to directly generate a torque, an electric motor (a rotary actuator) is often employed, where an angular displacement is delivered (e.g., to rotate the rudder left or right).

9.4.9 Open-Source Commercial Autopilots

Open-source commercial autopilots are widely used in the RC planes and quadcopters. There are a number of open-source, open-hardware autopilots available in the market. A UAV designer has the option to select the complete autopilot from a vendor and focuses on its assembly and integration. A few popular commercial autopilots are available in the market: Ardupilot, Micropilot, DJI WooKong, PixHawk, and CubePilot. These commercial autopilots are constantly miniaturized and modernized for a lighter, more effective, and higher performance. For Ardupilot, even the software codes are fully downloadable from their website.

Furthermore, the main elements of a commercial autopilot can be purchased from various vendors and manufacturers, and then incorporated and customized as one device. For instance, each of the following devices (each in one piece) can be purchased from different manufacturers and providers: avionics, flight controller, software code, electronic speed controller, data link (receiver and transmitter), power distribution board, and navigation equipment. For this case, the job of the UAV designer is to compile and upload the software into the controller board, mix and wire connect (and solder) all pieces, and finally match the characteristics.

As commercial examples for RC model planes, the Radiolink AT10II 2.4 GHz with 12 channels and a dual stick and Flysky FS-i6 2.4 GHz with 6 channels and a dual stick are equipped with both a radio transmitter and a receiver.

9.5 Sensors Supporting the Autopilot

In order for an autopilot to apply control command, the output of the aircraft (i.e., flight parameters) must be measured using measurement devices. Both the inner and outer loops of the control system require sensors that measure the current state of the air vehicle, so that the deviation from desired state can be determined. The basic things to be sensed include x and y coordinates, altitude, airspeed, and attitude angles and rates. In the following sections, altimeter, airspeed sensor, attitude sensors, and GPS are introduced. These sensors are often considered as the primary elements of a navigation system.

Avionics (short for aviation electronics) is a category of electronic equipment specifically designed for use in aviation. In the past decade, a number of mechanical/electric measurement pieces of equipment (such as a gyroscope and other inertial navigation devices) have been miniaturized and incorporated into integrated circuits (IC). Thus, modern avionics are much lighter, compacted, and integrated.

9.5.1 Altimeter

In a cruising flight, it is often important for the air vehicle to fly at a constant altitude and airspeed. To meet this requirement, or to provide an automatic leveling off, when a desired altitude is reached, a barometric sensing device is used in the altitude-hold mode. Three conventional altitude measurement devices are: (1) barometric (basic) altimeter, (2) radar altimeter, and (3) GPS receiver. In small air vehicles and modern UAVs, the GPS receiver is employed to determine flight altitude. In large UAVs, the radar altimeter is also employed, which measures altitude using a radio signal and radar.

A basic barometric altimeter consists of a pressure transducer (e.g., a hole in a tube) connected to a partly evacuated chamber, an amplifier, and a follow-up motor. The partially evacuated chamber is subject to changes in static pressure when the air vehicle changes altitude, causing it to expand or contract and move a pickoff element that generates an electric current proportional to the position of the pickoff element and hence the static pressure or altitude.

This current is amplified and sent to the pitch control channel to operate the elevator actuator and thereby restore the air vehicle to the desired altitude. The change in static pressure will also cause the follow-up motor to move the pickoff element in the opposite direction to reduce the error signal to zero.

9.5.2 Airspeed Sensor

A number of flight parameters such as flight range are functions of ground speed (i.e., the relative speed of the air vehicle with respect to the ground). However, in a number of flight missions and safety/control issues, airspeed (i.e., the relative speed of the air vehicle with respect to the

9.5 Sensors Supporting the Autopilot

surrounding air) is required and needs to be controlled. For instance, lift, drag, and aerodynamic moments are all functions of airspeed. Furthermore, for airworthiness reasons, the airspeed should not be allowed to fall below the stall speed. Note that the GPS is only able to measure the ground speed, not airspeed. Thus, the air vehicle must be equipped with a sensor to measure airspeed rather than ground speed.

Airspeed measurement devices often use a static pressure sensor in addition to a dynamic pressure sensor called "pitot tube," also known as pitot probe. There are two types of pitot tubes: (1) pitot static port (an anemometer to measure static pressure only) and (2) full pitot tube.

A static port only measures the static pressure, since the hole is perpendicular to the air flow, so the flow must turn 90 degrees to enter into the tube. In contrast, a pitot tube measures the dynamic pressure, since the hole is facing the airflow. When a pitot tube has a static port, it is often referred to as the pitot-static tube.

The only difference between barometric altitude sensors and airspeed sensors is that an airspeed sensor requires a differential between static pressure and dynamic pressure. The chamber, instead of being sealed, is open to a source of dynamic pressure (pitot tube) and static pressure is admitted to the sealed container in which the entire assembly is located.

The chamber expands or contracts from the pressure differential created by a change in airspeed. The rest of the system is identical to the altitude hold system, and the airspeed error signal is sent to the pitch axis and engine throttle control.

9.5.3 Attitude Sensors

A number of attitude angles must be measured in any flight mission: (1) angle of attack (α), (2) bank angle (ϕ), (3) heading angle (Ψ), (4) pitch angle (θ), and (5) climb angle (γ). In some modern UAVs, the measurement of the sideslip angle (β) is also desired.

The angle of attack sensor is of the wind vane type. Its sensing element is simply a small vane that is positioned in the direction of the free stream flow. The shaft of the vane senses the rotation of the vane when the UAV body or wing pitches up or down. Measuring the angle of attack is to ensure that the UAV does not stall during flight. Small UAVs and quadcopters are not equipped with an angle of attack sensor, to reduce the overall cost.

Some of the attitude angles (e.g., ϕ, γ, Ψ) are often measured using a mechanical device that maintains its attitude in inertial space, called a gyroscope (or just "gyro"). The gyroscope as a mechanical sensor is composed of one or more sets of nested wheels, each spinning on a different axis. It maintains the angular momentum and is utilized to measure the change in attitude and/or its rate of the UAV.

A gyro operates based on the *gyroscopic effect or law*, which states that any object rotating about one axis will react to any input about other axes. Thus, the reaction to any incoming change to its axis will be to create a force on another axis. A gyro uses this effect to track the orientation of a body with the aid of a gimbal-mounted flywheel that maintains constant orientation in inertial space. There are two groups of gyros: (1) attitude gyro (which measures attitude angles) and (2) rate gyro (which measures the rate of change of attitude angles).

The gyros associated with an autopilot are not generally used directly to measure attitude in inertial space. Rather, each gyro generally is used to measure the rate of change of attitude in one axis and the rate of change is integrated electronically to estimate the present attitude. Attitude-old is usually accomplished by measuring the rate of change in the air-vehicle attitude.

At a minimum, it is necessary to measure the pitch angle of the air vehicle (using the pitch gyro). A yaw gyro is added when a second axis is required (to measure the yaw angle), and a full, three-axis system adds a roll gyro (to measure the bank angle). Gyros add to the cost of a system but are necessary when precise attitude control is required for accurate target location.

Various indirect ways of measuring the direction of gravity are used to correct accumulated errors in gyros and to keep the estimate of yaw and roll from drifting too far from the Earth's inertial frame. In this application, the sensors are called "rate gyros". Rate gyros are not suitable for long-term navigation. If that function is required a higher quality inertial reference is required.

Mechanical gyroscopes add to the cost and complexity and accumulate errors. In the past decade, electronic integrated circuits have been developed to provide the same outputs and are integrated into the flight controller hardware package.

9.5.4 GPS

The global positioning system (GPS) is a satellite-based positioning/navigation system that is funded, owned, and operated [27] by the US government. It is based on a network of 33 satellites orbiting at 20,180 km that continuously transmit coded information. The information transmitted from the satellites can be interpreted by receivers to identify locations on Earth by measuring distances from the satellites. The GPS makes it possible to determine altitude, ground speed, linear acceleration, and air vehicle position based on satellite signals rather than the variety of mechanical sensors previously required.

However, the update rate for GPS measurements may not sometimes be sufficient to support the inner loop of the autopilot, so rate sensors are still required for that function. However, GPS provides an accurate long-term reference that can be used to avoid drift in the short-term estimates and for navigation.

As long as the GPS signal is available for the GPS receiver of a UAV, the GPS is a highly reliable reference for locating the air vehicle anywhere around the world. UAVs that are desired to be able to fly independently of GPS require an inertial navigation system (INS) using an internal inertial platform. Almost all UAVs in the world are currently equipped with a GPS receiver for navigation. However, some countries such as China and Russia have their own global navigation system.

The GPS uses simultaneous measurements of the range to three satellites (whose positions are precisely known) to determine the position of a receiver (e.g., a UAV) on the surface of the Earth and its atmosphere. If the range to four satellites is known, the altitude of the receiver also can be determined. Accuracies of 5 to 15 m are available in the restricted military version of the system, while accuracies of 100 m are available from the civilian version. The GPS receivers that use the L5 frequency band can have a high accuracy (about 30 cm).

Even higher accuracies are available if one or more supplemental ground stations are available whose positions are known precisely. These ground stations can be 100 km from the GPS receiver that takes advantage of their signal. Using the so-called "differential GPS" approach, the addition of ground stations allows accuracies of the order of 1–5 m, even for the civilian version of GPS.

The GPS signals from the satellites are transmitted in a direct spread-spectrum mode that makes them resistant to interference, jamming, and spoofing. (Direct spread-spectrum data communications are discussed in Part Five on Data Links.) Differential GPS could also use jam-resistant signal formats, although most present civilian systems do not do so.

At present, the only reasons for using any other form of AV navigation would be:

a) Concern about anti-satellite weapons used to destroy the GPS constellation during a war (much less of a concern today than it might have been a few years ago).

b) The susceptibility of GPS, particularly in its more accurate, differential form, to jamming.

9.6 Navigation and Target Location 199

While GPS is resistant to jamming or deception, it is not immune. If, as appears to be occurring, the military becomes highly dependent on GPS in areas ranging from navigation to weapon guidance, then GPS will become an attractive target for enemy electronic warfare.

9.5.5 Accelerometers

When a UAV is not employing a GPS receiver and an inertial system is used for navigation, accelerometers are the sensors to determine the UAV coordinates. An accelerometer is a mechanical measurement device to measure linear acceleration of a motion. The integration of the accelerometer output leads to the ground speed and a secondary integration of such signal produces the position of the flight vehicle (e.g., x). Three orthogonal accelerometers will allow a UAV to calculate the three coordinates (x, y, and z). In the past decade, miniaturization of sensors and employing new technologies developed new MEMS (Micro-Electro-Mechanical System) gyros and MEMS accelerometer sensors in an effective small measuring unit.

9.6 Navigation and Target Location

Another function that is conducted via autopilot is the navigation of AV, which also requires target location. Navigation is usually done using sensors such as gyros, accelerometers, altimeters, and the GPS.

Navigation is defined as the skill that involves the determination of position, orientation, and velocity of a moving object (e.g., UAV). To accurately determine the target location, it is first necessary to know the position of the AV. Navigation systems may be entirely onboard a UAV or they may be located elsewhere (which require the communication via radio or other signals with the UAV) or may use a combination of both methods.

The position of the AV over the ground must be known in order to carry out the planned flight path and to provide orientation for the use of the sensors. Furthermore, one common use for a UAV is to find some target of interest and then determine its location in terms of a map grid. The UAV sensor typically provides the location of the target relative to the AV. This information must be combined with knowledge of the location of the AV, in order to determine the target location on the map grid.

The AV Position is defined by three coordinates: (1) latitude, (2) longitude, and (3) altitude above sea level. The latitude of a place on Earth is its angular distance north or south of the equator. Latitude is usually expressed in degrees ranging from 0° at the equator to 90° at the North and South poles. Similarly, the longitude of a place on Earth is the angular distance east or west of the Greenwich meridian. Longitude is ranging from 0° at the Greenwich meridian to 180° east and west.

In many early UAV systems, the position of the AV was determined relative to a surveyed location of the MPCS data-link antenna, using azimuth and range data determined by the data link. This form of navigation has been replaced in most systems by onboard absolute position determination using systems such as the GPS. GPS receivers have become so inexpensive and small that it seems clear that they should be considered a standard navigation system for UAVs.

However, the AV position is determined, the remaining requirement, in order to determine the location of an object on the ground, is to determine the angles and distance that define the vector from the AV sensor to the target. The angles ultimately must be known in the coordinate system of the Earth, not of the AV.

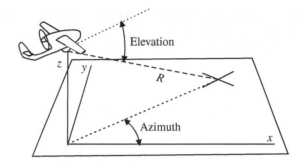

Figure 9.7 Geometry of target position determination

The first step in this process is to determine the angles of the sensor line of sight relative to the AV body. The geometry of this is shown in Figure 9.7. This normally will be accomplished by reading the gimbal angles of the sensor package. These angles must then be combined with information about the attitude of the AV body to determine angles in the Earths coordinate system.

The attitude of the air vehicle in the Earth's coordinated system will commonly be kept current by data from the GPS system, but the update rate for orientation information from the GPS may be too slow to provide accuracy during rapid maneuvers or in turbulent air. This can be dealt with using the onboard inertial platform that is required by the autopilot and must have enough bandwidth to support a control loop with roughly the bandwidth of the motion of the airframe.

The GPS provides the information needed to keep the high-bandwidth dead-reckoning of the onboard inertial system aligned with the Earth's coordinate system. The accuracy required for target location may be much greater than required for successful autopilot operation. Thus, the specification of the inertial platform for the AV may be driven by the target-location requirement, not the autopilot requirement.

Since the sensor is likely to be slewing relative to the AV body (even when it is looking at a fixed point on the Earth), and the AV body is always in motion, it is essential that the angles all be determined at the same moment in time. This requires either that the air vehicle is capable of sampling both pieces of data simultaneously, or that both are sampled at a high enough rate that the nearest samples of the two sets of angles will occur at a time interval that is short compared to significant motion of either the sensor or the AV body. Depending on the manner in which the data is sampled, it may need to be time tagged so that the data from two different sources can be matched when the calculation is performed.

The last element of the calculation of target location is the range from the AV to the target. If a laser range finder is provided or a radar sensor is in use, this range is determined directly. Again, it may need to be time tagged to be matched up with the appropriate set of angle data.

If the sensor is passive, the range may be determined by one of several approaches:

I) Triangulation can be used by measuring the change in azimuth and elevation angles over a period of time as the AV flies a known path and elevation. For relatively short ranges and accurate angle measurements, this approach may be adequate, although less accurate (and more time consuming) than use of a laser or radar range measurement.
II) If the terrain is available in digitized form, it is possible to calculate the intersection of the vector defined by the line-of-sight angle with the ground and find the position of a target on the ground without ever explicitly calculating the slant range from the AV. This calculation requires an accurate knowledge of the AV altitude. A less accurate variant of this approach is

to assume a flat earth and make the same calculation without taking into account terrain elevation variations.

III) A passive technique based on the principle of stadia range finding could be used (measuring the angle subtended by the target and calculating the range based on assumed target linear dimensions). In a UAV system, this process could be refined by allowing the operator to "snatch" a target image, define the boundaries of the target, and indicate the type of target, and then doing a calculation based on stored target dimensions for that type of target, rotating the stored target "image" as required to match the outline defined by the operator. While this is a labor-intensive process, it may be the only approach possible in a system that has no active range finder and does not have accurate altitude and attitude information.

If GPS navigation, to military accuracy, is used to locate the AV, passive triangulation may be able to provide sufficient accuracy to keep the overall errors within 50 meters.

9.7 Controlling Payloads

There are a great many different possible payloads for a UAV, as discussed in the next part of this book. For the purposes of this discussion, most of the possible payloads fall into one of a few generic classes:

- Signal relay or intercept payloads
- Atmospheric, radiological, and environmental monitoring
- Imaging and pseudo-imaging payloads

All of these payloads are discussed in some detail in Part Four of this book, and many of the specific control tradeoffs are presented there. We limit ourselves here to some generic characteristics of the classes of payloads that directly impact the issue of human control versus automation.

In general, controlling payloads is of a closed-loop type (see Figure 9.2); the main difference is the plant that is the payload (e.g., camera) instead of the UAV airframe (i.e., dynamics). For the case of the remote control of flight of a UAV and payload (Level 2), the controller is the ground pilot/operator, the plant is the AV and payload, the measurement device is the human eye, and the actuator is his/her hand via the GCS joystick. However, since two plants (UAV and payload) are controlled simultaneously, two feedbacks are present. Here we are mainly addressing the payload control.

The block diagram of a generic remote-control system of a payload (i.e., with one negative feedback) is shown in Figure 9.8.

This output is measured by a measurement device (i.e., operator eye) to produce the primary feedback signal, which is then compared with the reference input. The difference between the

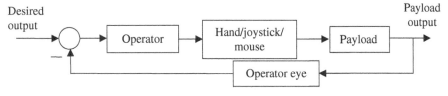

Figure 9.8 Block diagram of a remote generic closed-loop payload control system

reference input and the feedback signal (i.e., error) is the input to the operator. For a remote control, the controller and the comparator are in the operator brain. Then the controller generates a signal to the actuator based on a control law. The actuators are the joystick/mouse on the handheld control box, which is deflected/moved/clicked by the operator. The control function can just be to turn the payload on/off or to change the attitude (e.g., view angle) of the payload.

9.7.1 Signal Relay Payloads

These payloads are discussed in somewhat more detail in Chapter 12. In the context of this discussion, their primary characteristic is that their mission involves detecting electromagnetic signals and either (1) amplifying and retransmitting them or (2) analyzing and/or recording them.

In the relay case, the mission plan is likely to be very simple, consisting of orbiting at some position over the area to be supported by the relay and relaying some set of signals whose frequency and waveform is well specified. As long as this mission plan does not need to be changed, it is feasible for the UAS to operate with great automation, probably modified only with an "exception" reporting system for AV or payload failures and a capability for the operators to upload a new mission plan if their requirements change during the flight.

In the intercept case, it is likely that the mission plan also involves orbiting at some location and receiving signals in specified frequency bands and of specified waveforms, but there is a significant additional function that may be required in real time, which is to analyze those signals and exploit their content. How much of this can be automated is not public information and not known to the authors of this book. It is obvious on first principles that it is possible to classify at least some signals based on frequency and waveform. It is reported in the press that it is possible to scan voice intercepts for keywords. At some point, however, it is likely that human evaluation of an intercept is required in order to determine whether it should be forwarded to the "customer" for whom the UAS is performing the mission.

For an intercept mission of the type hypothesized here, it seems likely that some level of human involvement in the evaluation of the intercepts would be required if a real-time use of the intercepted information is part of the mission. This might not require any action in the UAS ground station, as it could be limited to downlinking the raw or processed signal to the ground station and an automatic passing of the downlinked signal information to the user of the information.

Therefore, a generic signal intercept mission probably can be highly automated with the same exception reporting and intervention provisions as a relay mission.

9.7.2 Atmospheric, Radiological, and Environmental Monitoring

Atmospheric, radiological, and environmental monitoring missions are similar to the signal intercept mission in the sense that they monitor information sensed by specialized sensors on the AV and downlink and/or record those readings as a function of time and location. If there is no requirement for real-time or near-real-time response to unusual readings, the mission plan consists of flying some specified flight plan while operating the sensors and, at most, monitoring the operation of the sensors. This type of mission can be fully automated with no more than exception reporting and intervention.

Some simple modifications to the mission plan might be automated. An example would be to watch for some reading, for example, a radiation level exceeding some threshold, and to insert a preplanned search pattern to map the readings over an area. A slightly more sophisticated response that could be automated would be to adapt the search pattern to the readings being acquired in an attempt to find the location at which the reading is at a maximum.

9.7 Controlling Payloads　　　　　　　　　　　　　　　　　　　　　　　　　　**203**

This would allow a highly automated operation with some automated changes in the flight plan. However, it seems likely that the UAS system designer would choose to consider any reading that triggered a change in the mission plan to be an exception and report it to the control station so that a human could become involved at some level in any change to the flight plan, even if that is simply to ratify an automatic "decision" to execute some type of mapping or search routine. In this case, the operator response might be in the form of an opportunity to veto the automated decision and the design might allow the automated choice to be executed if the veto was not received.

9.7.3　Imaging and Pseudo-Imaging Payloads

Imaging and pseudo-imaging payloads present a special challenge for automation of the operator function because the ability of the human eye–brain system to interpret images is not yet even nearly matched by any computer.

Of course, if the only function of the sensor is to downlink and/or record images of preplanned areas, with no real-time interpretation, then the function of the operator is simply to point the sensor in the correct direction and turn it on and off. Those functions are easy to automate and were fully automated in the reconnaissance drones of 50 to 60 years ago.

Similarly, there are some missions in which an imaging or pseudo-imaging system might be able automatically to detect objects of interest with reasonable reliability. In particular, if the sensor is a radar system or is augmented by a range-sensing subsystem, such as a scanning laser rangefinder, it may have a capability to reliably detect some special classes of objects. One of these would be to detect vegetation encroaching on the cleared right of way of a power line. Another important class that can reliably be detected by radar systems consists of objects that are moving across the ground, the surface of a body of water, or in the air.

There is a major area of ongoing research and development related to "automatic target detection." The objective of this effort is to develop a combination of sensors and signal processing that is capable of automatically finding some specified type of "target" when it is embedded in a noisy and cluttered background. If the target is moving and the sensors are capable of determining that, the problem is reduced to further characterizing the object that is moving.

In the discussion of target detection in Chapter 10, we will define a hierarchy of target characterizations that starts with detection (determining that there is some object that is of potential interest) and proceeds up to identification (determining that the object is the specific thing for which one is looking). Here, we will settle for "further characterization" and simply state that there has been some progress in achieving various levels of further characterization over the last 30 or 40 years, and some quite sophisticated approaches have been developed.

Many of these approaches are most effective when applied only to a small area that contains the object of interest and little or none of its surroundings. Therefore, the most successful present automatic target recognition approaches apply only to the "further characterization" that follows detection of "an object of potential interest." Unless the object has some signature that stands out clearly relative to the noise and clutter of the image or pseudo-image, a human operator is at least very useful in real time to use the uniquely powerful eye–brain system to detect the things that need to be looked at more closely.

The result is that in a generic imaging or pseudo-imaging situation, it is likely that the images need to be downlinked to a human operator in real time, and that if the sensor has variable magnification and pointing, as is common for imaging sensors, the operator needs to be able to control the pointing and magnification so that he or she can take a closer look at things that might be objects of interest and/or to zoom in to further characterize something that has been detected.

There is at least an implication that there needs to be a capability to alter the flight plan to look at something from a different angle or to allow more time to examine it. (To some extent the requirement to look again can be met by the ability to play back and freeze the images already acquired.)

As will be discussed in some detail in Chapter 10, the operator may need assistance from the computers in order to conduct a systematic search of a specified area. This need is created by the fact that the operator typically is "looking through a soda straw" with no peripheral images to allow him or her to retain orientation relative to the wider view. This creates a need for an assisted control mode if area searches are part of the UAS mission.

9.8 Controlling the Mission

We use "controlling the mission" to describe the direction of "what" to do, as opposed to "how" to do it. There is some unavoidable ambiguity about this distinction. In general, we will include most of the choices about the approach to accomplishing a task as part of the "what" and limit the "how" to the mechanics of implementing the approach. This amounts to assuming that the aircraft commander is a "micromanager" who makes most of the decisions for the next level down in the structure (the payload operator and the pilot). This is consistent with the assumption that the autopilot is provided with a detailed flight plan and that any change in that flight plan is a function of the aircraft commander.

One possible reason for a change in the flight plan has already been mentioned – loss of power. This is one of the more dramatic events that might be anticipated. Others include the following:

- Loss of the command uplink of the data link
- Loss of GPS navigation (if used)
- Payload malfunctions
- Weather changes
- Change in flight characteristics (possibly due to structural damage)
- Something that has been observed with the sensor payload that triggers a task that has a higher priority than the preplanned mission

Some of these events can be recognized in a straightforward manner by the computers on the AV. Loss of power, loss of data link, changes in flight characteristics, payload malfunctions, and loss of GPS are in that category. Others may or may not be easily determined in a fully automatic way. In particular, imaging and pseudo-imaging sensors may not be able to "notice" anything out of the ordinary without a human in the loop.

As mentioned in the discussion of payload control, there are exceptions. Sensors looking for chemical or biological agents or for radiation may be quite capable of detecting any of the things that they are looking for in a fully automatic manner. Once that detection has been made, a computer could look up rules about what should be done. The rules might tell the computer to interrupt its preplanned flight in order to map the distribution of whatever it is detecting over some specified area oriented around the initial detection.

It is easy to imagine software that could adapt the mapping process to the intensities measured in order to determine the geometry of the chemical, biological, or radiological contamination. This could lead to a significant amount of automation in the adaptation of the mission plan to accommodate unexpected situations, or, at least, anticipated possible situations that cannot be explicitly planned as it cannot be known in advance when and where they will occur.

9.8 Controlling the Mission

Another exception could apply to one of the types of missions often mentioned with regard to non-military applications of UAVs, flying along a power line looking for vegetation or other encroachments into the area meant to be kept clear along the right of way. It might be possible for an imaging or a radar sensor autonomously to recognize possible encroachments and take some simple action such as performing an orbit and getting data from all angles so that the full extent of the situation can be determined by later human review.

What many of these exceptions have in common is that they involve events or "targets" that involve relatively simple, "threshold-crossing" signatures, that are just about as easy for an electronic circuit to detect as for a human, and the response to the detection is a simple rote response that can be programmed without any need for a judgmental decision. In the two cases described above, the events to be detected are well defined when the mission is being planned and are limited in number. That is, there are one or two or three well-defined possibilities that need to be addressed, not a large number of poorly defined possibilities.

In general, it can be said that if a limited number of well-defined events are known in advance, and if those events can be detected by signal processing associated with the sensors of the AV, then it is possible, at least in principle, to provide a preprogrammed logic that specifies that if, say, event 4 has occurred and, perhaps, event 7 has also occurred but, perhaps, event 2 has not occurred, then event 4 becomes the highest priority and is to be responded to with some new mission plan. As can be seen by the complexity of the sentence needed to describe it, the logic can get very complicated, even for well-defined rules, as the number of distinct events that are to be addressed grows even a little larger than two or three.

The second element that can seriously complicate the problem of dealing automatically with unplanned situations is when the response is not simple or rote. Even if the event can be anticipated as a possibility and is easy to detect, it may be very hard to write software that can achieve an acceptable outcome based on information that is available to the computers on the AV.

An important example of this is loss of power. It will sometimes happen, must be anticipated, and is easy to recognize. The problem is that the response may depend on many factors that have to be balanced very quickly. The ability to test a large set of "rules" quickly is, of course, a strong point for computers relative to humans. Unfortunately, the computer may not be able to acquire some critical information that would be immediately obvious to a human pilot looking out of the cockpit.

The situation is somewhat simplified by the fact that the unmanned aircraft can be allowed to crash without injury to the ground pilot and payload controller. It is complicated by the fact that there probably will be less tolerance for any injuries or fatalities, or even serious damage caused by a UAV crash than there would be if they had resulted from the crash of a manned aircraft. Therefore, the location at which the UAV attempts a crash landing is important and it may be desirable for it to deliberately create a crash that minimizes the damage on the ground.

This might be achieved by diving steeply into a body of water or some open area in which there are no people, a choice less likely to be acceptable for a manned system. Diving into a schoolyard would be a bad choice, but a UAV may not be able to distinguish between a schoolyard and a large vacant lot, while a human operator would have a good chance of doing so under many circumstances.

All of this might be incorporated in a logic table that applies a series of tests to determine where the AV should attempt to glide to before it runs out of altitude and the autopilot probably can do a very good job of getting there. However, the rules would require information that is not available to the computers unless it is predetermined by the mission planner and incorporated in the mission plan.

The simple solution to this problem is an exception reporting and control system that alerts a human operator as soon as power is lost and allows the human to determine and direct the response. In the particular example of loss of power, the autopilot on the AV can immediately establish a minimum descent rate flight mode and the computers in the ground station can provide help in the form of determining the areas on the ground that the AV can reach without power and looking for possible "safe" crash sites within those areas.

The human can evaluate the information available, use the sensors to look at possible crash sites, and apply judgment to decide where to crash or crash land. Once that decision is made, piloting to the crash landing or deliberate dive into an open area can be accomplished with any of the levels of automation available in the UAS, depending on how that particular system is designed and how its operators are trained.

Each General Atomics MQ-1 Predator is operated remotely by a team of three, a pilot, a sensor operator, and a mission intelligence coordinator. The pilot is controlling the flight, while the sensor operator controls and monitors the performance of the various sensors/payloads. In 2018, SkyGuardian UAV successfully flew over the UK on the first ever historic transatlantic drone journey through civilian airspace. Moreover, ScanEagle, which is operated in autopilot mode for the entire flight, demonstrated the air vehicle's ability to successfully accept in-flight re-programming.

9.9 Autonomy

The leading edge of "remote control" issues for UAVs and all unmanned systems is a quest for autonomy. "Autonomy" is defined in dictionaries as the state of being self-governing or self-directing. Its basic meaning in the context of a UAV or UAS is that the system is capable of carrying out some function without the intervention of a human operator.

In terms of the aircrew functions, this might be thought of as replacing the aircraft commander with a computer exhibiting "artificial intelligence" while also delegating complete, unassisted, and unsupervised piloting of the aircraft to an autopilot that takes general directions from the computerized aircraft controller.

A significant degree of autonomy is already common in some fielded UAV systems. Even full autonomy, in the sense that a UAS might perform complex missions without human intervention, is not at all far-fetched as long as the missions can be fully planned in advance and do not require the ability to adapt to unplanned changes in the details of the mission.

Less than 15 years after the first flight at Kitty Hawk, the Kettering Bug could fly itself in roughly the right direction for some fixed time and then crash on a target with no human control during the mission. Reconnaissance drones of the 1960s could be programmed to fly over one or more targets and photograph them and return to a recovery point with no human intervention between launch and recovery.

The objective of the research in autonomy goes beyond this to attempt to make computers capable of making decisions that require something that might be called "intelligent judgment." As with the entire field of artificial intelligence (AI), one's opinion of how far we are from that capability, or even whether it ever will be achieved, depends on what one means by "intelligent."

Artificial intelligence (AI) is the simulation of human mind/intelligence in machines that are programmed to think/decide like humans and mimic human actions. Two active research areas under the topic of "artificial intelligence" are machine learning and deep learning. Machine learning refers to the concept that systems (e.g., machines) can analyze data, learn from experiments, identify patterns, solve problems, and make decisions with no or minimal human intervention. In deep learning, machine learning is carried out through the absorption of large amounts of unstructured data such as text, images, and video.

The well-known "Turing Test" for an "intelligent" computer requires the computer to be able to carry on a conversation (using text messages) that cannot reliably be distinguished from a conversation with a human. An equivalent test for a UAV might be put in terms of carrying out various elements of a mission in a manner that cannot reliably be distinguished from a manned system. Using this "UAV Turing Test" and our specific definition of the core control functions for a UAS, we can make some general observations.

It is feasible to have full autonomy for the pilot function, in the sense that a modern autopilot probably could fly an AV well enough that an external observer would not be able reliably to tell whether or not there was a human pilot at the controls. It may be true that it is not now possible to build an autopilot that is able to fool another pilot into thinking that it is a "great" pilot under extreme conditions, but it might be indistinguishable from an average pilot most of the time. Note, however, that this is true only because we have defined unplanned changes to the flight plan to fall under the responsibility of the aircraft commander.

All payloads could operate with full autonomy as long as there is no requirement for real-time or near-real-time response to what they "see" or detect. The cameras on unmanned drones were "autonomous" in that they carried out a preplanned sensor mission. If all that is required is to record some preplanned data and bring it home, then no real-time operator is needed.

Some modern payloads can automatically (autonomously) detect what they are looking for. Most of these payloads are measuring the level of some kind of signal, typically something like radiation or chemical contamination. In those cases, the sensor can operate autonomously, but if there is a requirement to alter the mission plan in response to a detection by the sensor, then an "exception" needs to be reported to an aircraft commander (which might be another computer module) that is capable of determining the appropriate response and altering the mission plan.

We conclude from this that the basic issues for system-level autonomy are as follows:

a) Real-time interpretation of sensor information
b) Response to exceptions that require alteration of the mission plan

These are the two areas that require some sort of "artificial intelligence," over and above what presently is generally available, in order to perform in a manner that cannot reliably be distinguished from what would be expected from a manned system or a UAV under human control.

As has already been mentioned, there is a great interest, in at least the military and security communities, in "automatic target detection, recognition, and identification." Furthermore, there is a significant ongoing research and development in the area of Intelligence, Surveillance, and Reconnaissance (ISR).

At first thought, these areas would seem to be easier than creating an artificial intelligence that can exercise human-like judgment with regard to complicated decisions, as we might think that it is a more mechanical function involving a synthesis of geometry and some other signature information ("hot or cold," "bright or dim," etc.).

However, the manner in which the human eye and mind process image information is so complex that it is yet to be shown that emulating that process in a computer is any less difficult than creating a set of rules that can be used to make an intelligent choice between complicated alternatives of any other kind.

Reference [28] divides autonomy into 11 (from 0 to 10) levels as a gradual increase in Guidance-Navigation-Control (GNC) capabilities: (0) remote control, (1) automatic flight control,

(2) external system independence navigation, (3) fault/event adaptive, (4) real time obstacle/event detection and path planning, (5) real time cooperative navigation and path planning, (6) dynamic mission planning, (7) real time collaborative mission planning, (8) situational awareness and cognizance, (9) swarm cognizance and group decision making, and (10) fully autonomous.

In Section 9.2 for modes of control, we introduced six levels of autonomy; at higher levels, the use of AI is a required tool. We will not attempt to address the broad area of artificial intelligence in this book, but will return to the question of autonomy briefly when we consider issues related to use of UAVs to deliver a lethal force.

There are a number of UAVs (Northrop Grumman RQ-4 Global Hawk and General Atomics MQ-9 Reaper) that are capable of partial autonomy using their payloads (i.e., are able to perform autonomous or preprogrammed missions). For instance, the Northrop Grumman RQ-4 Global Hawk covers the spectrum of intelligence collection capability with its sensors (OBC[1], HISAR[2], and SYERS[3]). Moreover, its Synthetic Aperture Radar (SAR) has three imagery collection modes: a 0.3-meter resolution spot mode, a 1-meter resolution wide area search mode, and a 4-knot minimum detectable velocity moving target indicator (MTI) mode. Furthermore, the MTI mode provides the position and speed of moving targets.

With the deployment of General Atomics MQ-9 Reaper (see Figure 1.5) in 2006, the US Air Force has moved from using UAVs primarily in intelligence, surveillance, and reconnaissance (ISR) roles to the precision strike role. This enabled the Air Force to launch a UAV from a remote field on the other side of the globe, and then pilot that aircraft from a base in the United States.

The Boeing Insitu ScanEagle – which operates in autopilot mode for the entire flight – is capable of the long-endurance necessary to complete an array of missions including ISR and communication.

Questions

1) What do FPV, RPV, AFCS, QFT, PID, LQR, IC, ISR, OBC, HISAR, SYERS, SAR, and MTI stand for?
2) What do MEMS, ADC, DAC, GNC, GPS, and AI stand for?
3) What are the remote air crew key functions in a UAS?
4) What is meant by "controlling the payloads"?
5) List six levels of control in a UAV.
6) What was the first ever historic flight of the SkyGuardian UAV In 2018?
7) Briefly describe features of "Level 1 – Partial remote control" of a UAV.
8) Briefly describe features of "Level 2 – Full remote control" of a UAV.
9) Briefly describe features of "Level 3 – Assisted remote control" of a UAV.
10) Briefly describe features of "Level 4 – Exception control" of a UAV.
11) Briefly describe features of "Level 5 – Full flight automatic control" of a UAV.
12) Briefly describe features of "Level 6 – Full flight and payload automatic control" of a UAV.
13) In dealing with obstacles, compare control levels of UAVs in terms of detecting and avoiding obstacles.

---

1 Optical Bar Camera.
2 Hughes Integrated Surveillance And Reconnaissance.
3 Senior Year Electro-optic Reconnaissance System.

Questions 209

14) Under normal circumstances, is an autopilot able to fly the aircraft better than the best human pilot?
15) Provide a block diagram to illustrate a generic closed-loop control system. Then write the relationship between elements.
16) List four basic elements in the control system of a plant in a closed-loop form.
17) Provide a block diagram to illustrate the relationship between three subsystems of an autopilot.
18) Briefly describe the video piloting.
19) In controlling the payloads of a UAV, what jobs are conducted by a remote pilot/controller?
20) For the case of the remotely controlled flight of a UAV (Level 1), identify the controller, the plant, the measurement device, and the actuator.
21) While drawing a block diagram, briefly describe the process of remote piloting a UAV.
22) Sketch the block diagram of a generic remote flight control system (i.e., with one negative feedback).
23) Sketch the block diagram of a generic remote-control system of one payload (i.e., with one negative feedback).
24) Identify the type of six signals in a Level-1 remotely controlled flight of a UAV (i.e., closed-loop).
25) Briefly describe how the remote piloting of a UAV, beyond the line of sight, is conducted.
26) Briefly describe the remote piloting in the autopilot-assisted control mode.
27) What typical tools in the GCS are available/used when autopilot-aided manual control is combined with autonomous navigation from waypoint to waypoint?
28) Briefly describe the features of the pure fly-by-mouse mode of UAV control.
29) Briefly describe features of an autopilot.
30) What are four main UAV autopilot functions?
31) Provide a block diagram to illustrate the relation between the control system, the guidance system, and the navigation system of an autopilot.
32) Define guidance.
33) What is the output of the guidance system?
34) Provide a block diagram to illustrate a general closed-loop control system.
35) Write four categories for UAV autopilots.
36) In general, there are three stability augmentation systems. What are they?
37) Is the stability augmentation required for quadcopters? Why?
38) Write at least five functions of an autopilot in navigation mode.
39) Write control tools, sensors, and navigation signals in longitudinal control of a UAV.
40) Write control tool, sensors, and navigation signals in directional control of a UAV.
41) Write control tool, sensors, and navigation signals in lateral control of a UAV.
42) Write two general types of servos in a control system.
43) What flight parameter is measured by an attitude gyro?
44) What flight parameter is measured by a rate gyro?
45) What flight parameter is measured by a pitot tube?
46) What flight parameter is measured by a compass?
47) What flight parameter is measured by an altimeter?
48) An automatic flight control system of a UAV has a single axis. What is usually the function of this system?
49) An automatic flight control system of a UAV has two axes. What is usually the function of this system?
50) What is a wing leveler?
51) What is the function(s) of the controller element in a closed-loop control system?

52) Write at least five control laws in linear systems.
53) Write at least four control laws in nonlinear systems.
54) What is the most widely used control law in industry?
55) Briefly describe features of a PID controller.
56) What is the function(s) of the actuator element in a closed-loop control system?
57) What are two basic groups of actuators?
58) Write at least four popular commercial autopilots available in the market.
59) Write at least five sensors supporting the autopilot.
60) What is avionic?
61) Write at least three sensors to measure altitude.
62) Briefly describe how a basic barometric altimeter is working.
63) What is the difference between airspeed and ground speed?
64) Briefly explain why airspeed is more important than ground speed in air vehicle flight control.
65) Briefly describe how an airspeed measurement device works.
66) What is the function of the Pitot static port?
67) What is the difference between a barometric altitude sensor and an airspeed sensor?
68) List at least five attitude angles.
69) Briefly describe how an angle of attack sensor works.
70) What is the gyroscopic effect?
71) What is gyroscope? Briefly describe.
72) What is the output of: (a) pitch gyro, (b) yaw gyro, and (c) roll gyro?
73) What is the difference between attitude gyro and rate gyro?
74) Briefly describe how the global positioning system is operating.
75) Is the GPS measuring airspeed or ground speed?
76) Discuss the accuracy of the GPS in navigation.
77) Are the GPS signals resistant to interference, jamming, and spoofing? Discuss.
78) Briefly describe how GPS is employed to determine the position of a receiver (e.g., UAV) on the surface of the Earth.
79) What is the function of an accelerometer?
80) Define navigation.
81) The AV position is defined by three coordinates. What are they?
82) Define longitude.
83) Define latitude.
84) Define elevation and azimuth when employed in the target position determination.
85) Name two devices that can determine the target range from the AV directly.
86) Name and briefly describe three approaches to determine the target range from the UAV, if the UAV sensor is passive.
87) List three generic classes of payloads.
88) Draw a block diagram of a remote generic closed-loop payload control system. Then briefly explain the relations between blocks.
89) Briefly describe the process of remotely controlling a signal relay payload of a UAV.
90) Briefly describe the process of remotely controlling an imaging payload of a UAV.
91) Briefly describe the process of remotely controlling an atmospheric and environmental monitoring payload of a UAV.
92) What is the objective of automatic target detection?
93) In controlling the flight mission of a UAV, there are dramatic events that might be anticipated and are reasons for a change in the flight plan. Write at least four such events.

Questions 211

94) Why do imaging payloads present a special challenge for automation of the operator function?
95) If the loss of power happens during a UAV regular flight mission, what could be the new flight plan which is conducted by autopilot?
96) Define "autonomy" in the context of a UAV.
97) What is artificial intelligence?
98) What is machine learning?
99) What is an "intelligent" computer required to do in a "Turing Test"?
100) Write the basic issues for a system-level autonomy.
101) Write at least two areas that require some sort of "artificial intelligence" for a UAS system-level autonomy.
102) Autonomy can be divided into 11 (from 0 to 10) levels as a gradual increase in Guidance-Navigation-Control capabilities. List the levels.
103) List payloads of the Northrop Grumman RQ-4 Global Hawk used for intelligence collection.
104) What does MTI mode of the Northrop Grumman RQ-4 Global Hawk provide?
105) With the deployment of General Atomics MQ-9 Reaper in 2006, what new capability has been provided to the US Air Force?
106) What is deep learning?

Part IV

Payloads

The term "payloads" is somewhat ambiguous when applied to a UAV. It sometimes is applied to all equipment and stores carried on the airframe, including the avionics and fuel supply. This definition results in the largest possible specification for "payload" capacity. However, using this definition makes it difficult to compare the useful payload of two different UAVs, since one does not know how much of the total "payload" capacity is dedicated to items that are required just to fly from one point to another. UAVs share this characteristic with manned aircraft.

The UAV payload is also defined as a piece of equipment for the purpose of performing some operational mission. The payload is either: (1) delivered or (2) employed for a flight mission. This excludes the flight avionics, data link, and fuel. It includes cameras, sensors, stores that perform the missions of reconnaissance, electric warfare, and weapon delivery.

In this book, all of the equipment and stores necessary to fly, navigate, and recover the AV are considered part of the basic UAV system and are not included in the "payload." The term "payload" is reserved for the equipment that is added to the UAV for the purpose of performing some operational mission – in other words, the equipment for which the basic UAV provides a platform and transportation. This excludes the flight avionics, data link, and fuel. It includes sensors, emitters, and stores that perform such missions as:

- Reconnaissance
- Electronic warfare
- Weapon delivery
- Scientific experiments
- Cargo/merchandise delivery

Using this definition, the payload capacity of a UAV is a measure of the size, weight, and power available to perform functions over and above the basic ability to take off, fly around, and land. This is a more meaningful measure of UAV system capability than the more general definition that includes flight-essential items in the payload. However, it must be understood that some tradeoffs are available between the "mission" payload and the more general definition of payload. For instance, it may be possible to carry a heavier "mission" payload if the fuel supply is limited, and vice versa.

A system designer must be aware of the ambiguity about payload capacity and be careful to use a definition that is appropriate to the particular situation being considered.

Introduction to UAV Systems, Fifth Edition. Paul Gerin Fahlstrom, Thomas James Gleason, and Mohammad H. Sadraey.
© 2022 John Wiley & Sons, Inc. Published 2022 by John Wiley & Sons, Inc.
Companion website: www.wiley.com/go/fahlstrom/uavsystems5e

214 | *Part IV Payloads*

The ScanEagle payloads include electro-optical and infrared sensors, biological and chemical sensors, laser designators, and a magnetometer for identification and locating magnetic anomalies.

In Part Four, we discuss system issues related to several types of mission payloads. Reconnaissance and surveillance are fundamental missions for UAVs and are addressed in the first chapter (i.e., Chapter 10) in this section. The second chapter (i.e., Chapter 11) addresses the issues involved in using UAVs to carry and deliver lethal warheads (i.e., weapons such as a missile, rocket, and bomb), which has become a major driver for worldwide proliferation of UAVs in the last decade. The last chapter (i.e., Chapter 12) provides some discussion of a variety of other possible UAV mission payloads.

10

Reconnaissance/Surveillance Payloads

10.1 Overview

Reconnaissance payloads are by far the most common used by UAVs and are of the highest priority for most users. Even if the mission of a UAV is to gather some specialized information, such as monitoring pollution, it is often essential that it is able to locate specific "targets" on the ground for the purpose of collecting data in the vicinity of those "targets."

Reconnaissance/surveillance payloads rely on electromagnetic waves in order to perform their functions. The electromagnetic waves are the output of electromagnetic radiation, which refers to waves of the electromagnetic field, propagating through space, and carrying electromagnetic energy. It includes spectrum of radio waves, microwaves, infrared waves, light, ultraviolet light, X-rays, and gamma rays. The difference between these waves is mainly due to their wavelength and frequency range. Figure 11.4 in Chapter 11 illustrates the spectrum of electromagnetic waves.

As the name implies, electromagnetic waves are synchronized oscillations of electric and magnetic waves. All types of electromagnetic waves including light are traveling with the speed of 300,000 km/s.

Visible light has wavelengths in the range of 400–700 nanometer (nm), between the infrared and the ultraviolet. However, infrared waves (also known as thermal radiation) are the spectrum with wavelengths above red visible light. Infrared light extends from the nominal red edge of the visible spectrum at 780 nm to 1 mm.

These payloads, or sensors as they often are called, can be either passive or active (see Figure 10.1). Passive sensors – for example, (1) a daytime camera and (2) a night-vision camera – do not radiate any energy and do not provide their own illumination of targets. Photographic and video/television (TV) daytime cameras are examples of passive sensors. Passive sensors must rely on energy radiated from the target; for example, a heat wave in the case of an infrared (IR) sensor or reflected energy, such as the sun, star, or moon light for an EO camera. Since daytime cameras are utilizing electric energy and daylight, they are also referred to as an electro-optic (EO) camera.

On the other hand, active sensors transmit energy to the object to be observed and detect the reflection of that energy from the target. Radar is a good example of an active sensor. Both passive and active sensors are affected by the absorbing and scattering effects of the atmosphere.

Introduction to UAV Systems, Fifth Edition. Paul Gerin Fahlstrom, Thomas James Gleason, and Mohammad H. Sadraey.
© 2022 John Wiley & Sons, Inc. Published 2022 by John Wiley & Sons, Inc.
Companion website: www.wiley.com/go/fahlstrom/uavsystems5e

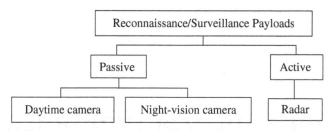

Figure 10.1 Reconnaissance/surveillance payloads general classification

The two most important kinds of reconnaissance sensors will be discussed in detail in this chapter:

1) Daytime (i.e., EO) photo/video camera
2) Night-vision (i.e., IR) photo/video camera

A camera is an imaging sensor for recording or capturing images, which may be stored locally and/or transmitted to another location. The purpose of these sensor payloads is to search for targets and, having found ("detected") possible targets, to recognize and/or identify them. Additionally, in conjunction with other sensors, such as rangefinders, and the UAV's navigation system, the sensor payload may be required to determine the location of the target with a degree of precision that depends on the use to which the information will be put.

For all sensors, the ability to detect, recognize, and identify targets is related to the individual target signature, the sensitivity and resolution of the sensor, and environmental conditions. Design analysis of these factors for imaging sensors (both EO and IR) follows the same general procedure, described in detail in the following sections.

10.2 Imaging Sensors

A reconnaissance/surveillance payload collects data from a real target and often generates an image. A sensor is described as "imaging" if it presents its output in a form that can be interpreted by the operator as an image/photo/video of what the sensor is viewing. A photo is a single image while a video is a series of images played in sequence at a specified frame rate. A video file is played by a software (if digital), while a video tape is played by a hardware (if analog). Thus, the image is still, while the video has movement.

In general, images can be classified into four groups: (1) real color image (for an EO camera), (2) real monochrome image (for near-IR camera), (3) real monochrome image (for mid- or far-IR camera), and (4) synthetic images (for radar).

In the case of an EO sensor, the meaning of an "image" is straightforward. It is a photo/picture of the scene being viewed. If the camera operates in the visible portion of the spectrum, the picture is just what everyone is accustomed to seeing from a color or black and white TV.

In the case of an IR sensor, the camera operates in the near-IR, the picture (almost always monochrome in this case) has some unfamiliar characteristics related to the reflectivity of vegetation and terrain in the IR (e.g., dark green foliage may appear to be white, due to its high IR reflectivity), but the general features of the scene are familiar.

If the IR sensor operates in the mid- or far-IR, the image presented represents variations in the temperature and emissivity of the objects in the scene. Hot objects appear bright (or, at the option of the operator, dark).

10.3 Target Detection, Recognition, and Identification

Caddx Ratel (EO Camera)   Adafruit MLX90640 (IR Camera)

Figure 10.2 Two small reconnaissance sensors

The scene presented to the operator still has the gross features of a picture, but interpretation of the details of a thermal scene requires some familiarization and training. Some intuitive impressions based on lifelong experience with what things look like in the visible spectrum can be deceptive when viewing an IR image. Various interesting effects appear in a thermal scene, such as a "shadow" that remains behind after a parked vehicle moves (due to cooler ground that had been shaded from the sun while the vehicle was parked).

Some radar sensors provide synthetic images, often including "false" colors that convey information about target motion, polarization of the signal return, or other characteristics that are quite distinct from the actual color of objects in the scene. While the synthetic image is usually designed to be intuitively interpreted by the operator, training and experience are even more important when dealing with radar images than with thermal images. Figure 10.2 illustrates two small reconnaissance sensors: (1) Caddx Ratel 2 EO camera with a mass of 5.9 grams and (2) Adafruit MLX90640 IR camera with a mass of 3.5 grams.

The EO cameras are electro-optical instruments to record or capture monochrome/color image/video, while infrared (IR) cameras measure the differences between heat radiations of a target from the surrounding environment. An EO camera detects light that is reflected off surfaces in either the visible or near-IR bandwidth, while an IR camera detects mid- or far-IR waves that are being emitted from objects. Electro-optical images may be individual still photos or sequences of images constituting videos/movies. Both can be stored locally and/or transmitted to another location. An IR camera in a thermal sensor (imager) operates in the infrared bandwidth of the light spectrum, so depicts thermal images of targets.

RQ-4 Global Hawk (see Figure 1.4) payloads include: (1) the synthetic aperture radar (for both spot images and wide-area search images), (2) the electro-optical and infrared sensor system, (3) an active electronically scanned array radar (part of the Air Force airborne signals intelligence (SIGINT) platform), and (4) a ground movement target indicator.

10.3 Target Detection, Recognition, and Identification

Imaging sensors are tools to provide the autopilot (in the UAV) or remote operator (in the GCS) with the means (i.e., image) to perform further analysis and operations. An imaging sensor takes an image/video of a point of interest or the target area (e.g., object, natural scenery, human, vehicle, farm, tower, bridge, and house). A remote operator or autopilot will employ these images to perform a number of operations including target detection, recognition, and identification.

This section is dedicated to a methodology for performance analysis of imaging sensors in target detection, recognition, and identification. The discussion applies primarily to EO and IR images. The factors that influence the performance of these two types of imaging sensors are very similar, and the methodology used for predicting their performance is almost identical. Imaging radar systems share some characteristics with optical and thermal-imaging systems, but are sufficiently different to require separate treatment. Imaging sensors are used to detect, recognize, and identify targets. These three key terms used to describe the operation of the sensor are briefly defined as follows:

- *Detection*: Defined as determining that there is an object of interest at some particular point (i.e., target area) in the field of regard of the sensor.
- *Recognition*: Defined as determining that the object belongs to some general class, such as a truck, a tank, a small boat, or a person.
- *Identification*: Defined as determining a specific identity for the object, such as a *dump* truck, an *M1* tank, a *cigarette-class* speedboat, or an *enemy* soldier.

The successful accomplishment of these tasks depends on the interrelationship of a few factors: (1) sensor resolution, (2) target contrast, (3) transmission through atmosphere, (4) target signature, and (5) display characteristics. The means of image transmission to the remote operator (a data link) is also an important factor. In the following sections, we discuss these six factors and then provide a methodology for sensor performance analysis (including a few considerations and pitfalls).

10.3.1   Sensor Resolution

The quality and resolution of an image taken by an imaging sensor will impact the target detection process including the probability of detection. Resolution of an image is represented by the number of pixels in an image. Pixel – short for "picture element (i.e., cell)" – is a small square; in fact, the computer picture's smallest element. The ideal is that each pixel has only one color. However, there are three colors in each pixel with different brightness levels. The number of pixels represents the resolution of an image. For instance, the ScanEagle high-resolution uncooled thermal imager provides 640×480 pixels with a 25-micron pitch.

Sensor resolution usually is defined in terms of scan lines across the target's dimension. It would seem reasonable to use the maximum dimension of the target when discussing resolution. However, most imaging sensors have a higher resolution in the horizontal direction than the vertical, and experience has shown that, unless the target has very elongated proportions, reasonable results are obtained by always comparing the vertical resolution of the sensor to the vertical dimension of the target. This convention is used in the most commonly applied models of imaging sensor performance.

Sensor resolution is specified in resolvable lines or cycles across the target dimension. A line corresponds to the minimum resolution element in the vertical direction, while a cycle corresponds to two lines. A cycle is sometimes called a "line pair". Lines and cycles can be visualized in terms of a resolution chart having alternating white and black horizontal bars. If the lines of a TV display were perfectly aligned with these bars, a TV could, in principle, just resolve the bars when each white or black bar occupied exactly one line of the display.

It should be noted that the discrete nature of the image sampling as one goes from one TV line to the next leads to some degradation of resolution. However, this effect is relatively small and often is not explicitly considered in sensor analyses. Figure 10.3 illustrates the lines and cycles of

10.3 Target Detection, Recognition, and Identification

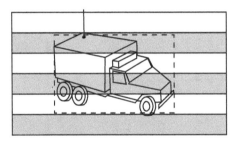

Figure 10.3 Target with resolution "lines" superimposed

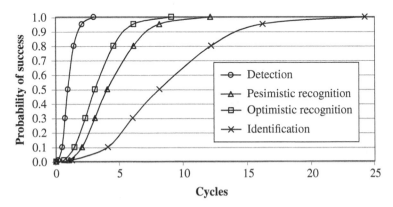

Figure 10.4 Johnson Criteria for the probability of success

resolution across a target. In the case shown, the target is a truck and from the aspect at which it is being viewed it spans four lines, or two cycles, in the vertical direction.

The well-known "Johnson Criteria" established that about two lines across a target are necessary for a 50% probability of detection. Additional lines are required to increase this probability as well as determine more detailed features of the target, that is, to recognize or identify it. Figure 10.4 presents curves for the probability of detection, recognition, and identification as a function of cycles of resolution across the target. Note that two curves are presented for recognition, representing optimistic and conservative criteria for determining whether the sensor will be able to perform this function.

Electro-optical (EO) sensors operate in the visible, near-IR, and far-IR, at wavelengths ranging from about 0.4 μm to 12 μm. The theoretical resolution available from reasonable optical apertures (5–10 cm) at these wavelengths is very high. The diffraction-limited angular resolution of an optical system with a circular aperture is given by Equation (10.1), where θ is the angle subtended by the smallest object (see Figure 10.5) that can be resolved, D is the diameter of the aperture, and λ is the wavelength used by the sensor:

$$\theta = \frac{2.44\lambda}{D} \tag{10.1}$$

For example, if $\lambda = 0.5$ μm and $D = 5$ cm, then $\theta = 24.4$ μrad.

The actual resolution of most EO sensors is determined by the characteristics of the detector used in the sensor system (vidicon, charge-coupled device (CCD), or IR detector array). All of these detectors have fixed numbers of resolution elements (TV lines or rows of individual detector elements).

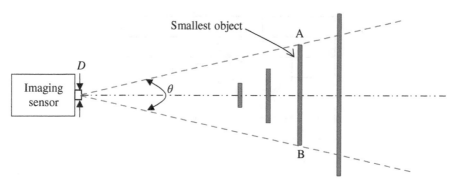

Figure 10.5 Angle subtended by the smallest object for an imaging sensor

For instance, an IR imaging focal plane array with 480 horizontal rows, each containing 640 individual detectors would be described as a "640-by-480" detector array and would ideally have 480 "lines" of resolution and 307,200 (i.e., 640 × 480) pixels.

The old-fashioned standard vidicon had 525 lines of resolution. Silicon focal-plane arrays now available can have 10 million or more pixels arranged in various ways, depending on their aspect ratio (width to height).

The EO sensor of the Boeing Insitu RQ-21 Blackjack (see Figure 17.7) is capable of providing the GCS operator sufficient visual resolution to support classification of a 1-meter linear sized object from 3,000 feet altitude above ground level and a sensor depression angle of 45 degrees, resulting in an assessment at a slant range of 4,242 feet. Moreover, its IR sensor is capable of classifying a 3-meter sized linear object from 3,000 feet above ground level and slant range of 4,242 feet.

The angular resolution of the sensor system is determined by dividing the angular field of view (FOV) by the number of resolution elements of the detector in that dimension. Thus, a TV with a 525-line resolution and a vertical FOV of 2 degrees would have an angular resolution of 262.5 lines per degree. More common units for resolution are lines or cycles per milliradian (mrad). Since 1 mrad is equivalent to 0.057 degrees and 2 degrees equals 34.91 mrad, the resolution given above is 7.5 lines/mrad, or 3.75 cycles/mrad.

Notice that 7.5 lines/mrad is equivalent to 0.133 mrad/line, which implies an angular resolution of about 133 μrad. This resolution is much worse than the diffraction-limited optical resolution of 24.4 μrad calculated from Equation (10.1), illustrating the common situation in which the actual sensor system resolution is limited by the detector, not by diffraction.

If the vidicon were replaced by a 10 megapixel array with a 1:1 aspect ratio, having a little over 3,000 lines of resolution, the detector-limited resolution would be about 22 μrad and would about match that of the optics. However, there would be serious problems in attempting to transmit a 10 megapixel video to the ground in real time, as described in the discussions of data-link issues later in this book. As a result, despite the availability of very large detector arrays, it remains common for the resolution of imaging systems to be limited by the detector rather than the diffraction limits of the optics.

In addition to the basic limitation on resolution due to the sampling structure of the sensor element, further limitations may be imposed by:

- Blur due to linear or angular motion or vibration of the sensor;
- Attenuation of high frequencies in the video amplifiers used with the sensor or the display system (increasingly not a problem as all-digital imaging becomes dominant in UAV payloads as in all other imaging areas);
- Distortion due to the manner in which the image is processed and transmitted by the data link.

10.3.2 Target Contrast

Another parameter that also has an important impact on the target detection quality is the target contrast. It affects the ability of the autopilot (in the UAV) or remote operator (in the GCS) to detect a target. The diffraction- or detector-limited resolution calculated above assumed a large signal-to-noise ratio in the image. If the signal-to-noise ratio is reduced, it becomes harder to resolve features in the image. The signal level in an image is specified in terms of the contrast between the target and its background. For sensors that depend on reflected light (i.e., visible and near-IR), contrast (C) is defined as

$$C = \frac{B_t - B_b}{B_b} \tag{10.2}$$

where B_t is the brightness of the target and B_b is the brightness of the background. For thermal-imaging sensors operating in the mid- or far-IR, contrast is specified in terms of the radiant temperature difference between the target and its background:

$$\Delta T = T_t - T_b \tag{10.3}$$

The combined effects of resolution and contrast are expressed in terms of a "minimum resolvable contrast" (MRC) for visible and near-IR sensor systems and "minimum resolvable temperature difference" (MRT, MRTD, or MRΔT) for thermal sensors. These parameters are defined in terms of multibar resolution charts.

MRC is the minimum contrast between bars, at the entrance aperture of the sensor system, that can be resolved by the sensor, as a function of the angular frequency of the resolution chart in cycles per unit angle.

MRT is the minimum temperature difference between bars, at the entrance aperture of the sensor system, that can be resolved by the sensor, as a function of the angular frequency of the resolution chart in cycles per unit angle. Two things must be emphasized about the MRT and MRC:

1) They are system parameters that take into account all parts of the sensor system from the front optics through the detector, electronics, and display, to the human observer, and including the effects of blur caused by vibration and/or motion of the sensor FOV across the scene being imaged.
2) The contrast or ΔT that must be used with the MRC or MRT is the effective contrast at the entrance aperture of the sensor system, after any degradation due to transmission through the atmosphere.

Both the MRC and MRT are curves versus angular frequency, not single numbers, although they are sometimes specified in terms of one point (or a few points) of the total curve. Thus, an MRC is a curve of contrast versus angular frequency and MRT is a curve of ΔT versus angular frequency. Figures 10.6 and 10.7 show typical, generic MRC, and MRT curves.

The actual calculation of an MRC or MRT curve is beyond the scope of this book. It involves a detailed determination of the modulation transfer functions (MTF) related to optics, vibration, linear or angular motion, and displays; gains and bandwidths of the video circuits; signal-to-noise levels in the detector and display subsystems; and factors related to the ability of the operator to perceive objects in the display.

From the standpoint of a system designer, the MRC or MRT curve is the logical starting point for analysis of sensor performance. Once the appropriate curve is known, provided by the designer of the sensor subsystem, operational performance can be predicted by a relatively simple "load line"

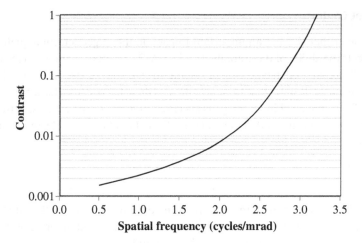

Figure 10.6 Generic MRC curve

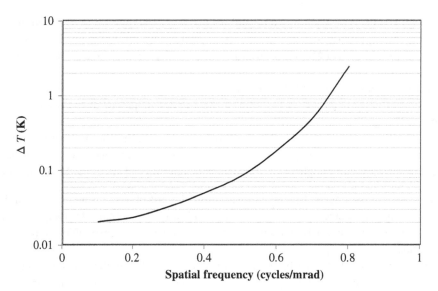

Figure 10.7 Generic MRT curve

analysis that determines the range at which the expected target contrast or ΔT is at least equal to the MRC or ΔT of the sensor system.

10.3.3 Transmission Through the Atmosphere

Other than sensor resolution and target contrast, the atmosphere will impact the target detection process, since the image is transmitted through the atmosphere. The atmosphere is a dynamic system that could be different every day in terms of parameters such as meteorological visibility, haze, fog, rain, and humidity.

The contrast at the sensor aperture is determined by the inherent contrast of the target (at zero range) and the contrast transmission of the atmosphere. For thermal radiation, the contrast

10.3 Target Detection, Recognition, and Identification

transmission is equal to the ordinary transmission at the wavelength in use, since T_t and T_b are both reduced, so that the effective ΔT at the sensor is given by

$$
\begin{aligned}
\Delta T(R) &= \tau(R)T_t - \tau(R)T_b \\
\Delta T(R) &= \tau(R)(T_t - T_b) = \tau(R)\Delta T_0
\end{aligned}
\tag{10.4}
$$

where $\tau(R)$ is the transmission to range R and ΔT_0 is the thermal contrast at zero range.

The situation is more complicated for visible and near-IR sensors. At these wavelengths the atmosphere may scatter energy from the sun or other sources of illumination into the sensor. This energy produces "veiling glare" that further reduces the target contrast beyond the basic reduction caused by attenuation. The effect of this glare is characterized by the "atmosphere to background" ratio (A/B), where A is the radiance of the atmosphere along the line of sight (LOS) to the target and B is the radiance of the background. The contrast at range R (i.e., $C(R)$) is then given in terms of the contrast at zero range (C_0) by the equation

$$
C(R) = C_0 \left[1 - \frac{A}{B}\left(1 - \frac{1}{\tau(R)} \right) \right]
\tag{10.5}
$$

Unfortunately, the ratio A/B is rarely known with any confidence. It is not commonly reported in weather and atmospheric databases. A rough estimate of the magnitude of A/B can be obtained by setting A/B equal to the inverse of the background reflectivity for overcast conditions and A/B equal to 0.2 times the inverse of the background reflectivity for sunlit conditions. Since the background reflectivity in the visible spectrum is on the order of 0.3–0.5, we can use $A/B \cong 2$–3 for overcast conditions and $A/B \cong 0.4$–0.6 for sunlit conditions.

To apply either Equation (10.4) or (10.5), one must know the value of the transmission through the atmosphere, $\tau(R)$. For the visible and near-IR, atmospheric attenuation of electromagnetic radiation is almost entirely due to scattering. Under these conditions, $\tau(R)$ is given by Bier's law:

$$
\tau(R) = e^{-k(\lambda)R}
\tag{10.6}
$$

where $k(\lambda)$ is called the "extinction coefficient" and depends on both the state of the atmosphere and the wavelength at which the sensor operates.

It turns out that for ordinary haze, $k(\lambda)$ can be estimated to a very good approximation in the visible and near-IR by using the empirical equation

$$
k(\lambda) = \frac{C(\lambda)}{V}
\tag{10.7}
$$

which relates k to the meteorological visibility (V) through a constant that takes into account the wavelength of the radiation. The constant $C(\lambda)$ is plotted in Figure 10.8. For example, at a wavelength of 500 nm, in the green portion of the visible spectrum, $C(\lambda) = 4.1$. Therefore, if the visibility were 10 km (a fairly clear day), $k(\lambda)$ would be 0.41 km^{-1}. Applying Equation (10.6), the transmission at a range of 5 km would be about 0.13.

In the mid- and far-IR, atmospheric attenuation mechanisms include scattering and absorption. Absorption in the atmosphere is largely due to water. The water may be present as a vapor, rain, or fog. The attenuation mechanisms are different for these three cases. Water vapor primarily absorbs the IR energy. Rain primarily scatters energy. Fogs both absorb and scatter. Haze scatters mid- and far-IR energy just as it does visible and near-IR radiation. However, the effects of haze are less at the longer wavelengths, since scattering efficiency decreases rapidly as a function of wavelength for wavelengths longer than the characteristic dimensions of the scatterers.

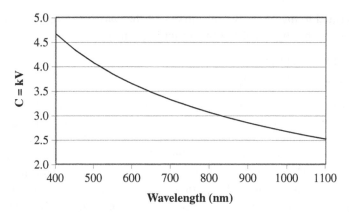

Figure 10.8 Extinction coefficient versus visibility

Attenuation of mid- and far-IR radiation generally is assumed to follow Bier's law (Equation (10.6)). Bier's law is based on the assumption that the fractional absorption of energy over a unit distance is constant over the entire path length. If there is significant atomic or molecular absorption in "absorption lines" that have a narrow wavelength extent, it is possible for all the energy in some highly-absorbed lines to be absorbed over a short path length, so that less absorption occurs per unit distance for the rest of the path. This may need to be taken into account in some circumstances, but usually is neglected at the levels of attenuation that are consistent with successful operation of imaging sensors.

Following the common approach, we will use Bier's law to estimate attenuation. The extinction coefficient that must be used is the sum of several separate extinction coefficients that address each of the processes described above. The total IR extinction coefficient (k_{IR}) is given by

$$k_{IR} = k_{H_2O} + k_{fog} + k_{haze} + k_{smoke/dust} \tag{10.8}$$

Only the H_2O term is subject to the caveat mentioned above, as all of the other terms relate to broadband attenuation phenomena that are not subject to the saturation effects that can lead to the breakdown of Bier's law. It is beyond the scope of this book to discuss all of these extinction coefficients in detail. A brief description of each of the components of the total extinction coefficient is as follows:

- Water-vapor absorption is determined by the density of water vapor in the atmosphere, in g/m^3. This, in turn, depends on the temperature and relative humidity, or, equivalently, temperature and dew point. Note that it is the absolute density of water vapor that matters, so a cool, humid ("clammy") day is much better than a hot, humid day for use of IR sensors.
- Rain scattering is in addition to the absorption due to water vapor in the air through which the rain is falling. The extinction constant due to rain can be calculated (or looked up) if one knows either the visibility (at visible wavelengths) or the rain rate in millimeters per hour.
- Fog both absorbs and scatters. Furthermore, fog tends to have a strong variation in density as a function of height above ground level. The IR extinction coefficient in fog has been related empirically to the visible extinction coefficient, and there are models for the vertical structure of typical fogs. Two general types of fog are recognized that have significantly different effects on IR radiation: wet fogs (in which condensation occurs on surfaces at ambient temperature such as windshields) and dry fogs (in which such condensation is not observed). As one might expect, attenuation is greater in a wet fog than in a dry fog, for the same visible range.

10.3 Target Detection, Recognition, and Identification

- Haze scatters radiation. Although it has more effect in the visible than in the IR, it turns out that Equation (10.7) still holds, with $C(\lambda) = 0.29$ at a wavelength of 10 µm. Note that this is much smaller than the values of $C(\lambda)$ in the visible and near-IR given in Figure 10.8.
- Extinction coefficients for smoke and dust depend on the integrated density of material along the LOS, expressed as the "cL" product, where c is the density of material at any point along the LOS (in g/m^3) and L is the length of the LOS. If, as is typical, the density, c, varies along the LOS, then cL must be calculated by integrating $c(s)$ ds over the LOS (where s is the position along the LOS). Once cL is known, $k_{\mathrm{smoke/dust}} = \alpha cL$, where α is a constant with units of m^2/g that characterizes the particular kind of smoke at the wavelength of the sensor. The effect of dust is nearly independent of wavelength over the entire visible to far-IR spectrum and α for dust is 0.5 m^2/g.

As was alluded to with regard to fog, most atmospheric attenuators vary in density as a function of height above ground level. Models are available to describe this variation and allow calculation of effective extinction coefficients for slant paths from the sensors of a UAV down to the ground. In most cases, looking down at a relatively steep angle, as is typical of a UAV, has advantages over attempting to look over the same range in a near-ground path, because the attenuation drops off rapidly with altitude. Some fogs and low clouds are the obvious exceptions to this rule.

10.3.4 Target Signature

The final parameter needed to predict imaging sensor performance is the target signature. The definitions of visible/near-IR and thermal signatures, contrast, and ΔT have already been discussed. The determination of actual signatures is quite complex. The target signature for an EO sensor is primarily a function of "reflective contrast," while for an IR sensor is primarily a function of "thermal contrast." The "reflective contrast" depends not only on the surface properties of the target (paint, roughness, etc.) and background (materials, colors, etc.) but also on lighting conditions.

In some cases, the contrast to be used in the analysis will be specified as part of the system requirement. For general systems analysis, it is common to assume a contrast of about 0.5. Lower values might be used to explore worst-case conditions. It is reasonable to assume that most targets will have contrasts in the range from 0.25 to 0.5. However, it must be understood that some targets will have essentially zero contrast, some of the time. Those targets will not be detected.

The thermal contrast at which the payload must operate may be specified. If not, it is reasonable to use a ΔT in the range from 1.75 °C to 2.75 °C as a nominal value for systems analysis. Actual targets tend to have localized hot spots with contrasts much higher than these nominal values of ΔT. However, unless such hot spots are known reliably to be present, they usually should be considered to contribute to a margin of performance for the system, rather than being used to support basic performance predictions.

10.3.5 Display Characteristics

One other point is worth mentioning, although it relates to the ground station rather than the payload. If the display screen is too small, then the remote operator's eye will not be able to take advantage of the full resolution of the sensor. This effect should be included in the MRC or MRT curve. However, the curves provided for sensors often are not taken with the actual display to be used in the system.

It can be shown [29] that a target embedded in clutter must subtend at least 12 minutes of arc (1/5 degree) at the operator's eye, in order to have a high probability of being detected. If one were trying to find a target that had two lines of a 500-line display across its height, then it would fill $1/250 = 0.004$ of the vertical height of the display. Suppose that the operator's eye is about 24 inches in front of the display screen. Then, the height (not diagonal measurement) of the screen would need to subtend 250 times 1/5 degree (50 degrees) at a distance of 24 inches. This requires a screen height of about 22 inches. This would require a screen with a diagonal measurement of 37 inches if the display used the common 4:3 aspect ratio!

In fact, many sensor images may have less than 500 lines of resolution. At 350 lines of resolution, a 25-inch diagonal screen would be adequate. However, many tactical displays are 12 inches or less. The operator can always move his or her head closer to the screen, but this should be considered in the design of the operator station if it is going to be required. Distances less than 20 inches may lead to eyestrain if used for long periods of time. There turns out to be a good technical justification for using large, high-definition display screens in a ground station when there is room for them.

10.3.6 Range Prediction Procedure

Once the MRC or MRT, atmospheric extinction, and target signature are known, it is possible to predict ranges for detection, recognition, and identification using a simple, graphical procedure.

The first step in this procedure is to convert the cycles per mrad axis of the MRC or MRT to range. This is accomplished by using the relationship:

$$R = \frac{hf_s}{n} \tag{10.9}$$

In Equation (10.9), R is the range to the target, h is the height of the target, n is the number of lines of resolution required to perform the desired task with the desired probability of success, and the spatial frequency (f_s) is expressed in lines per rad, assuming that R and h are both in the same units (m, km, etc.). For instance, if the target of interest has a height (projected perpendicular to the LOS) of 4 m ($h = 4$ m), and the desired task performance is a 0.5 probability of detection, then two lines across the target height are required ($n = 2$). Now, Table 10.1 can be constructed, which maps lines or cycles per mrad (the likely units of the horizontal axis of the MRC or MRT) directly to range to the target.

A table of this sort can be used to create a new horizontal axis for the MRC or MRT with range to target as the parameter. Note that this axis applies only to a particular value of h and task (value of n).

Table 10.1 Lines or Cycles versus Range

$h = 4$ m, 50% detection probability			
R (m)	Lines/rad	Lines/mrad	Cycles/mrad
500	250	0.250	0.125
1,000	500	0.500	0.250
1,500	750	0.750	0.375
2,000	1,000	1.000	0.500
2,500	1,250	1.250	0.625

10.3 Target Detection, Recognition, and Identification

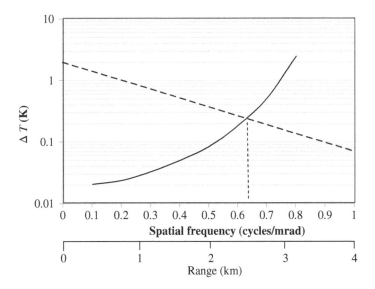

Figure 10.9 Load-line analysis

Once this mapping is performed, the x axis of the MRC or MRT curve can be relabeled from spatial frequency to range, as shown in Figure 10.9 by placing a second horizontal axis below the original spatial frequency axis.

To establish the maximum range at which the task can be performed with the required probability, we must determine the contrast available at the sensor as a function of range. The plot of contrast versus range is often referred to as a "load line." For this example, a thermal contrast at zero range of 2 °C is assumed and used as the y-intercept of the load line. The slope of the load line is calculated using Bier's law. For example, a total IR extinction coefficient of 0.1 km^{-1} is assumed.

If a semilogarithmic scale is used, as in Figure 10.9, the target contrast versus range becomes a straight line (the dotted line sloping downward from the left to right). The target contrast line intercepts the MRC or MRT at the maximum range for which the available contrast is greater than or equal to the required contrast. Therefore, the value of range at the interception of the target contrast line and the MRC/MRT curve is the predicted maximum range at which the sensor system will be able to perform the task being assessed (i.e., the task for which the value of n was selected, against the target whose dimensions were used in Equation (10.9) to convert from spatial frequency to range).

In the specific example used in Figure 10.9, a generic MRT curve is used to estimate the range at which a 4 m high target can be detected with a probability of success of 0.5. The MRT curve is plotted versus cycles per mrad on semilogarithmic scales. Equation (10.9) is used to convert cycles per mrad to range in kilometers for this task and target, as tabulated above. A second horizontal axis shows the range corresponding to each spatial frequency.

The load line intercepts the MRT curve at 0.63 cycles/mrad, which is equivalent to about 2.5 km. At ranges less than or equal to 2.5 km, the available contrast (plotted in the load line) exceeds the required contrast (plotted in the MRT). At longer ranges, the available contrast is less than the required contrast. Therefore, we estimate that the maximum range at which there is a 0.5 probability of detecting a 4-m target with this sensor is 2.5 km.

The methodology described above is the standard approach to prediction of detection, recognition, and identification ranges for imaging sensors. It can be carried out in either direction, using a given MRC or MRT curve to predict performance or using required performance to generate a required maximum value of the MRT or MRC at each spatial frequency.

Example 10.1

Consider a target with a height of 5 m, at a distance of 1,500 m from a UAV. In using a surveillance payload, we are looking to have a 90% probability of detection. Determine:

a) Spatial frequency (in lines/rad)
b) Resolution (in cycles/mrad)

Solution:
From Figure 10.4, when the probability of detection is 90%, the number of cycles is 1.8.

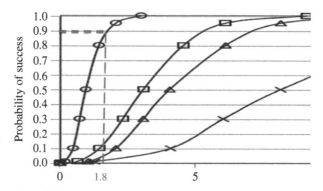

Then the number of lines (i.e., cycles) is 1.8, with $n = 2$ cycles $= 2 \times 1.8 = 3.6$. Now we use Equation (10.9) to calculate the spatial frequency (f_s):

$$R = \frac{hf_s}{n} \rightarrow f_s = \frac{Rn}{h} = \frac{1,500 \times 3.6}{5} = 1,080$$

Each rad has 1,000 milli rad and each cycle has two lines. Thus, the resolution is

$$n = \frac{f_s}{1,000 \times 2} = 0.54 \text{ cycles / mrad}$$

10.3.7 A Few Considerations

While the methodology – presented in Section 10.3.6 – is standard, and can be carried out with great precision, it is important for a system designer to understand that it produces only an estimate of the performance that actually will be achieved. The estimate has proven reasonably accurate in practice, and the use of a standard methodology results in considerable confidence when comparing two similar sensors – nonetheless, it is no more than an estimate. A number of considerations should be kept in mind when using this estimate:

- The estimate is made for a probability of success when averaged over a large number of trials. Since it includes the performance of the operator, the averaging must include data from several operators, some of whom will perform better than the average and some of whom will perform worse.

10.3 Target Detection, Recognition, and Identification 229

- The estimate is for a particular target contrast and set of atmospheric conditions. Both contrast and the atmosphere vary with time and location. This means that one is unlikely ever to get a large sample of test data that is taken under uniform conditions that exactly match those used in the estimate.
- Both the target signature and the atmospheric conditions actually present along the LOS from the UAV to the target at a particular instant are difficult to measure and rarely are measured at the same time and from the same aspect angles as used by the UAV. Therefore, it is rarely possible to specify all of the factors affecting a particular data point in a field test.
- If the task to be performed is detection of a target, the level of clutter in the scene affects the probability of detection in a complicated manner (discussed later) and is not included in the estimation methodology given above.

The result of these considerations is that it is difficult to make any precise comparison of the estimated performance to the actual performance measured in a field test. This is not an uncommon situation in systems engineering. As in other cases, it must be dealt with by designing for a level of robustness in estimated performance that ensures that the actual operational performance will meet the needs of the user. Although it is impossible to make any "scientific" proof of the robustness of designs developed using the methodology described above, the fact that this methodology has survived and become a standard in the design of imaging systems gives some assurance that it is sufficiently conservative in its predictions to result in an acceptable performance margin.

One of the strengths of the standard methodology is that it can be used with considerable confidence to compare similar sensor systems. For instance, if two image sensors with different MRC curves are predicted to have a maximum detection range of 2 and 2.2 km for some specific target and atmospheric conditions, it may be that they actually would achieve ranges of 2.5 and 2.75 km under those conditions (or 1.8 and 2 km), but it is likely that the difference in range actually is about 10%, and it is virtually certain that the sensor predicted to have a longer detection range actually does have a longer-range capability under any conditions that are approximately the same as those used in the calculation. Therefore, this methodology can be used with considerable confidence in making tradeoffs between similar sensor designs.

If the sensors are not similar, however, considerable care must be exercised in using the methodology to compare their performance. This is particularly true when comparing an EO to a forward-looking IR (FLIR) imaging device. There is considerable anecdotal evidence that FLIRs perform better than the model predicts for detection, and some evidence that they perform worse than predicted for identification. Reasons why this might be true can be hypothesized, but the authors are not aware of any "definitive" study that offers proof of any such hypothesis. It is particularly likely that an FLIR will be able to detect a high-contrast target at ranges greater than predicted by the Johnson Criteria.

A "hot" target can appear as a beacon in the scene, equivalent to a fire or flare in a TV scene. Thermal contrast can be tens of degrees for some targets, such as engine exhausts. This can allow detection with less than one resolution line across the target. Since hot objects are likely to be of some interest, the ability to easily detect their location at long range can lead to effective target detection at long range in many scenarios.

On the other hand, an EO sensor uses only reflected light and target contrast can never exceed 1.0. There may be many small, high-contrast clutter items in an EO sensor scene, so that it may be difficult to separate a possible target from the clutter, even if it has the maximum possible contrast, until there is enough resolution in the image to tell something about the target shape and for the patch of high-contrast (light or dark) to be big enough to call attention to itself.

These arguments are, of course, very qualitative. They depend on generalizations about the types of clutter present in the scene and the types of targets that are of interest. However, they illustrate the hazards of attempting to compare an EO sensor to an FLIR with the standard methodology. Some relative "calibrations" of the model in the particular situation that is of interest for the comparison are required before such a comparison is attempted. For instance, one might find that the criterion for detection should be reduced to one line across the target dimension for an FLIR against some classes of targets.

Fortunately, the limiting class of targets for most systems is relatively cool. If thermal contrast does not exceed a few degrees, the performance of an FLIR is better represented by the standard model. Therefore, when designing to deal with the minimum target signatures required in a system specification, the standard model is not likely to be too pessimistic.

A special case of comparing two similar sensors that is very important to a system engineer is the determination of sensitivities to changes in system design. This is an area in which the standard methodology should work very well. For instance, the system engineer can have considerable confidence that the predictions of fractional degradation due, say, to an increase in vibration will be reasonably accurate.

10.3.8 Pitfalls

Having already, perhaps, caused some dismay with regard to the accuracy of the performance estimations when the input information is accurate, it is, unfortunately, necessary to point out some pitfalls with regard to the input information.

Potentially the most serious problem is in ensuring that the MRT or MRC curve that is used truly is a system-level curve. In particular, the effects of AV motion and vibration and the effects of the data link must be included in the curve if the predictions are to be used to predict system performance. Even the displays in the control station may turn out to limit performance and must be included in the system MRT or MRC. The designer often is provided with an MRT or MRC measured in the laboratory, with the sensor firmly supported on a laboratory bench, coaxial cables connecting the sensor to the display, and a high-quality display being observed by an experienced operator. This curve may be very optimistic, compared to the actual operational configuration.

At the very least, the system designer must determine whether the MTF of the sensor vibration, the data link, and the display have features that are likely to have a significant impact on the total system curves. If so, they must be folded into the total curve. This can be done analytically, using procedures described in the literature but beyond the scope of this book. As an alternative, the MRT or MRC can be measured using the data link and actual ground-station displays. Unfortunately, it is usually very difficult to introduce realistic sensor vibrations and motion into an MRT/MRC measurement, so these factors probably will have to be introduced analytically if a preliminary analysis indicates that they will degrade the system curve.

Another major pitfall, already alluded to, is the difficulty in carrying out a field test that produces results that are comparable to the system-design calculations. Of course, if the system "works" in an operational scenario, then one might say that this is not a problem. However, it often is difficult to define just what is meant by "working" in an operational sense. Most systems will have specifications of a range at which they must detect, recognize, and/or identify certain targets. It is likely that the organization responsible for accepting the system will want to test it against these specifications. The system engineer must be aware that this is not an easy test to perform.

It is hard to provide targets that have the specified contrast, very hard to provide an atmosphere that has a well-characterized transmission over the actual LOS from the UAV to the target, and almost impossible to ensure that these conditions are constant long enough to get a statistically significant sample over several operators. The atmosphere is particularly difficult to provide and characterize if the specification calls for some moderately limited visibility (say 7 km). Very clear atmospheres are relatively easy to find at desert test sites. Limited visibility may be common at some test sites, but is likely to be highly variable over time and over different LOS.

The result of this difficulty is that it usually is necessary to build the best possible model of the sensor, using the methodology described above, and then validate the curves at a few points that may not be very close to those stated in the specification (clear air, high-contrast targets). The "validated" model is then used to prove compliance with the specification "by analysis." This situation needs to be understood by those who write specifications as well as by those who must plan and budget for system testing.

10.4 The Search Process

The analysis methodology described above deals only with the static probability of being able to detect, recognize, or identify a target, given that it is present within the display. This is the essential first step in design of an imaging sensor system for a UAV, but does not address the critical issue of searching for a target over an area many times the size of a single FOV.

The mission requirements for using a UAV to search for something are conveniently discussed in terms of military or pseudo-military applications (such as police or border patrol), because those are the missions for which UAVs have been most often employed up to this time. Civilian search applications generally will fall into the same categories as illustrated by some examples cited in the discussion, so the conceptual framework developed by the military provides a good way to organize the discussion. A number of factors impact the search process: (1) type of search, (2) field of view of the imaging sensor, (3) search pattern, (4) search/stare time. These four factors are briefly discussed in the following sections.

10.4.1 Types of Search

One of the most common missions for a UAV is reconnaissance and/or wide-area surveillance. These missions require the UAV and its operator to search large areas on the ground, looking for some type of target or activity. An example might be to search a valley looking for signs of an enemy advance. There are three general types of search: (1) point, (2) area, and (3) route. It is important to understand how the fundamental characteristics of a UAV and its imaging payload affect the ability of the UAV system to perform these three types of searches.

A "point" search requires the UAV to search a relatively small region around a nominally known target location. For instance, an electronic interception and direction-finding system may have determined that there is a suspected command post located approximately at some grid coordinate. However, the uncertainty in the location determined from radio direction finding at long range is often too great to allow effective use of artillery to engage the target without very large expenditures of ammunition to blanket a large area. The mission of the UAV would be to search over a region centered at the nominal grid location of the command post and extending out to the limits of the uncertainty in the actual location, perhaps several hundred meters in each direction.

An "area" search requires the UAV to search a specified area looking for some type of targets or activity. For instance, it might be suspected that artillery units are located somewhere in an area of several square kilometers to the east of a given road junction. The mission of the UAV would be to search the specified area and determine the presence and exact location of these units. A civilian equivalent might be to search some specified area looking for stray livestock.

A "route" search can take two forms. In the simplest case, the mission is to determine whether any targets of interest are present along a specified length of a road or trail, or, perhaps, whether there are any obstructions along a section of a road. A considerably more difficult task is to determine whether there are any enemy forces in position to deny use of the route. The second type of route reconnaissance actually is more like an area search over a region that centers along the road, but extends at least several hundred meters to either side of the road, to include tree lines or ridges that would provide cover and a field of fire that includes the road.

There have long been proposals to use civilian UAVs in applications that resemble the simple route search. One such application is to maintain surveillance on transmission lines or pipeline rights of way to find potential problems, such as trees too near the power lines, so that they can be dealt with before they lead to failures.

The NASA Airborne Science Program provides a set of NASA supported aircraft that benefit the earth science community. In 2016, NASA Ames has delivered an unmanned aircraft system (UAS) to the University of Kansas for analyzing glaciers in Greenland (i.e., "Area") associated with sea-level rise around the globe. The UAV – named Viking 400 (Figure 10.10) – has a maximum range of 600 nautical miles and endurance of 11 hours, and is designed for data collection missions in remote or dangerous environments. The UAV with a gross take-off weight of 520 lb is powered with one 1 kW 24 V DC electric motor. It can carry up to 100 pounds of scientific instruments and sensors as payload.

The attraction of a UAV for these missions in the military world is based on its ability to fly almost undetected in hazardous airspace with greater survivability than a manned aircraft, as well as the UAVs relative expendability, since it does not carry a human crew. For civilian applications, the hoped-for advantages are mainly related to reducing cost by eliminating the flight crew and using smaller aircraft with lower operating costs. The price that is paid for leaving the human operators behind is that the visual perception of the operator is limited to the images that can be provided by a sensor payload.

10.4.2 Field of View

The relevant basic limitation of imaging sensors is resolution, which is closely related to Field of View (FOV). As we have seen, if the sensor provides 500 lines of resolution there is a fixed

Figure 10.10 NASA Viking 400 (Source: NASA)

10.4 The Search Process

relationship between the dimensions of the FOV and the maximum range at which there is a reasonable probability of being able to detect the presence of a target within the FOV. If we require two lines across a 2-m target for detection (assuming that there is sufficient contrast) and the sensor has a total of 500 lines available, then the FOV cannot cover more than 500 m (one line per meter).

Consider a case where we require two lines across a 2-m target for detection and the sensor has a total of 500 lines available. What is the maximum length that the FOV cannot cover?

For any look-down angle, the slant range to the far edge of the FOV will be greater than the range to the near edge of the FOV (or at the center of the FOV if the UAV sensor is looking straight down). In general, the sensor will not be looking straight down and the geometry will be as shown in Figure 10.11. The figure assumes an altitude of 1,500 m, reasonable for immunity from small arms fire, and a nominal look-down angle of 45 degrees.

A fairly routine EO sensor could have a good probability of detecting a 2-m target out to a slant range of about 2,200 m, if it had an FOV of about 7 degrees. A 7-degree × 7-degree FOV would cover the keystone-shaped area (i.e., trapezoid) on the ground shown in the figure. Taking into account the fact that most TVs actually have a four to three aspect ratio for their FOVs, the actual area of the FOV on the ground would be about 350 m × 350 m, still with the keystone shape. Recall that, $\sqrt{1{,}500^2 + 1{,}695^2} = 2{,}263$ and $2{,}263 \times \tan(7 \text{ deg}) = 278$. Moreover, $\sqrt{1{,}500^2 + 1{,}327^2} = 2{,}003$ and $2{,}003 \times \tan(7 \text{ deg}) = 246$.

Smaller look-down angles would lead to greater depth for the area on the ground, but the sensor would not be able to detect a target in the upper portion of the scene. If the system uses a simple, manual search process, there is a danger that the operator will not be conscious of the detection limitations of the sensor system and will manipulate the pointing angles of the sensor in such a way that much of the scene viewed in the display will be at slant ranges greater than the maximum effective detection range of the sensor.

It may then appear to the operator that he or she has "searched" large sections of terrain in which, in fact, he/she would not have detected targets, even if they were out in the open. While training and experience can reduce this problem, an operator looking at a screen in a control station may have difficulty in making effective use of the sensor unless provided with information over and above the raw imagery.

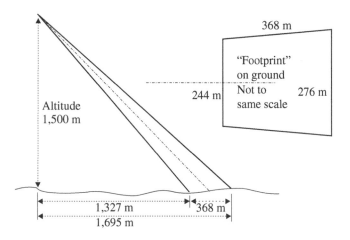

Figure 10.11 Geometry for a typical UAV field of view on the ground

A simple form of additional information that can be provided by the system is a line across the scene that indicates the "detection horizon" of the sensor. This line indicates the perimeter on the ground at which the slant range from the sensor exceeds the nominal detection range for the class of targets being sought. Its position in the scene can be computed based on the look-down angle of the sensor and the altitude of the AV. This will allow the operator to confine the search to ranges at which there is a reasonable chance of detecting targets.

10.4.3 Search Pattern

An important factor in the design of a search process is the pattern by which the search process is conducted. The importance of confining the search to ranges at which success is possible is related to the fact that searching the ground with an EO camera or thermal-imaging system is rather like looking at the world through a soda straw. As we have seen, the FOV of a sensor capable of detecting targets with dimensions of a meter or two at ranges of about 2 km is only a few hundred meters on a side. This forces the operator to search a succession of small areas. Assuming the nominal field on the ground shown in Figure 10.11 (i.e., about 350 m × 350 m), and allowing some overlap required to fit a set of keystone shapes (i.e., trapezoid) into a square and ensure that no part of the square is unexamined, it would take about 12–15 separate "looks," at a minimum, to cover a square kilometer. This is illustrated in Figure 10.12.

At no time would the operator be able to see the entire square kilometer unless he/she switched back and forth between the 7-degree search FOV and a much larger "panoramic" FOV. This leads to a significant problem related to searching large areas – it is difficult for the operator to manually carry out a systematic search that covers the entire area in an efficient manner. An observer looking out of the window of a manned aircraft uses peripheral vision over a large FOV to retain orientation to the ground and carry out a systematic search.

An operator looking at a display in a control station has no peripheral vision. Each patch of terrain is seen in isolation. Unless some automatic system is provided to keep track of what part of the ground has been looked at and to guide the operator in selecting the next aim-point for the sensor, he/she is likely to search a relatively random sample of the total area without even realizing that he/she has not looked at all parts of the assigned region.

This problem can be addressed by training the operator to move the FOV in some systematic pattern. However, this approach is likely to require significant overlap between "looks" in order for the operator to be able to retain a sense of how the next look is related to the previous look. If the area is many FOVs wide and the operator tries to perform a raster scan or line-by-line scanning

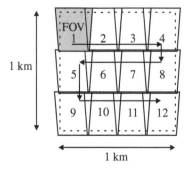

Figure 10.12 Automated search pattern

10.4 The Search Process 235

(i.e., across the bottom, then move up one FOV and go back across the area, etc.) it is very difficult to keep the scans parallel and slightly overlapping unless there are conveniently-spaced linear features in the scene to use as references.

In fact, it is likely that the only way to perform a thorough and efficient search of an area that is much larger than a single FOV is to provide an automatic system that uses the navigation and inertial reference systems of the UAV to systematically move the FOV over the area with some reasonable degree of overlap and at a rate that allows the operator to look at each scene long enough to have a good probability of detecting any targets that may be present.

Since slewing the sensor continuously introduces some blur, masks target motion, and requires a high data rate to transmit a constantly changing scene, it is best to use a "step/stare" approach to the search. In this approach, the sensor is rapidly moved to an aim-point in the center of each desired FOV on the ground and is then stabilized at that point for some period while the operator searches a stationary scene. Then the sensor slews rapidly to the next FOV.

If target motion relative to the scene is relatively small during one stare period, then there is little benefit from motion clues and there is a need to transmit only one frame of video per FOV on the ground. Whether this situation applies depends on the rate of target motion and on the length of the stare. In any case, if the data rate is limited, as it often is, it may be necessary to forego any possible advantages of target-motion clues and settle for one "still" picture per FOV on the ground.

10.4.4 Search Time

Search time is an important factor in the search process; we would like to design a process at which the time is minimized. The required search/stare time can be estimated from experimental data on the ability of an operator to detect targets in video scenes versus the time that is allowed to look at each scene. The experimental data has the interesting characteristic that the cumulative probability of detecting a target, if it is present, climbs rapidly for the first few seconds that an operator looks at the scene. The curve then flattens out, and much longer look times result in only slight increases in the probability that a target will be detected. There is some evidence that the elapsed time before the curve flattens is correlated to the probability that there is a target in the scene.

This could be explained by a "discouragement factor" that leads to reduced attention by the operator if he or she does not find a target in the initial scan of the scene. The discouragement factor would be increased if the operator were scanning many scenes in succession and finding that most of them did not contain any targets. In other words, if the operator does not expect to find a target in the scene, he/she will give up serious examination of the scene more quickly than if he/she thinks that there is a good chance that there is something there.

Reference [30] applies a methodology from Reference [31] to calculate search times for a scene of a video for three levels of clutter. The results are shown in Table 10.2. The "congestion factor" is defined as the number of clutter objects per eye fixation by the operator within the scene. It requires about 15 fixations to search a typical video display, so a congestion factor of 3 corresponds to about 45 clutter objects in the scene. A "clutter object" is defined as any object whose size and contrast approximates a target of interest, so that it must be examined with more than a passing glance in order to distinguish it from a target.

One of the authors of previous editions of this book had experience with the Aquila system that suggests that these times may be somewhat longer than optimum if the "discouragement factor" is taken into account. This subject is not well documented and would be a fruitful area of research for human-factors organizations. It is an area that must be of concern to a system designer, since the implications of these numbers are that a UAV has relatively limited capability for large-area searches.

Table 10.2 Single-frame display search time

Level of Clutter	Congestion Factor	Search Time (s)
Low	<3	6
Medium	3 to 7	14
High	>7	20

For instance, if the mission were to search an area 2 km × 5 km in extent, using the 7-degree FOV discussed above, and the clutter were high, one would need about 15 FOVs (scenes)/km^2 at 20 s/scene, plus about 1 s/scene for the sensor to slew and settle at its new aim-point. This works out at 320 s/km^2, or 3,200 s (53.3 min), to search the assigned 10 km^2 area. A 1-hour search would consume much of the on-station endurance of many small UAVs. Furthermore, if the targets were moving, the low search rate might allow them to move through the area without being seen, since only a very small fraction of the total area would be under surveillance at any given time.

By comparison, a manned helicopter or light aircraft would probably search the same 2 km × 5 km area in a few minutes, making a few low-altitude passes up and down its length.

Naturally, the search time would be shorter if the target were larger or more prominent. If the object to be found were a sizeable building, the UAV might be able to use its widest FOV and a rather small look-down angle to search the entire area in only a few FOVs. On the other hand, if the target were personnel (e.g., smugglers backpacking drugs across a border or guerrillas moving through open terrain), then the UAV would have to use a smaller FOV or get closer to the ground (which reduces the size of the footprint of the FOV on the ground) and might take many times longer to search the same area.

If there are many individual targets operating in concert, finding any one target may lead to finding them all. For instance, finding one person would lead to a closer examination of the surrounding area. Finding a few more people would confirm that a group of people were present in the area to be searched. This situation increases the probability of detecting the array of targets, since many can be missed as long as one is detected. However, it does not have a major effect on the time needed to search an area, since that time is dictated by the FOV and the time it takes to search one FOV. The FOV depends on the size of a single target.

Having ten tanks in the FOV will not make it possible to detect any of the tanks with a larger FOV (less resolution) than is required to detect any one of the tanks alone. There may be a minor effect on detection due to being able to accept a lower probability of detecting any one target and/ or due to the clue provided by several tiny spots moving together. However, this effect is likely to be small and is hard to predict when fixing the system design. As with many other marginal effects that might favor the system, it is best considered as part of the system margin, not to be counted on when setting the system specifications.

The conclusion that a UAV, with presently available sensors and processing, needs long times to search areas that might be searched rapidly by a manned aircraft is supported by experience with Aquila and other UAVs that have been tested in this mode. It is possible that automatic target recognizers, when they become available, will allow a much more rapid search by reducing the dwell time on each scene. Until that time, UAV system proponents and designers need to be cautious in claims related to searching large areas, particularly if the individual targets are small and there is significant clutter.

On the other hand, point and route searches can be performed within reasonable times. If the location of the target is known to within about 500 m, then only about 1 km^2 need be searched. Even in heavy clutter, this would take no more than about 5 min. As with the large-area search, however, it is probably necessary to provide an automated search system to ensure that the area is completely and efficiently covered by the FOV.

A route search of a road or highway and its shoulders can be performed by a single row of FOV positions strung out down the road. Depending on the targets and whether they may be hiding in tree lines along the shoulder of the road, the clutter may be anywhere from low to high. If looking for a convoy actually on the road, it may be possible to simply scan the FOV down the road as the UAV flies over it. Even if a more thorough, step/stare search is required, it should be possible to move down the road at a rate of no less than about 1 km/min.

If a full route reconnaissance is required, including possible ambush positions, then the task is essentially an area search for an elongated area centered on the road. Search time estimates can be made using the same methodology as for an area search.

It should go without saying, but sometimes seems to be forgotten, that a UAV brings with it no magic way to see through trees. In fact, the lower resolution, lack of peripheral vision, and slow progress of the search, all make a UAV system even less likely than a manned aircraft to detect targets moving under forest cover.

The IR sensor of Global Hawk (Block 10) provided a search rate of at least 12,000 km^2/h at an average NIIRS (National Image Interpretability Rating Scale) of 4.7 or better on actionable imagery.

10.5 Other Considerations

The topics discussed above all relate to the static performance of the imaging sensor subsystem of the payload, even though the mechanical motion of the system is inherent in the calculations through such terms as the MTF. However, it is important to have some explicit understanding of the mechanical factors that govern payload (here, imaging sensor) performance. To have an acceptable performance for imaging reconnaissance and surveillance payloads, one needs to also carefully address the following factors: (1) sensor location, (2) sensor installation, and (3) stabilization of the line of sight.

10.5.1 Location and Installation

Imaging sensors (both EO and IR) must be located such that they have the best view of the target without any obstacle. This requirement – for most UAV configurations – necessitates the sensor location to be naturally under the fuselage nose. Such a location further requires the engine(s) not to be positioned at the fuselage nose. This is the case for most fixed-wing UAVs such as Northrop Grumman RQ-4 Global Hawk (with a pusher turbofan engine) and General Atomics MQ-9 Reaper (with a pusher turboprop engine).

Moreover, the imaging sensors should be installed at a location that has the least amount of structural vibrations. Reconnaissance/surveillance payloads are a kind of delicate instrumentation that should be free of severe vibration. The UAV airframe is highly prone to stiffness problems such as aeroelasticity and induced vibration. Since the UAV structural elements are elastic, they are influenced by atmospheric disturbances such as gust.

The structure will experience the impact of gust; the effect will become severe as you move away from the air vehicle center of gravity. For such impact, the worst location is the wingtips, which will constantly vibrate in a gusty environment. Thus, the recommendation is to install the imaging sensors: (1) close to the UAV center of gravity and (2) away from wingtips – as much as possible.

The EO/IR cameras should be covered such that they produce the least amount of drag, while they can rotate about three axes. In general, there are four alternatives: (1) faired fuselage (e.g., Global Hawk and RQ-11 Raven), (2) nose roll-tilt (e.g., Aeronautics Orbiter), (3) belly pan-tilt (AeroVironment RQ-20B and Lockheed Martin Stalker), and (4) nose pan-tilt faired (e.g., ScanEagle and RQ-11 Raven). Most VTOL UAVs (including quadcopters) have no aerodynamic cover for their cameras; they sacrifice drag to gain lower cost and weight.

10.5.2 Stabilization of the Line of Sight

An EO/IR camera takes still imagery, so it is necessary to limit its angular movement – within an acceptable range – when attempting to take a picture/video. The stabilized platform of the payload (e.g., an EO/IR camera) is of central importance, because of the need for an accurate EO camera output and a high-quality image/video. Imaging systems mounted in UAVs cannot be rigidly fixed to the airframe because the airborne platform generally cannot maintain a constant orientation such as angular attitude (e.g., pitch angle, heading angle, and angle of attack) necessary for many missions. Two reasons for this problem are: (1) engine propeller/fan/shaft angular rotation and (2) atmospheric disturbances such a gust.

After a gust hits a stable air vehicle, the airframe will oscillate about its trim point for a few seconds and sometimes for a few minutes. The inherent stability will try to nullify the gust input by opposing its change and returning to the UAV initial trim point. During this transition period, the fuselage nose and wingtips will oscillate. If the imaging sensor is rigidly fixed to the airframe, it will also oscillate during this period. To get rid of such oscillation, we need to stabilize the imaging sensor and its line of sight. The objective would be to keep the LOS variations below 0.02 degrees or 0.4 mrad. In another words, in order to keep the minimum resolution required to effectively detect a target, the maximum allowable LOS variation is 0.02 degrees.

For example, for a UAV flying at an altitude of 2 km, a 3-km slant range would give the platform a 4.4 km circle of coverage on the ground. A sensor system would require about 0.4-mrad resolution to maintain one resolution cycle (two lines) over the dimensions of a tank (2.3 m) at 3 km, which is the minimum resolution required to detect that target. It would require much better than 0.02 deg (0.4 mrad) mechanical stability in order to maintain a reasonable MRC at 0.4 mrad/cycle (2.5 cycles/mrad).

A UAV airframe cannot maintain an angular stability approaching 0.4 mrad, so the sensor must be suspended by a stabilized platform that can support it with minimum angular motion. Maintaining image quality, tracking targets, or pointing the sensor accurately are accomplished through a "multiple axis gimbal-gyro set." The ability to hold the optical axis of the sensor about a nominal line – in the presence of disturbance – is called the line of sight (LOS) stability. It is measured in terms of the root-mean-square (RMS) deviation from a desired pointing vector and is usually described in units of mrad. The LOS stability required depends on the mission – high stability usually meaning high cost and high weight.

10.5.3 Gyroscope and Gimbal

Stabilization of the line of sight is provided by employing a gyroscope/gimbal set. In this section, the functions of gyroscope and its combination with gimbal is discussed.

10.5 Other Considerations

Gyroscope (or gyro) is (see Figure 10.13a) a spinning disk or wheel (i.e., rotor) that reacts to an input based on the gyroscopic law. The input is the deflection (see Figure 10.13b) of its spin axis (e.g., rotation about the x-axis) and the output is a force that causes a rotation about the other axis (e.g., y). Indeed, all spinning objects have gyroscopic properties: (1) rigidity in space and (2) precession.

This gyro reaction – referred to as the precession – is a force and is mainly a function of two parameters: (1) the disk mass moment of inertia and (2) angular rotation. There will therefore be a greater reaction when: (1) the disk is heavier, (2) the disk has a larger diameter, and (3) the angular speed is higher. The rotation of the disk is often provided with an electric motor. The law of the gyroscope is based on Newton's second law: The sum of the applied torques is equal to the rate of change of angular momentum. Since the disk has a fixed shape, the angular momentum of the disk is $I\omega$, where I is the mass moment of inertia of the disk and ω is the angular speed of the disk. Thus, by application of the law, we obtain:

$$T = \frac{d}{dt}(I\omega) = I\frac{d\omega}{dt} \tag{10.10}$$

The applied torque (T) is the only factor that causes the input deflection to the gyro disk.

A gyroscope has another performance when it is installed within one, two, or three gimbals. The first gimbal is connected to both ends of the shaft (see Figure 10.13c) of the disk. Then, this gimbal can be connected to another gimbal that is perpendicular to the first gimbal. Moreover, the first gimbal could be free to rotate inside the second gimbal. Then, the second gimbal can be connected to a third gimbal that is perpendicular to both the first and second gimbals. Thus, each gimbal can freely rotate about one of the axes of the gyro. To allow a free rotation, the connection between the gimbals and between the first gimbal and the disk is via ball bearings that have negligible friction.

A gimbal is a base/ring/frame that is mounted on an axis of a spinning disk and allows the disk to freely deflect about the other two axes. This output of the gyroscope would not be possible without the presence of a gimbal. This function of the gyro-gimbal set has two applications: (1) to measure the input deflection and (2) to maintain the initial spin axis. The first application is covered in Chapter 5 as the angular attitude measurement device (e.g., pitch/roll/yaw angle sensor). The second application is a means for stabilization of the line of sight, when the gyro spin axis is aligned along the LOS. Such a gyro/gimbal set is a stabilized platform for a payload.

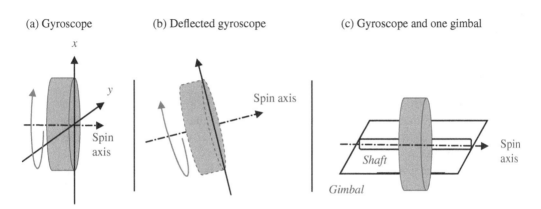

Figure 10.13 Gyroscope and gimbal

In the past few decades, a number of non-pure-mechanical devices with gyroscopic properties have been developed. Examples are: (1) microelectromechanical systems (MEMS) gyroscope, (2) ring laser gyroscope, (3) coriolis vibratory gyroscope, (4) fiber optic gyroscope, and (5) digital gyroscope.

10.5.4 Gimbal-Gyro Configuration

An imaging sensor must be suspended by a stabilized platform that can support it with minimum angular motion. A stabilized platform requires a stiff structure and careful placement of a gyro. The stabilization platform for a UAV typically uses either a two or three gimbal configuration (two-axis rotation and three-axis rotation) to offset the angular rotations of the aircraft to ensure that the sensor's line of sight remains pointed in the desired direction.

The first consideration in a gimbal design is the configuration; for example, the number of gimbals. While a two-gimbal mount may be satisfactory for some missions, it will not allow operations over a complete field of regard (i.e., there are directions in which the sensor cannot be pointed). A four-gimbal mount eliminates the notorious "gimbal lock" in which the gimbal reaches a limit in some direction beyond which it cannot go, but are heavy and large. Two- or three-gimbal systems meet most UAV mission requirements.

Some systems are designed with the IR receiver and laser range receiver located on the stabilized gimbal. As an alternative, a stabilized mirror with the sensors located off the gimbal can be used.

The primary advantages of the stabilized sensor configuration are that:

- The volume for the mirror is eliminated.
- The LOS is directly stabilized in inertial space without the need for half-angle correction (reflections from a tilting mirror move through twice the angle by which the mirror tilts).
- There is no need to compensate for image rotations that are introduced at mirrors.
- A smaller aperture is required.

Disadvantages of this configuration are that:

- There is a need for larger torque motors to drive the higher gimbal inertia associated with more mass moving on the gimbal.
- There may be a need to compensate for sensor-generated torque disturbances due to a variety of causes that are discussed below.
- There may be a requirement for a higher stabilization-loop bandwidth to reject the on-gimbal disturbances.
- A more complex gimbal structure may be required to support the higher bandwidth.
- There will be tighter balance requirements on individual replaceable components that mount on the gimbals.

Through use of structural modeling and careful control loop design, these problems can be solved and the required stabilization performance achieved, but the tradeoff to determine how much of the sensor system is on the gimbal is a key part of the initial design of the system. Figure 10.14 shows two- and three-gimbal configurations.

A gyro with three gimbals is a spatial memory that is able to maintain the spin axis, despite: (1) any atmospheric disturbance and (2) any intentional change of the UAV attitude angles (e.g., pitch/bank angle).

10.5 Other Considerations

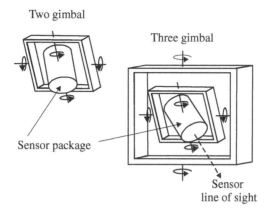

Figure 10.14 Two- and three-gimbal configurations

10.5.5 Thermal Design

Implicit in the design of the gimbal is the need to dissipate heat generated by the sensor and gimbal control electronics without undue impact on stabilization or structure. A number of concepts have been used for thermal management. Liquid cooling could provide the necessary heat transfer, but requires tubes to cross the gimbals with attendant torque disturbances. Other drawbacks to liquid cooling include higher weight associated with the plumbing and coolant, low reliability because of potential leakage and corrosion, and more difficult maintainability. For these reasons, ram air is the most common choice for thermal management.

10.5.6 Environmental Conditions Affecting Stabilization

In order to take advantage of the benefits offered by a stabilized sensor, LOS stabilization must be accomplished in the presence of wind/gust loads and sensor-generated torque disturbances, primarily compressor vibration from the cooling system for an FLIR. If the detection system involves a mechanically scanned detector array, which was common in early IR systems and still might be seen in some cases, the gyroscopic reaction associated with applying torque to the rapidly-spinning scanning mirror may be significant in sizing of gimbal torque motors. The need to deal with these issues implies the need for high-bandwidth, low-noise servo loops and a stiff gimbal structure.

A significant mechanical effect is caused by wind/gust loads on the exterior of the sensor housing. The exposed portion of the housing generally is spherical, but often has flat surfaces at the optical windows through which the sensor looks (and the laser beam emerges, if a rangefinder or designator is included in the payload). This choice is dictated by cost, since spherical windows are expensive, and their optical properties significantly complicate the design of a sensor system. Gust-loading effects often depend on the orientation of the sensor relative to the AV body, which determines the orientation of the flat optical window relative to the airflow around and past the AV.

Measurements of the disturbances generated by wind/gust loads and compressor vibration must be made in order to determine the required stabilization bandwidth. Wind/gust load measurements can be obtained by monitoring torquer current during flight tests of prototype payloads mounted on manned aircraft.

Vibrations from both rotary and linear cryogenic compressors used with FLIRS are another source of disturbances to LOS stabilization. These disturbances can be compensated by active systems that measure compressor current and estimate the vibration inputs from internal compressor motion to the stabilization system. This estimate can be used as a feed-forward correction directly to the gimbal torquers to compensate for the disturbance. Using this technique, compressor-generated LOS disturbance effects can be reduced by over 50%.

10.5.7 Boresight

"Boresight" refers to keeping the LOS of one sensor aligned with that of another sensor or with the beam of a laser that is being pointed using the sensor. For some applications, it is necessary to keep a laser beam aligned within a few hundred microradians of a sensor LOS. This is a tolerance that is small enough that it may require special test instruments to perform the initial alignment and to determine whether it has shifted.

Boresight accuracy must be maintained during aircraft maneuvers that apply inertial loads to the structure and over the full temperature range at which the payload will operate. While their shifts can and should be measured experimentally to confirm their magnitudes, the design to minimize them will be accomplished using finite-element modeling. This is done before any hardware is built and tested, with the success of the design determined by testing after prototype hardware is available.

The analysis of boresight shifts due to inertial loads (or thermal loads) is accomplished by defining these shifts algebraically as a function of the displacements and rotations of the system's optical elements, such as mirrors, lenses, cameras, etc. These equations are entered into the software being used to model the total structure and the boresight shifts computed directly by the program.

The boresight relationships may also be used to estimate LOS jitter caused by vibration. This is accomplished by applying the appropriate vibration input to the model. Jitter is then calculated directly by the boresight relationships in the model.

The details of this process are beyond the scope of this book, but it is important to be aware of the fact that maintaining a tight boresight requirement, as may be necessary for systems that use a sensor to point a laser very precisely at a target, is a significant challenge for the mechanical designers. It may require sophisticated mechanical structures that incorporate passive compensation for thermal effects so that the effects of thermal expansion of one component are cancelled out by the expansion of another. These approaches are complicated by requirements for low mass and small volume, so it is important to consider all of these tradeoffs early in the system-design process if a new payload package is anticipated.

An option is to select an existing payload package and then design to provide the space, weight, power, and other environmental factors required by the selected payload. It is likely to be risky to assume that it is possible to make even fairly small "improvements" in the performance of the selected payload package without risk of weight or volume growth and/or cost and schedule slippage.

10.5.8 Stabilization Design

Gimbal resonances must be at a high enough frequency (at least 3–4 times the loop bandwidth) so that notch filtering of these resonances does not decrease stability margins. This requires a stiff structure and careful placement of the gyro. The resonant modes of vibration that affect servo performance are those that respond to inputs from the servo torque motors and that excite the gyro sensor. Torsional modes, which are twisting modes involving rotation of the structure, generally are easily excited by the torque motors and are the most damaging to stability. Bending modes are linear and do not respond to torque inputs.

To achieve the required torsional stiffness, attention must be given very early in the design to building gimbal structures that have good torsional load paths. Closed torque-tube-like sections can be designed into the structure in some circumstances.

One can see from the foregoing that the stabilized-platform design is just as important and often as complicated as the design of the optical portions of the sensor subsystem, and the selection of a particular system should not solely depend on MRC or MRT curves. It should be realized that MRC/MRT curves usually contain total MTF information, which in turn includes the LOS MTF. Therefore, the gimbal performance may be embedded in the curve. Extreme caution should be used when selecting systems using only information reported by the sensor manufacturer, which may not include the effects of platform motion and vibrations. One must always make every effort to understand what is included in the published data.

Questions

1) What do FLIR, EO, IR, LOS, FOV, MTF, MRTD, mrad, and nm stand for?
2) Briefly describe electromagnetic waves.
3) What is the wavelength range for visible light?
4) What is the speed of electromagnetic waves?
5) What is the wavelength range for IR light?
6) Briefly describe target detection, recognition, and identification.
7) Briefly define the UAV payload.
8) Define MRT and MRC.
9) List main groups of civil payloads.
10) Write four groups of images and provide one sensor for each group.
11) Name two groups of reconnaissance/surveillance passive payloads.
12) Briefly describe the differences between EO and IR cameras.
13) Name one example of the reconnaissance/surveillance active payload.
14) What is a synthetic image?
15) Define detection.
16) Define recognition.
17) Define identification.
18) Define pixel.
19) What operations are reconnaissance/surveillance payloads capable of doing?
20) List payloads of a RQ-4 Global Hawk.
21) How is the sensor resolution specified?
22) Briefly describe Johnson criteria for probability of success.
23) Based on "Johnson Criteria," how many lines across a target are necessary for a 50% probability of detection?
24) What is the wavelengths range for electro-optical (EO) sensors?
25) What does the number of pixels represent?
26) How many "lines" of resolution and pixels are there in a "640-by-480" detector array?
27) How is the angular resolution of a sensor system determined?
28) Write equivalent in mrad for one degree.

29) Write equivalent in degree for one mrad.
30) What items can impose limitations to a sensor element? Write at least three items.
31) How is contrast defined for sensors that depend on reflected light (i.e., visible and near-IR)?
32) How is contrast defined for thermal-imaging sensors?
33) How are the combined effects of resolution and contrast expressed for visible and near-IR sensor systems?
34) Briefly describe the resolution of the EO sensor of the Boeing Insitu RQ-21 Blackjack.
35) Briefly describe the resolution of the IR sensor of the Boeing Insitu RQ-21 Blackjack.
36) How are the combined effects of resolution and contrast expressed for thermal sensor systems?
37) How is MRC (minimum resolvable contrast) illustrated?
38) How is MRT (minimum resolvable temperature difference) illustrated?
39) Briefly describe how contrast at the sensor aperture is determined for thermal radiation.
40) Briefly describe how contrast at the sensor aperture is determined for visible and near-IR sensors.
41) What is the "atmosphere to background" ratio? How can it be obtained?
42) What is the value of "atmosphere to background" ratio for: (a) overcast conditions and (b) sunlit conditions?
43) How can "extinction coefficient for ordinary haze" be estimated?
44) What is the visibility (in 10 km) for a fairly clear day?
45) How much is the "extinction coefficient" at wavelength of 500 nm for a fairly clear day?
46) How much is the value of the transmission at wavelength of 500 nm for a fairly clear day when the range is 5 km?
47) Briefly compare the atmospheric attenuation mechanisms for water, when it is in the form of: (a) vapor, (b) rain, and (c) fog.
48) What law is used to estimate attenuation of mid- and far-IR radiation?
49) Which one is much better for use of IR sensors: (a) a cool, humid day or (b) a hot, humid day?
50) Briefly compare wet fog and dry fog.
51) Haze scatters radiation. Which one gets more effect: (a) visible sensor or (b) IR sensor?
52) What is the reasonable contrast range for most targets?
53) What factors impact the target signature for (a) an EO sensor and (b) an IR sensor?
54) What must a target – embedded in clutter – subtend at the operator's eye, in order to have a high probability of being detected? Assume that the operator's eye is about 24 inches in front of the display screen.
55) What is the common aspect ratio of a display in a GCS?
56) Briefly describe the graphical procedure to predict ranges for detection, recognition, and identification.
57) Consider a target of interest has a height (projected perpendicular to the LOS) of 4 m and the desired task performance is a 50% probability of detection. How many lines across the target height are required?
58) What plot is often referred to as a "load line"? How is the slope of this plot calculated?
59) What factors impact the search process?
60) Write three general types of search. Describe each type briefly.
61) Write: (a) gross take-off weight, (b) payload weight, and (c) engine power, of NASA Viking 400.
62) Write: (a) maximum range and (b) endurance, of NASA Viking 400.
63) Consider an EO sensor could have a good probability of detecting a 2-m target out to a slant range of about 3,000 m. If it had an FOV of 6 degrees, determine the actual area of the FOV on the ground.

Questions 245

64) Consider an EO sensor could have a good probability of detecting a 2-m target out to a slant range of about 2,000 m. If it had an FOV of 8 degrees, determine the actual area of the FOV on the ground.
65) What is a detection horizon?
66) Consider the actual area of the EO sensor FOV of a UAV on the ground would be about 250 m × 250 m. How many separate "looks" by an operator – looking at a display in the GCS – are needed to cover a square kilometer?
67) What is a raster scan?
68) What is a "step/stare" approach to the search?
69) What is a discouragement factor in the operator search process?
70) Define "congestion factor."
71) Define "clutter object."
72) Consider a UAV mission to search an area of 2 km × 5 km in extent, using a 7-degree FOV EO sensor, and the clutter is high. How long does it take to search this area?
73) With presently available sensors and processing (i.e., not having automatic target recognizers), which one is faster in searching an area for a target: (a) UAV and operator in a GCS or (b) manned aircraft?
74) With presently available sensors and processing (i.e., not having automatic target recognizers), which one is faster to detect targets moving under forest cover: (a) UAV and operator in a GCS or (b) manned aircraft?
75) Briefly describe the search rate (in km^2/h) of the IR sensor of Global Hawk (Block 10).
76) The inter-relationship of what factors will impact the successful accomplishment of target detection, recognition, and identification?
77) List the main atmospheric phenomena that impact the total IR extinction coefficient.
78) What factors impact the search process using imaging sensors?
79) Discuss the imaging sensor installation concerns and their solutions.
80) Discuss the best locations for the imaging sensors.
81) Why cannot imaging systems mounted in UAVs be rigidly fixed to the airframe? Discuss.
82) Why is the stabilized platform of an EO camera of central importance? How can it be provided?
83) What is the maximum allowable imaging sensor LOS variation needed in order to keep the minimum resolution required to effectively detect a target?
84) Explain how maintaining image quality, tracking targets, or pointing the sensor accurately are accomplished in installing imaging sensors.
85) Define the line-of-sight stability. How it is provided?
86) What is a gyro?
87) Write two gyroscopic properties of all spinning objects.
88) Define gyro precession. What factors have an impact on it?
89) How can one increase the gyro precession?
90) Write the law of gyroscope.
91) Briefly describe gyroscope structure and its properties.
92) Briefly describe how adding a gimbal to a gyroscope can provide a stabilization platform for the payload line of sight.
93) Discuss the gyroscope performance when it is installed within three gimbals.
94) What is the role of a gimbal in a gyroscope?
95) Write two applications of a gyro-gimbal set.
96) List four non-pure mechanical devices that have gyroscopic properties.
97) What is gimbal lock? How is it eliminated?

98) Write primary advantages of the stabilized sensor configuration.
99) Write primary disadvantages of the stabilized sensor configuration.
100) What is a spatial memory? How is it provided?
101) Write drawbacks of liquid cooling for thermal management of gyros.
102) What is the most common choice for thermal management of gyros?
103) Write UAV internal/external sources of disturbances to LOS stabilization. How can their effects be reduced?
104) What is boresight? What is a typical tolerance for this parameter?
105) Explain a method to maintain a tight boresight requirement.
106) What are the recommended gimbal resonances?
107) Write two groups of resonant modes of vibration that affect servo performance.
108) Consider a target with a height of 6 m, at a distance of 1,800 m from a UAV. In using a surveillance payload, we are looking to have a 95% probability of detection. Determine: (a) spatial frequency (in lines/rad) and (b) resolution (in cycles/mrad).
109) Consider a target with a height of 5 m, at a distance of 1,500 m from a UAV. In using a surveillance payload, we are looking to have a 90% probability of detection. Determine: (a) spatial frequency (in lines/rad) and (b) resolution (in cycles/mrad).
110) Write options for camera payload aerodynamic coverage to reduce drag.
111) Why have most VTOL UAVs (including quadcopters) no aerodynamic cover for their cameras?
112) What is the resolution of the ScanEagle high-resolution thermal imager?

11

Weapon Payloads

11.1 Overview

This chapter is dedicated to weapon payloads, which is mainly a gun/missile/bomb, or just the explosive or ammunition delivered by a UAV to a target. Ammunition is the material fired, dropped, or detonated from any weapon or weapon system. Ammunition is both expendable weapons (e.g., missile and bomb) and the component parts of other weapons that create the effect on a target (e.g., warhead and bullet). When the warhead makes physical contact with the target, or in the vicinity of the target, the explosive material is detonated. Groups of lethal warheads or weapon payloads can be classified as: (1) missile, (2) warhead, (3) bomb, (4) gun/machine gun, (5) electronic countermeasures suit, and (6) active susceptibility reduction payloads. The weapon payloads with an example for each group are tabulated in Table 11.1.

A missile is a guided airborne ranged weapon capable of self-propelled flight, usually by a jet engine or rocket motor. A machine gun is an autofiring, rifled long-barrel autoloading firearm employed for sustained direct fire with a full cartridge. The entire missile (payload) is delivered to the target, while only a bullet (from a gun payload) is fired at a target.

An electronic countermeasure (ECM) is a dummy object, or an electronic device, or a communication operation (e.g., anti-jamming) aimed to deceive or fool enemy detection systems. The deceive operation may include: (1) make an object (e.g., UAV) disappear to an eye of a radar of an incoming missile, (2) make one object as many separate targets to the enemy, and (3) make an object seem to move about randomly. The topic of Electronic Warfare is further discussed in Chapter 12.

Armed UAVs carry weapons to be fired dropped, launched, or exploded, or fire a bullet. "Lethal" UAVs carry explosives or other types of warheads and may be deliberately crashed into targets. We distinguish between three classes of unmanned "aircraft" that may deliver some lethal warhead to a target:

1) UAVs that are designed from the beginning to operate in an intense surface-to-air and air-to-air combat environment as a substitute for the present manned fighters and bombers.
2) General-purpose UAVs that can be used for civilian or military reconnaissance and surveillance but can also carry and drop or launch lethal weapons.
3) Single-use platforms such as guided cruise missiles that carry a warhead and blow themselves up either on or near the target in an attempt to destroy that target.

Introduction to UAV Systems, Fifth Edition. Paul Gerin Fahlstrom, Thomas James Gleason, and Mohammad H. Sadraey.
© 2022 John Wiley & Sons, Inc. Published 2022 by John Wiley & Sons, Inc.
Companion website: www.wiley.com/go/fahlstrom/uavsystems5e

Table 11.1 Weapon payloads

No.	Weapon Payload	Example
1	Gun	
2	Missile	
3	Bomb	
4	Ammunition, warhead	
5	Electronic countermeasures suit	Jammer
6	Active susceptibility reduction payloads	
		Flares, chaff, other decoys

Source: For Missile Example, Sr. Airman Theodore J. Koniares / Wikimedia Commons / Public Domain

The description of an unmanned combat air vehicle (UCAV) is ambiguous, since any unmanned flying vehicle that is used in any sort of combat might "earn" that title. We follow what we believe is more or less standard usage and apply it primarily to the first class of UAVs.

The third class of system we consider to be guided weapons, mostly not UAVs. There can be a significant overlap between guided weapons and UAVs, as described in the history of lethal unmanned aircraft in the next section of this chapter. Guided weapons are not addressed in this chapter except for historical reasons, and to contrast them with UAVs in some cases, because the system tradeoffs for expendable flying objects that transport an internal warhead are different from those for a UAV intended to return to a base to be recovered and used over and over again.

The main subject of this chapter is the process of integrating the carriage and delivery of weapons on to what might be called "utility" UAVs, in analogy to the "utility" class of helicopters, to which a variety of weapons were added in the 1960s and which remain important combat systems to this day. We discuss UCAVs in a qualitative manner in the introduction to this book, and most of the issues related to carrying and delivering weapons with a utility UAV apply to them as well, but UCAVs are designed around the weapons from the beginning and the design issues and tradeoffs for them are often different from those discussed here for utility UAVs. A number of nontechnical issues that may be important for a UAV designer are mentioned, but this book does not attempt to deal with them in any form other than to discuss the practical technical factors that may have an impact on how the nontechnical issues are addressed.

This chapter attempts to address some of the more universal issues related to weapons carriage and delivery from a UAV/UAS. Integration of any particular weapons system and its fire-control elements into a UAS will involve many details that are unique to the type of weapons system (e.g., semiactive laser-guided missile or imaging IR homing missile) and/or to the specific weapons

11.2 History of Lethal Unmanned Aircraft

There is nothing new about using a UAV as a weapon. As described in the introduction to this book, many of the early applications of pilotless aircraft were as flying bombs. The Kettering Aerial Torpedo (or "Bug") and the Sperry Curtis Aerial Torpedo both were relatively conventional aircraft with early forms of "autopilots" intended to fly to a point over a target and then crash to the ground, detonating an explosive charge carried onboard.

The post-World War I British Fairy Queen biplanes equipped for remote control were used as a target for training air-defense gunners, but had the potential to be used in the same manner as the two US "aerial torpedoes."

Subsequent efforts through the end of World War II largely concentrated on the use of drones or other forms of pilotless aircraft as targets, not as weapon delivery systems. However, there were some very notable exceptions. These included German radio-controlled glide bombs and the V-1, which was an autopilot-controlled aircraft, launched from a rail and powered by a pulsejet engine. As with the early aerial torpedoes, it flew for a fixed time, then cut off its engine and crashed to the ground.

On the Allied side of World War II, Joseph Kennedy, the older brother of President-to-be John Kennedy, died while piloting a B-24 heavy bomber that was rigged for remote control for use as a large aerial torpedo. It required a minimum crew to take off and get up to altitude, after which the crew bailed out and it was remotely controlled by a pilot in an accompanying aircraft. The modified bombers were heavily loaded with explosives and were intended to be used to attack super-long-range artillery positions and other high-value, point targets in German-occupied France.

Subsequent to World War II, the arrival of guided missiles of various types largely supplanted the concept of a conventional aircraft without a crew as a way to deliver explosives on a battlefield. This culminated in the fielding of long-range, penetrating, guided cruise missiles that are, in concept, a modern version of the Kettering Aerial Torpedo.

A continuous competition between air power and air-defense capabilities was a major feature of the Cold War. As a result of improvements in air-defense technology, any airspace accessible to first-class enemy air-defense systems became very dangerous.

The arrival of precision-guided weapons in the late 1960s led to a gradual decline in the importance of bomber aircraft in the tactical arena, as it became possible to destroy most targets with a small number of guided bombs or tactical guided missiles. With the size of the bomb load less important, the tactical air strike came to depend on attack aircraft, sometimes called fighter-bombers because they often were capable of both roles. The combination of relatively small, fast, and agile attack aircraft with precision-guided weapons preserved the effectiveness of manned aircraft in the ground attack role, but survivability was of great concern, particularly after experience in Mid-Eastern wars, using the air-defense systems of the two great powers against the latest in attack aircraft, showed that the frontline air defenses could be very effective against aircraft that pressed home an attack against ground forces.

In the same time frame, the use of low-performance observation aircraft in an intense air-defense environment became increasingly risky, leaving a gap in reconnaissance, surveillance, and fire-direction capability that led to a renewed interest in UAVs.

The first systems fielded in the new wave of UAVs were capable of providing observation and adjustment for lethal artillery fires, and some were intended to be able to provide laser designation for the precision delivery of tactical laser-guided weapons, including bombs, missiles of the Hellfire class, and guided artillery projectiles such as the Copperhead. They were intimately connected to the delivery of weapons, but did not actually carry and launch the weapons, which had to be provided by manned attack aircraft, armed helicopters, or field artillery rockets or howitzers. The use of UAVs was driven by the desire to allow the manned aircraft to stay as far as possible away from their targets and to delegate the "close in" work to the unmanned systems.

There was at least one program for a lethal pilotless aircraft during the Cold War, called the Harassment Drone. It was a relatively small UAV equipped with a seeker that could home on radar emissions and a warhead that detonated either on contact or on close approach to the radar. The concept was that it would orbit over an area that contained air-defense radars and when a radar transmitter was turned on, it would home on it and try to destroy it. If the radar turned off again before the drone reached it, the drone could climb back up to altitude and orbit waiting for that radar or some other radar to turn on and provide it with a new target.

As further evidence of the overlap between cruise missiles and UAVs, when the US military initially began serious development of modern UAVs in the 1980s, the management of that effort was assigned to an organization whose primary mission was to manage cruise missile development. Furthermore, when the time arrived to put missiles on a UAV so that they could be launched against a target on the ground, legal questions were raised about whether or not doing so would turn the UAV into a ground-launched cruise missile, a type of weapon system that was banned by treaties negotiated and signed during the Cold War.

Starting in the 1990s the world political and military situation changed as the Cold War ended. Since then, combat has been dominated by so-called "asymmetric" conflicts in which advanced military powers fight insurgents or relatively poorly equipped military forces, and in which "combat" operations often spill across national boundaries without great attention being paid to the formalities of an open war between two sovereign powers.

In this context, the existing UAV resources started being used in a semi-covert manner to try to locate terrorist forces so that they could be attacked using conventional air power or long-range cruise missiles (which are, as we have seen, the modern equivalent of the old aerial torpedoes). This led to situations in which the desired targets were "in the crosshairs" of an imaging sensor on a UAV, but there was nothing onboard the UAV with which to shoot at them. The time lag to bring in conventional air power or a cruise missile was too great to allow success against a fleeting target.

In addition, the use of manned aircraft carried with it the risk of losing an aircraft with the crew either killed or captured, often in a country with which there was no war in progress and no authorization by the government to allow the aircraft and crew to be operating over their territory. It seemed that if something had to be lost, a UAV with no crew was far better than a manned aircraft. This had been the rationale for using cruise missiles in earlier strikes.

Based on this perception, a program was begun in the United States to arm a medium-sized, general-purpose UAV with a precision-guided weapon so that there would be a capability to engage a target immediately if one were located and identified.

The first publicly revealed UAV that could deliver a weapon under remote control was the Predator UAV armed with Hellfire laser semiactive guided missiles, shown in Figure 11.1. It achieved considerable success and a great deal of publicity and is largely responsible for the public's present perception of the nature of armed UAVs. On February 4, 2002, the first UAV military strike was conducted in history using a Predator A in a targeted killing via a Hellfire missile in Afghanistan.

Figure 11.1 Armed Predator, showing missiles on launch rails and optical dome for sensors and laser designator (Source: General Atomics Aeronautical Systems Inc.)

In parallel with this, however, beginning in the latter part of the 1990s, there had been a growing interest in UCAVs in the air forces of a number of countries, partly based on experience with the use of UAVs as unarmed reconnaissance systems during the first Persian Gulf War in the early 1990s and the use of the Predator and other UAVs in a similar role in the areas of the former Yugoslavia that were subject to international peacekeeping efforts later in the 1990s.

Once the legal and psychological barriers to arming a UAV had been broken by the armed Predator, this interest increased rapidly, and major air forces began to speak openly about the possibility that future generations of fighters and bombers might be unmanned.

The process of developing the first systems that would be true UCAVs in the sense of being designed from the start as fighters, attack aircraft, interceptors, or bombers is now underway. As stated in Section 11.1, this chapter does not address "true" UCAVs directly, although the issues raised apply to them as well as to the arming of general-purpose utility UAVs.

Various light and expendable "vehicles" that might be described as UAVs – such as the AeroVironment Switchblade 600 (see Figure 2.5) – have been described in the historical notes or ultra-light flying objects. They are currently under development – these might be hand, mortar, or rocket launched and carry a hand grenade or small rocket-propelled grenade (RPG) class of internal warhead – and are properly considered as guided weapons.

Moreover, there are a number of new research projects to develop advanced munitions and weapon systems for UAVs. For instance, DARPA challenged researchers in 2015 to develop single-use, propulsion-less, aircraft-deployed UAVs, with a range of 100 miles, which lands within 30 feet of its target. These UAVs are made of electrical components, disintegrate after completion of the mission, and destroy themselves on a molecular level.

They typically have to survive long storage in weapon bunkers, rough handling in transit, and nearly instantaneous activation when used. All of these factors have a significant effect on the system designs and design tradeoffs associated with the weapons. Much of the material in this book applies to them as well as to small UAVs, but their special requirements are not addressed.

11.3 Mission Requirements for Armed Utility UAVs

The mission requirement for a UAV to carry weapon payload may be one of the following items: (1) to deliver the entire weapon payload (e.g., a missile) to a target, (2) to deliver part of the weapon payload (e.g., a few bullets from a machine gun) to a target, and (3) to deliver the entire UAV (including weapon payload) to a target. All of these UAVs may be loosely referred to as armed utility UAV. However, we may assign a specific title for each group that accurately represent the type of UAV and mission of the flight.

For instance, the third group of UAVs are in fact a precision guided loitering munition (e.g., AeroVironment Switchblade (see Figure 2.5)) and have a low collateral damage. Another example is the Aeronautics Orbiter 1K Kingfisher with a fuselage adapted to carry a 2 kg (4.4 lb) explosive payload, turning it into a loitering munition.

The loitering munition concept is a relatively new concept and technology that blurs the line between UAV and missile. Moreover, these UAVs can re-engage the target, or find new targets, or self-destruct safely. This type of UAV is designed as a "Kamikaze," which implies that it is able to crash into its target with an explosive warhead to destroy it. The title "expendable attack UAV" may be assigned for these groups of UAVs. The General Atomics MQ-9 Reaper is sometime is referred to as the hunter UAV.

The mission for an armed utility UAV is much like that presently being performed by the General Atomics MQ-1 Predator (see Figure 11.1) and MQ-9 Reaper (see Figure 1.5). It might be described as "medium surface attack." It involves the delivery of relatively small tactical weapons, mostly precision guided, that are suitable for attacking vehicles up to and including heavily-armored tanks, small-to-medium boats/ships, small groups of personnel (or individual personnel), small buildings, and many other "point targets" such as a particular room in a large building (perhaps by entering through a window to that room) or the entrance to a bunker or cave. This type of mission could also include delivery of anti-submarine weapons to the water in the vicinity of a detected submarine.

The description "relatively small" is intended here to indicate that the weapons to be delivered are small and light enough to be carried on small-to-medium UAVs, which would tend to include anything that can be carried and delivered from an attack or utility helicopter. This definition covers a very large range of weapon sizes and weights, ranging from a few pounds to perhaps as much as a few hundred pounds. It should be noted that an AV suitable for this mission would almost certainly need to be able to carry more than one weapon so that, for instance, a 200-lb weapon in a minimum quantity of two would require twice the weapon payload weight demonstrated by the Predator. The exception to this rule might be for a homing torpedo carried by a ship-based UAV, where carrying and delivering a single torpedo might be sufficient and the torpedo might, therefore, weigh as much as two or more missiles or bombs.

Special operations versions of armed utility UAVs might have stealth features to suppress radar, IR, and acoustical signatures and might be designed for longer-range/endurance to allow operations further from the locations at which they were based.

In 2021, Textron Systems won a contract [3] to engineer and convert the US Army's Shadow RQ-7B unmanned aircraft system to an updated configuration, which includes new features designed to increase the unmanned vehicle's engine power, reduce its acoustic signature, and support communications and high-definition video processing.

11.4 Design Issues Related to Carriage and Delivery of Weapons

There are a number of design issues related to carriage and delivery of weapons. In this section, the following topics are discussed: (1) payload capacity, (2) structural issues, (3) electrical interfaces, (4) electromagnetic interference, (5) launch constraints for legacy weapons, (6) safe separation, (7) data links, and (8) payload location.

11.4 Design Issues Related to Carriage and Delivery of Weapons

11.4.1 Payload Capacity

The first requirement for an armed UAV is that the AV be capable of taking off (or be launched) with a useful load of weapons. The US Air Force chose a US Army missile (e.g., Hellfire) for integration on the Predator/Reaper for the simple reason that all Air Force air-to-ground missiles (e.g., AGM-88 HARM) were too big for the Predator/Reaper to carry.

The tactical air-to-surface missiles in the inventory of most nations typically are sized to make them effective against at least medium armor. When combined with a seeker of some sort (e.g., radar or IR sensor), electronic processing, control systems and actuators, and a rocket motor, the net prelaunch weight of even those missiles that are intended for launch from the shoulder tends to add up to tens of pounds.

Partly in response to the arming of UAVs, a number of smaller munitions have either been developed or adapted for UAV use. These include small laser-guided bombs, shoulder-fired anti-armor missiles adapted for use from a UAV, and some unpowered "faller" munitions that were originally designed to be dispensed from larger "busses" that were, themselves, rockets, missiles, or bombs. Many of these munitions weigh around 50 lb and some even less.

Shoulder-fired surface-to-air missiles have also been adapted for launch from helicopters and UAVs as air-to-air missiles. These include the US Stinger surface-to-air missile. They may be useful for self-protection and may be required if there is some, but not too much, air-to-air threat to the system. If that threat is anything more than from armed helicopters, however, the environment may no longer be acceptable for the armed utility UAV as opposed to a true UCAV.

Weapons payload requirements are driven almost entirely by the mission. For all missions, the weight of an individual weapon is driven by the targets that must be engaged and the required standoff range at launch. The number of individual weapons that must be carried also is driven by the mission. The required standoff range at launch may be something that can be traded off to some extent. The number of weapons required may also be something that can be traded off against lower cost, longer range, longer endurance, and perhaps some aspects of launch and recovery that are sensitive to AV size and weight.

It usually is possible to trade off fuel for mission payload. This is a common operational tradeoff for all types of aircraft and applies equally to UAVs. Thus, it may be possible to achieve a larger maximum operating range or endurance if it is permissible to carry less than the maximum number of weapons when achieving the maximum range or endurance.

Hellfire, with a length of 163 cm, weighs about 100 lb and the Predator A UAV can carry up to two of them at takeoff. The Hellfire missile stands for helicopter launched fire and forget missile. However, it can be carried by Predator and Reaper UAVs. A General Atomics MQ-1 Predator with two hardpoints has provisions to carry 2 AGM-114 Hellfire, or 4 AIM-92 Stinger, or 6 AGM-176 Griffin air-to-surface missiles.

11.4.2 Structural Issues

In Chapter 7, we discussed basic concepts and fundamentals of air vehicle structures. In this section, structural issues related to carriage and delivery of weapons are briefly presented. In general, there are four techniques to structurally mounting a weapon payload to the UAV hardpoint structure: (1) launch rail for missiles, (2) rack for bombs, (3) tube for holding weapon payloads, and (4) internal bay.

Carrying and dropping or launching weapons requires provisions for mounting the weapons on so-called "hard points" under the wings and/or fuselage of the AV or for internal storage with provisions for bomb racks and/or launch rails. External storage is almost certain to be simpler and less expensive, but internal storage may be required if any significant degree of radar signature reduction is desired or to reduce drag for maximum range and endurance.

In either case, the airframe must be designed to provide mounting points capable of supporting the launch rails or bomb racks through all flight regimes, including maximum-g maneuvers and hard landings. If arrested landings or net recoveries are required, the forces associated with those processes must be considered in setting the specifications for the hard points. The mass supported by the hard points includes both the weapons and their launchers or racks. It is almost always necessary to design for landing/recovery with a full weapons load as it is not always possible to jettison weapons. This is significant as the loads on the structure that must be designed for to avoid failure during a hard landing are significant.

If internal storage is selected for rockets or missiles, there must be provisions for a clear path for the launch and for the rocket motor exhaust at launch. This may be possible by using "clam shell" enclosures that open up and leave the rocket or missile exposed outside the fuselage, but may require a launcher that moves out into the airstream after the weapons-bay doors are opened. This problem has been addressed in some manned aircraft with rotary launchers that drop far enough into the airstream to expose one missile and then rotate subsequent rails and missiles into that position for sequential launches.

Figure 11.2 illustrates the concept of a rotary launcher. When the weapons-bay doors are closed, it is entirely contained within the fuselage skin and contributes no radar signature or drag. When the doors are opened, it can be extended to place one missile in the airstream with enough separation for a safe launch. Successive missiles can be rotated into the launch position.

In addition to vertical forces due to gravity and hard landings, the hard points and launchers must be able to hold the weapons against lateral and longitudinal forces due to maneuvers. In particular, there may be high decelerations during landings, ranging from breaking, thrust reversals, or deployment of drag parachutes up to arrested landings or net recoveries. Rocket and missile launch rails provide a "hold-back catch" whose function is to prevent the rocket or missile from moving down the rail until the force exceeds some value that is set high enough to keep the weapon from sliding forward off the rail under the highest expected deceleration.

Hold-back is also required to allow the thrust of the launch motor to build up to a high enough level for the rocket or missile to accelerate to an airspeed that results in aerodynamic stability as it leaves the rail. The hold-back release force for a launcher designed for use on a helicopter, which may not be moving forward when it launches the missile or rocket, must be higher than for a

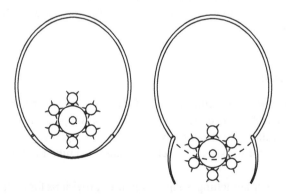

Figure 11.2 Rotary launcher retracted and extended

11.4 Design Issues Related to Carriage and Delivery of Weapons   **255**

fixed-wing aircraft, for which the rocket or missile starts out with the forward airspeed of the aircraft and has a head start toward aerodynamic stability. This must be kept in mind if the UAV can hover or if the weapons and launcher being used on a fixed-wing UAV were originally designed for use on a hovering aircraft.

The launching system must safely withstand the propulsive forces of the weapon. The forces applied to the hard points to which the bomb rack or launcher are mounted are passed on to the structure of which they are a part, so if there are going to be weapons mounted on a wing the basic structure of the wing must be adequate to deal with the additional forces generated by the presence of the weapons. The forces on a wing due to hold-back are in a direction that experiences little stress in small aircraft that do not have engines located under the wings, so special attention may be required in this area.

Most nations that have advanced air forces and some international alliances, such as NATO, have standard interfaces to allow weapon carriage on many different aircraft. This may extend to standard launchers that can launch more than one missile or rocket. If the standard launchers are designed for manned aircraft, they may be heavy enough to cause problems in integrating them on a small UAV and there may be a need to do a tradeoff between those problems and the possibility of a new or modified launcher that is better adapted to the small UAV applications. One relatively small modification would be to reduce the number of individual rails on a multirail launcher in order to match the limited weapons payload of the UAV.

11.4.3 Electrical Interfaces

As with the mechanical interfaces, many countries and some alliances have various standards related to the electrical interfaces from a platform to a standard weapons station or launcher and from the weapon station or launcher to the weapon itself.

Most guided missiles have some sort of electrical interface to their launch platform. This is accomplished by an "umbilical" connector that plugs into the weapon and comes free when the missile is launched. The types of information that may be transmitted over the umbilical connection include the following:

- Arming and "power-up" signals from the platform to the missile to make it ready to launch.
- Results of a "Built-In Test" (BIT) performed by the missile upon powering up to determine if it is functioning correctly and ready to launch.
- Laser pulse code information that allows a laser-guided missile to select the correct laser signal on which to home.
- Selection from among two or more different flight modes that might be implemented by the weapon for different target scenarios and are selected by the operator prior to launch. An example of this would be to choose between a steeply diving end-game trajectory intended to attack the roof armor of a vehicle, typically thinner than front and side armor, or a flat trajectory intended to attack bunkers or tunnel entrances.
- Imagery from an imaging seeker that can be used by the operator to lock an image auto-tracker on to the target that is to be attacked.
- Control signals from the operator that are used to center the imaging seeker's "track box" on the desired target and tell it when the part of the image that is in the "track box" is the thing that the operator wants it to hit.
- Lock-on and track signals indicating that the seeker believes that it is tracking the selected target.
- A launch signal to the missile that fires squibs that light the launch motor.

This list does not cover all possible prelaunch communications, but illustrates the general nature of the information to be handled over the interface. Much of it consists of flags and short, numerical messages, such as the code numbers for any errors detected by the BIT, but some of it may require high bandwidth and may not be very tolerant of delays and latency. Examples of the latter class of data are the images sent from the missile to the platform and the operator commands used to move the track box to the desired target, which may be moving within the images being sent to the operator. The undesirable effects of delays and latency in these signals upon the ability to perform tasks such as locking on an image auto-tracker are discussed in some detail in connection with data links elsewhere in this book and any delays or latencies introduced by the interface from the platform to the weapon contribute to those effects.

The interfaces required for specific weapons are specified by the weapon system designer. It is quite important to know what weapons will be carried on any particular UAV as early as possible in the design process, as adding additional wiring to each weapon station after the design is complete can be very expensive. This is a general problem on all types of aircraft that carry weapons and is actively being addressed by efforts to establish standards for all interfaces. In the absence of guidance from standards, it is prudent to try to provide a complete set of data interfaces at every weapons station, to include high data bandwidth of some sort. As more and more of the total electronic domain becomes digital, it may be adequate to provide high-bandwidth digital lines to all weapons stations and then multiplex and de-multiplex at the weapon station as needed to accept all of the data from the particular weapon and deliver all of the data from the platform, including the video that has been converted to a digital video if it was not in that form initially.

As an exception to the extensive interfaces often needed by a guided missile, many laser-guided bombs have no electrical interface at all to the platform. They are powered up after being released, based on a mechanical switch activated by their separation from the toggles on which they are hung prior to being released. In order to operate without an interface, all of the essential information is provided by mechanical switches and/or links or jumpers that are set during weapon preparation on the ground before the bomb is loaded on an aircraft. This results in simple integration on the AV, but may add requirements for additional personnel on the ground.

11.4.4 Electromagnetic Interference

Electromagnetic interference (EMI) is a general issue for any system that combines many different electronic subsystems and has to operate in the vicinity of radars or wireless communications systems. The environment in which military aircraft operate can be very difficult, particularly onboard naval vessels. Military ships have a high concentration of radar systems and the confines of a ship mean that the AV may often be very near the radar transmission antennas.

When we combine two electromagnetic waves, as they will interfere with each other, depends on the phase difference between the two waves: (1) the waves may reinforce each other leading to a larger amplitude (constructive interference) or (2) may partially or fully null each other out, leading to a smaller or even zero amplitude (destructive interference).

This problem becomes particularly acute when the AV is armed with rockets or missiles. There were incidents during the war in Vietnam in which the transmitted signals from radars on aircraft carriers coupled into the electrical system of rockets or missiles mounted on aircraft awaiting launch on the flight deck of an aircraft carrier and caused their motors to light, leading to rocket

11.4 Design Issues Related to Carriage and Delivery of Weapons 257

launches while the aircraft were still on the deck of the ship. The rocket or missile then struck other armed aircraft, starting fires that led to additional unintentional weapon launches and severe damage with casualties.

As a result of these incidents, the US Navy developed specifications and requirements related to Hazards of Electromagnetic Radiation to Ordnance (HERO). Presumably there are similar sets of requirements in other countries. While these requirements apply most directly to the weapons themselves, it is essential that the aircraft electronic system does not generate false arming and launch signals and that the wiring within the AV does not act as an antenna to couple harmful interfering signals into the weapon.

11.4.5 Launch Constraints for Legacy Weapons

Many of the UAVs now being armed are relatively small and have a very limited capability to carry any payload. The weapons must be small and light, compared to those routinely delivered from fixed-wing attack aircraft. It is desired that the weapons have a high probability of success with a single shot, so that the UAV does not have to carry very many of them. These requirements add up to a need for small, light, precision-guided weapons.

To avoid the cost of developing new weapons specifically for UAVs, many of the weapons being used are existing systems that were designed to be delivered by helicopters, such as the Hellfire missile, or are man-portable and even shoulder-fired. If the UAV is going to deliver these weapons from moderate-to-high altitude, there can be issues with regard to the "delivery basket," which is a volume in space to which the weapon must be delivered in order to acquire and home on a target. In many cases, it is desired that the weapon locks on to a target before being launched so that the operator can confirm that it is the target that it is intended to engage and also to reduce the probability of wasting a weapon that does not acquire a target. In those cases, the "basket" that matters is an "acquisition basket," which is the volume in space within which the weapon sensor can see the intended target.

The baskets are defined by at least two constraints: the minimum and maximum angles at which the weapon sensor can look up or down, relative to its axis as mounted on the UAV, and a maximum range to the target at which the target signature is detectable. All of this is very dynamic for a fixed-wing UAV flying toward, and eventually over, a target location.

If the weapon was designed to be pointed directly at a target by a gunner who rests the launch tube on his shoulder, it may have little capability to look above or below its axis and it may be necessary either to provide articulated launch rails or tubes that can be pointed downward or to modify the weapon to allow its sensor to be pointed in the direction of the target. The latter can be an expensive proposition, so the burden of pointing the FOV of the weapon sensor down toward the ground may fall largely on the UAV system integrator.

The ability of various weapons to achieve an acceptable acquisition basket within the required flight envelope and other physical constraints of the UAS is a critical factor in selection of which existing weapons are suitable for adaptation to UAV delivery.

11.4.6 Safe Separation

The term "safe separation" refers to the ability to launch a weapon from an aircraft with a very low probability that the weapon will end up striking the aircraft a glancing blow, or worse, as it separates from its launch rail, tube, or bomb rack. Here, a glancing blow refers to a missile when it hits the target at an angle (usually at a shallow angle) rather than directly from the front. The full force of the hit is not received by the target as the hit partly misses the target.

There are almost always complicated airflows in the immediate vicinity of the fuselage and wings of a fixed-wing aircraft, or in the rotor downwash of a helicopter. During the first few moments after the weapon is disconnected from the aircraft, it is critical that it is not carried by those airflows into contact with the aircraft structure. In many cases this requires restrictions on the flight conditions under which the weapon may be launched.

The main purpose is to ensure that the weapon safely clears the AV. Moreover, launching must occur at the optimum moment so that the weapon may function effectively. It must be highly reliable in order for the weapon to achieve the kill probability for which it is designed. The physical distance between the weapon on the hard point and the wing should be optimized such that the weight of the rail/rack is minimized.

The responsibility for safe separation is a system-level function, but both the AV and weapon designers clearly must be aware of the need to ensure that it occurs, and it should be among the issues considered when selecting a weapon for integration on to a UAV.

11.4.7 Data Links

Data links are the subject of Part Five (Chapters 13 through 16) of this book and the discussion there addresses the issues of operating in a hostile electronic environment. The presence of lethal weapons on a UAV heightens the importance of security against jamming, deception, and interception and exploitation of downlinked imagery, but does not qualitatively change the factors that are important in each of these areas.

11.4.8 Payload Location

The UAV center of gravity (cg) plays a crucial role in the stability and controllability during flight operations. The most aft location of the UAV cg is restricted by the longitudinal stability requirements, while the most forward location of the UAV cg is restricted by the longitudinal controllability requirements. The location of payload contributes to the location of UAV cg. When the payload is released, its contribution is also removed, so the UAV will have a new cg location.

The weapon payload should be located such that, after its delivery, the UAV cg does not fall out of the desired cg limit. For a fixed-wing UAV, the desired location of the cg range is about 15% to 30% of wing MAC. It is recommended that the payload is placed such that its cg is matched with the UAV cg. This ensures that the UAV cg will not be changed after the payload delivery.

11.5 Signature Reduction

Some degree of "stealth" is useful for any UAV that is to be used in a military role and is carrying a weapon payload. A stealth AV is required to avoid detection using a variety of technologies that reduce its signatures. The signatures that might be reduced include the following: (1) acoustic, (2) visual, (3) IR, (4) radar, (5) emitted signals, and (6) laser radar.

The last, laser radar, is included in this list for completeness, but is generally of low importance, because most laser "radar" systems are unable to perform a search function over a hemisphere and have to be cued by some other form of search system. The cued laser radar can then be provided a high-quality track of the AV, but if the other signatures are suppressed, the AV will not ever be detected, and no cue will be available to allow the laser radar to be pointed close enough to the AV to allow tracking.

All the other signatures are commonly exploited in military applications (either emitted or reflected) and may need to be suppressed to ensure the effectiveness and/or survivability of the UAV. The details of how to do so are well beyond the scope of an introductory text, but some conceptual information and general terminology can usefully be conveyed here.

11.5.1 Acoustical Signatures

11.5.1.1 Fundamentals

Any aero-mechanical system including UAV is generating some levels of sound while operating/ moving/flying. The first and most self-evident way to improve the stealth feature of a UAV is to reduce its acoustical signature (i.e., sound/acoustic intensity). Sound is a mechanical wave, while light is an electromechanical wave. Sound needs a media (e.g., air) to propagate, while an electromechanical wave does not need any media and propagates in space/vacuum. The sound intensity level is the level of the intensity of a sound relative to a reference value.

Most of us are familiar with the manner in which the sound from an aircraft often is the first and only clue that leads us to look up and see that one is passing overhead. A battlefield may often be a noisy place that will mask that sound, but many military or police applications of UAVs may involve surveillance of rural areas in which the ambient level of noise may be quite low. Even on a noisy battlefield, the sound of an aircraft overhead may become noticeable during brief lulls in the general noise. Therefore, the acoustical signature of a UAV may become the primary cue for its detection.

Before doing so, we will digress for a moment to introduce the basics of a very useful engineering practice – the expression of quantities in terms of their logarithms. Engineers sometimes express dimensionless ratios in decibels, usually abbreviated dB. For instance, if the ratio of two quantities is given by "R," and the ratio does not have any dimensions, then the ratio can be expressed as dB. In other words, when two quantities have the same dimensions – such as power – their dimensions cancel when the ratio is taken. For the sound intensity, the standard representation of the magnitude of R in dB is 10 log $|R|$, where the base of the logarithm is 10. The unit used in this representation of the magnitude is the dB:

$$R \text{ (in dB)} = 10 \log_{10} (R) \tag{11.1}$$

A variation in this procedure is to include some denominator that has units in the definition of a particular ratio. An example is the ratio of any voltage (V) to a reference value of 1 mV. Then any V can be expressed in dB as a number given by 10 $\log_{10}$ (V/1 mV). Sometimes, but not always, the reference unit is specified in the "name" of the dB ratio. For the case of a 1 mV reference level, this would be done by writing "dBmV."

In the particular situation presently being discussed, the sound levels, which are measured as variations in air pressure, are specified in dB with the reference level being 0.0002 microbar (i.e., mb). The reference level is an agreed nominal level for the lowest pressure variation that can be heard by a human. The noise levels from aircraft are specified in dBA, which is a value that takes into account the acoustical frequency content of the noise and the variation in sensitivity of the human ear over the same frequencies. Thus, a noise level stated in dBA is adjusted to match the results when using a human ear as the detector.

Example 11.1

A small UAV with a piston engine is causing variations in air pressure with the amount of 400 microbar at sea level. Determine the noise level in dBA.

Solution:

The reference level for variations in air pressure is 0.0002 microbar. We use Equation (11.1) to determine the noise level in dB:

$$\frac{\text{AP variation}}{\text{Ref. } AP} = \frac{400\,\text{mb}}{0.0002\,\text{mb}} = 2 \times 10^6$$

$$R = 10\log_{10}\left(2 \times 10^6\right) = 63.01\,\text{dB}$$

To discuss signature reduction techniques for UAVs further, we consider two general UAV configurations: (1) fixed-wing aircraft and (2) rotary-wing aircraft. This section will be ended with a brief discussion on automated detection of the noise from the AV.

11.5.1.2 Fixed-Wing Aircraft

In a fixed-wing aircraft, the propulsion system is the primary source of acoustical energy and signature. As presented in Chapter 6, conventional engines for UAVs range from reciprocating (i.e., piston) engines to turbine engines to electric motors. Piston, turboprop, and electrine engines are equipped with one or more propellers to convert engine power to thrust, while turbofan and jet engines have fans and nozzles for directly generating thrust.

The open literature investigates the concept of a "Silent Aircraft" in focusing on the airframe and engine. The goal of a silent aircraft concept is to be no louder than the background noise in a typical urban environment. The airframe and propulsion system configuration play important roles in the total aircraft noise. The challenge of manufacturing a silent aircraft lies within the development, integration, and optimization of efficient airframe and engine technologies. Three design examples of: (1) broad deltas, (2) blended wing bodies, and (3) joined wing airframe configurations when integrated with innovative propulsion systems have potentials in generating a silent aircraft.

The aircraft noise during takeoff is dominated by the turbulent mixing noise of the high-speed exhaust gas. However, it is the airframe and landing gear that generate the noise during approach and landing. Simple mufflers and other forms of baffles can significantly reduce the level of sound emitted by electric motors, which are effectively silent and are becoming more common in UAV applications (particularly in quadcopters). They are a possible choice for cases in which maximum acoustical stealth is required, such as in a reciprocating engine. Turbine engines are more difficult to "silence," but if silence is important enough, there are design tradeoffs that can be considered to reduce the noise created by the exhaust from the turbine.

As a practical matter, acoustical detection can be made quite unlikely for a small-to-medium UAV by combining some engine-noise suppression with high-altitude operations. The Predator A (with one reciprocating engine), for instance, is widely reported not to be audible on the ground, even in a relatively quiet environment when operating at altitudes of the order of 10,000–15,000 ft. This indicates that a medium-sized UAV with a reciprocating engine producing about 100 hp can be silent enough to operate covertly from an acoustical standpoint. It is not clear how much effort

11.5 Signature Reduction

went into muffling the engine noise of the Predator A, but it is likely that the techniques used are similar to those used for other reciprocating engines, including those for automobiles.

We can make a simple quantitative estimate of the acoustical signature of a small-to-medium UAV using generally available data and very basic physical principles related to the propagation of sound.

According to the FAA, the noise level [32] for airplanes – with a takeoff weight of up to and including 1,320 lb – to be certificated under FAR Part 36 should be under 68 dBA. If the takeoff weight is over 3,300 lb, the noise level should be under 80 dBA. For the aircraft with a takeoff weight between these two values, you may take a linear variation for the noise level. These regulations are provided for manned aircraft; we may assume similar values for UAVs.

The noise level on the ground from a light, single-engine aircraft with a reciprocating engine of about 100 hp, flying overhead at an altitude of 120 m (394 ft), has been reported to be about 65 dBA. This is comparable to the noise from an old air conditioner at a distance of 100 ft and a factor of 10 dB below the noise from a vacuum cleaner at 10 ft. The quietest fixed air conditioners emit a noise level between 25 and 46 dB, which is less than the average desk fan.

Neglecting any absorption of sound energy and assuming that the aircraft is effectively a point source of acoustical energy, one would expect the noise level to drop off as the square of the distance, which is equivalent to a drop of 6 dBA for every doubling of the distance. This is illustrated in Figure 11.3. At 8,000 ft, the sound level has dropped to about 38 dBA, which is about the level of noise for a "quiet urban background," or, from another source, "a quiet room." This very crude estimate supports the expectation that a UAV of the Predator A class, particularly if some effort has been made to muffle engine noise, would be hard to hear if operating 10,000–15,000 ft overhead.

Using data from the same sources, a small twin turboprop transport (4–6 passengers) is about 10 dB louder than the light single-reciprocating-engine aircraft, and a light twinjet corporate aircraft is about 15 dB louder than the twin turboprop. The perceived sound levels on the ground for these two aircraft are also shown in Figure 11.3. For the twin turboprop at 15,000 ft, the level heard on the ground is about 47 dBA. This is quieter than an old air conditioner at 100 ft, but louder than

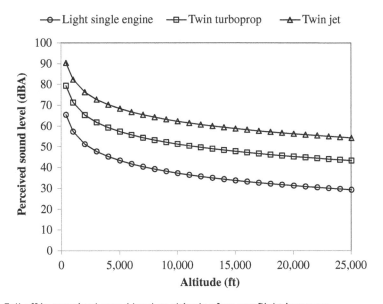

Figure 11.3 Fall off in perceived sound level as altitude of an overflight increases

a "quiet urban daytime." Increasing the altitude could reduce the sound level at the ground if doing so were consistent with the mission.

However, the curve is flattening out at 15,000 ft, and to get down to 40 dBA would require nearly a threefold increase in altitude, to about 45,000 ft. Adding another 15 dBA for the small twinjet clearly makes it unreasonable to expect to hold the perceived sound on the ground below 40 dBA.

These estimates are mathematically correct but illustrate a difficulty that is relatively common when attempting to calculate quantities like "detectability" that involve humans as part of the system whose performance is being estimated. The probability that humans on the ground will hear an aircraft flying overhead is extremely sensitive to the ambient noise level at the observer's location and to whether the observer is concentrating on listening for an aircraft or other aural indications of some activity about which he or she is concerned.

Common sense and our personal experience tell us that an observer standing near a busy highway or traveling in an off-road vehicle is much less likely to notice any but the loudest sounds than would be a lookout on a mountain peak. The sensitivity to ambient background noise is common to both automatic and human detection, but tends to lead to larger performance variations for human observers than for automatic signal-processing systems.

The question of how much attention is being paid to the observation task is not present for an automatic system but is very important for a human observer. The occupants of a moving vehicle deal with many sensory distractions in addition to the background noise and may be carrying out conversations, driving, studying maps, or otherwise involved in something other than listening for an AV. Even for a dedicated observer, the level of motivation tends to decline as the time between detections grows longer.

If an observer spends a long time listening for a faint noise and none is heard, he or she tends to become less alert. This is a general characteristic of human observers that applies to any boring task, and a search that "never" finds what is being looked or listened for certainly is a boring task. In the context of a really boring job, like listening for a sound that is not occurring, "a long time" may be half an hour, particularly in the middle of the night.

In addition, analysis of the probability of detection by a human observer is complicated by the fact that (1) the performance of any one observer will vary from event to event to a greater extent than would be seen for an automatic system and (2) the variations in performance between different observers may be greater than the variations for any one observer. This creates difficulty in determining the statistics of the process and leads to large standard deviations in any probability distribution that may be determined. It also requires large numbers of trials with large numbers of observers in order to make good experimental measurements of observer performance for any specific task.

For all of these reasons, calculating the perceived sound level, as done above, is only the beginning of the process of answering the question of whether the aircraft is detectable unless the environment of the observer is known. The engineer attempting to answer the complete question either must be given a perceived sound level on the ground that is considered acceptable "most of the time in real-world situations" or must address all of the observer uncertainties identified above.

What we can say based on these estimates is that making a UAV in the class of a light single-engine aircraft "hard to hear" should not be too difficult if it can operate at 10,000 or 15,000 ft, but that doing the same thing for a larger turboprop or jet AVs is likely to require significant investments in quieting technology.

11.5.1.3 Rotary-Wing Aircraft

The discussion up to this point has concentrated on engine noise, which is almost certain to be the dominant noise for a fixed-wing (piston or turbine engines) or ducted-fan aircraft. A rotary-wing

11.5 Signature Reduction 263

aircraft (e.g., helicopter UAVs), on the contrary, often have the characteristic "chop, chop" sound caused by shock waves from the rotor tips as they approach the speed of sound.

The shock wave generated by the rotor tips propagates perpendicular to the leading edge of the blade near the tip, which results in it scanning across any position on the ground once per revolution of the blade (when the blade is moving forward relative to the direction of flight of the helicopter, which results in the highest tip velocity versus the air mass).

There are two fairly simple ways to reduce the noise from the blade tips. Sweeping the tip of the blade backward reduces the component of blade velocity perpendicular to the blade leading edge and reduces the noise from the tip. Use of more blades in the rotor can allow the blades to be shorter, which reduces their tip speed and reduces the chopping noise. This technique has been employed in multicopters (e.g., quadcopters), due to having more engines and more and shorter blades.

11.5.1.4 Automated Detection

This discussion has been oriented toward human detection of the noise from the AV. There has been increasing interest in recent years in various systems that use acoustical detectors and computer processing to detect and locate threats of various types, particularly including snipers. An obvious extension of this would be to design the software to search for the sounds of an aircraft engine or helicopter rotor. This would be a new implementation of amplified acoustical detectors that were widely used in World War II for the direction of anti-aircraft artillery. It is likely that these approaches can significantly increase the probability of acoustical detection in many situations, although that probability will always be limited by the ambient noise background presented to the detector, whether it is a human or a signal-processing system.

11.5.2 Visual Signatures

Basically, two primary features of an aircraft contribute to its visual signatures: (1) size and (2) paint/color. Of course, the best way to make a UAV hard to see from the ground is to make it small. This is one of the inherent advantages of a UAV, which does not have to be sized to carry a pilot or other crew.

The typical approach to making an aircraft difficult to see is to paint it in a color scheme that blends into the background. If viewed from below, the background is either blue sky or white cloudy. A light gray or light "sky" blue is often used for the underbelly of military aircraft. If attack from above is anticipated, the upper surfaces of the aircraft may be painted in a blue, tan, or green shade, depending on whether the surface is water, desert, or vegetation. If night operations are anticipated, a dark color with a little gloss may be preferred.

There have been attempts to use active methods to reduce the contrast of an aircraft against the sky. As counterintuitive as it may seem, is has been suggested that having a bright light mounted on the aircraft can make it less visible against a bright sky.

Glints off reflective surfaces are a major source of visual clues for ground vehicles and helicopters operating near the surface. A typical example is sun glint off the windshield of cars. This depends on the geometry, requiring the sun to be behind the observer or in other preferred locations. Rounded reflective surfaces, such as cockpit canopies, can lead to glints with less restrictions on the relative positions of the sun, the aircraft, and the observer.

UAVs will not, in general, have cockpit canopies, but may have other rounded surfaces, such as the nose of the aircraft, and if those surfaces are shiny, they might become a source of glint. Glint is most easily avoided by using a flat paint or by making the surfaces flat at angles that will minimize

11.5.3 Infrared Signatures

The most important IR signatures are due to hot surfaces. Unlike visual signatures, which are passive and depend on there being some illumination from the sun, moon, stars, or artificial sources, IR heat signatures are internally generated and are present regardless of any ambient illumination.

IR signatures are by far the most common signatures used by small air-to-surface missiles, making them a significant issue for survivability in military systems. Until relatively recently only military organizations were likely to have IR imaging equipment, but that has changed during the first decade of this century and it is now possible that terrorists, partisans, or criminals may have IR viewers that could be used to detect UAVs used for surveillance or to deliver lethal attacks. Proliferation of man-portable surface-to air missiles also has reached the level of insurgents and might reach the better-funded criminal organizations.

It is difficult to avoid significant waste heat from an aircraft engine. For a purpose-designed UCAV, there are various approaches for hiding this heat from missile seekers and IR search systems that have been developed for use on manned stealth aircraft. The open literature discusses the concepts of using a "dog leg" in the input to a turbine engine to keep the hot portions of the turbine from showing in a frontal view of the aircraft. At the exit from a jet engine, it is possible to mix cold air with the hot exhaust to reduce the emission of the jet plume.

For a utility UAV that is not designed from the beginning as a "stealth" system, the most likely approach would be to put engine input and output apertures on the top of the airframe or wings so that the hottest sources cannot be seen from below. This is particularly effective for piston engine aircraft, where the exhaust system can be cooled by ambient airflow, and other views from below of any hot portions of the engine can be shrouded.

If the objective is to prevent detection, not just to prevent homing by IR missiles, it is important to note that a temperature difference of only a few degrees above the background is sufficient to lead to a prominent target when viewed with an IR imaging system. On a clear night, the "temperature" of the sky is near absolute zero, as what is being looked at is outer space. Under these circumstances, the skin of the UAV probably is many degrees warmer than the background and it may look like a bright light.

Fortunately, the resolution of most IR imaging systems is limited by the technology of IR detectors to something like 500–1,000 pixels within the height and width of the FOV. To use an IR viewer to search the sky for UAVs, one would need to have a fairly larger FOV, say 7.5 degrees (typical of 7×50 binoculars, which have often been used for similar applications in the visible). If the resolution of the IR system is 1/1,000 of 7.5 degrees, then a pixel would have angular dimensions of 0.13 mrad.

For useful search and detection of UAVs, one would need to be able to detect them long before they were directly overhead, so for a UAV at 15,000 ft, roughly 5,000 m, a slant range to detection of at least 20 km would be highly desirable. At that slant range, a 0.13 mrad would have linear dimensions of about 2.6 m (i.e., tan (2.6/20,000) = 7.5/1,000 deg = 0.13/1,000 rad). For a roughly head-on view of a medium-sized AV, this would not meet the Johnson Criterion of two lines on the target for detection. However, as explained in Chapter 10, a "hot" target can often be detected even if it fills significantly less than one pixel of a thermal detector, as long as it is hot enough to make that pixel stand out against the surrounding pixels.

11.5 Signature Reduction

For the worst case of a clear night sky background even the skin of the AV might be warm enough relative to the sky to make at least one pixel bright. The thermal emissivity of the skin can be reduced by using appropriate paints or even polished metal surfaces. The abundant airflow over the skin of the vehicle should keep its temperature roughly at that of the ambient air at altitude and prevent any significant heating up due to reduced radiation cooling.

Engine exhaust and radiators will have dimensions well below 2.6 m and thus will fill only a small part of a pixel, but are likely to be hot enough to create a bright pixel even when their contribution is averaged over the total area of that pixel. They can be hidden and/or cooled as described earlier in connection with heat-seeking air-defense missiles.

11.5.4 Radar Signatures

Radar signatures result from the reflection of electromagnetic waves off the structure of the AV. Before further discussing these signatures, we provide a short introduction to the basic features of the electromagnetic spectrum, reminding the reader of the terminology used in describing radio and radar waves.

11.5.4.1 Electromagnetic Spectrum

The electromagnetic spectrum from 1 MHz to 300 GHz is shown in Figure 11.4. This omits the "long wave" bands in the low kHz frequency region used for some broadcasting and for long-range, non-line-of-sight communications and the optical bands at extremely high frequency in the ultraviolet, visible, and IR. Radio frequency (RF) ranges from around 20 kHz to around 300 GHz and is roughly between the upper limit of audio frequencies and the lower limit of infrared frequencies.

Electromagnetic waves are characterized by frequency, wavelength, and polarization and travel at the speed of light, 186,000 miles (300,000,000 m) per second.

In the RF portion of the spectrum, it is common practice to describe an electromagnetic wave in terms of its frequency, although these two parameters are essentially interchangeable via the relationship shown, and frequency and wavelength are often mixed in expressions such as "microwave frequencies." The frequency (or wavelength) of an electromagnetic wave impacts the shape, size, and design of the antenna, the ability of the wave to propagate through the medium separating the transmitter and receiver, and the nature of the reflection of the wave off objects on

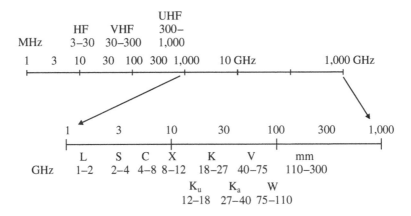

Figure 11.4 Electromagnetic spectrum

which it is incident. Wavelength (λ) and frequency (f) of a signal are related by the following equation:

$$f = \frac{c}{\lambda} \tag{11.2}$$

where c is the speed of electromagnetic waves (i.e., light).

Example 11.2

A UAV is transmitting an electromagnetic wave with a wavelength of 3 m. Is this wave within the RF band?

Solution:
Radio-frequency (RF) ranges from around 20 kHz to around 300 GHz. We need to determine its frequency first:

$$f = \frac{c}{\lambda} = \frac{300,000,000}{3} = 1 \times 10^8 \text{ Hz} = 100 \text{ MHz}$$

The frequency of this wave falls within the range. Therefore, yes, this wave is within the RF band.

There is a general tendency for atmospheric transmission to decrease with increasing frequency due to molecular absorption. Transmission is excellent at the longer wavelengths (lower frequencies) up to the top of the X band and not too bad out to the upper end of the K_u band (12–18 GHz). There is a local peak in absorption between 20 and 30 GHz, although the transmission does not get bad enough to prevent short-range use of the K band frequencies. Then there is a "window" of better transmission in the K_a band. Transmission is poor in the V band, but it may be desirable for some short-range applications where it is useful for the signal not to travel too far. Finally, there is a window of better transmission around in the center of the W band. The transmission in the K_a and W bands is significantly worse than in the L through K_u bands. Nonetheless, there has been a movement to use of the K_a band for narrow-beam radars where the shorter wavelength translates directly into smaller antenna sizes.

The nature of radar reflections depends strongly [33] on the ratio of the radar wavelength to the dimensions of the object from where the wave is being reflected. If the wavelength is comparable to or smaller than the dimension of the target normal to the radar beam, the reflection is much like the reflection of light off macroscopic objects. That is, a flat surface will tend to reflect the radar beam much like a mirror would reflect light. Since the radar wavelength is of the order of a few mm at least, any flat surface will appear to be smooth and reflect like a mirror, unlike optical reflections where the wavelength is of the order of a millionth of a meter and many surfaces are rough at the scale of the wavelength and produce diffuse reflections.

In this "specular reflection" regime, the radar signature is very sensitive to: (1) the geometry of the object and (2) the orientation from which the radar is looking at the object. As an illustration, a large, flat metal plate would have a very large signature when viewed at normal incidence but a very small signature when viewed at an angle well off the normal incidence, just as a flashlight beam shined on a mirror at normal incidence will be reflected right back to the source (the flashlight), but if the mirror is tilted, none of the light from the flashlight will be reflected back toward the source (if the mirror is clean and very smooth).

11.5 Signature Reduction

If the dimensions of the object are small compared to the wavelength of the radar, the energy is scattered in all directions instead of being reflected in a specular manner and the amount reflected is not as sensitive to the details of the shape of the object.

This short description of radar reflections is very simplified, but is sufficient to allow the discussion of the general source of the radar signature of an AV and of the ways in which those signatures might be reduced.

11.5.4.2 Radar Cross-Section

The radar signature of an object is expressed as a "radar cross-section (RCS)." The cross-section is defined as the cross-sectional area (in m^2) of a perfectly reflecting sphere (Figure 11.5), which would produce the same return signal in the direction of the radar receiver. A radar cross-section is the measure of a UAV's potential to reflect radar signals in the direction of the radar receiver. A common set of units is dBsm (decibels referenced to a **s**quare **m**eter), which is the ratio of that area to 1 m^2. The unit conversion is indicated in the following formula:

$$\sigma \text{ (in dBsm)} = 10 \log_{10} (\text{RCS} / m^2) \tag{11.3}$$

where σ is the RCS in dBsm. Recall that a circle with the area of 1 m^2 has the diameter of 113 cm.

Three parameters are impacting the RCS: (1) The projected cross-section, (2) reflectivity, and (3) directivity. The RCS is highest at the side-view of the fuselage due to the large physical area observed by the radar and perpendicular aspect. The next highest RCS area is the nose/engine/tail, largely because of reflections off the fuselage nose, engine inlet, and propellers.

In the non-specular regime (wavelength comparable to the dimensions of the target), the scattering is a function of the electrical properties of the target on a macroscopic scale (is it a conductor or an insulator) and the ratio of the wavelength to the dimensions of the target in three dimensions (along the beam, in the direction in which the radar beam is polarized, and the direction perpendicular to the polarization). There is not much that can be done to reduce the radar cross-section other than to tailor the electrical properties of the surface of the AV surface to cause as much as possible of the incident electromagnetic energy to be absorbed.

The most common approaches to treating the surface of an aircraft to increase radar absorption are to use radar-absorbing material (RAM) or radar-absorbing paint (RAP). RAM – often made of composite materials – often consists of tiles that are applied on top of the surfaces of the AV and can be replaced if damaged. This can have consequences for airflow, particularly on airfoils, and is likely to add weight to the AV. Its advantage is that the tiles can have significant thickness and this can allow a better absorption of the incident energy.

RAP is applied as a paint, which makes it less expensive to apply than tiles and may result in less effect on the airflow over the surface. It also may add less weight than tiles. RAM and RAP formulations are closely guarded by the organizations that use them, but some are available on the open market.

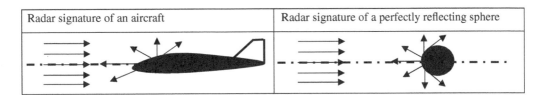

Figure 11.5 Concept of a radar cross-section

In the specular or semispecular regime, when the radar wavelength is less than the characteristic dimensions of the AV, the radar return in any given direction depends on the shape of the AV. This is the most likely regime when dealing with modern tracking radars. The most important basic principles of shaping to reduce radar cross-sections are as follows:

- Avoiding 90-degree dihedral and trihedral geometries
- Avoiding curved surfaces
- Orienting flat surfaces to locate all large radar signatures in a few directions off to the sides of the aircraft

Dihedral and trihedral geometries at 90 degrees are illustrated in Figure 11.6. A 90-degree trihedral, often called a "corner cube," will return all the energy in a collimated radar beam that strikes the inside of the corner in the direction of the radar transmitter/receiver, creating a return that is orders of magnitude larger than the return from a sphere of the same area. This return is called a "retroreflection" as it returns along the reverse of its incoming path. A corner cube is one type of retroreflector. The retroreflection occurs regardless of the orientation of the corner relative to the radar beam axis as long as the beam enters into the "corner." For this reason, trihedral corner cubes are often mounted on buoys to make them easier to detect by ship's radars in foggy weather.

A 90-degree dihedral, also illustrated in the figure, has an effect similar to that of a trihedral, but only in two dimensions. That is, if the radar beam is normal to the line along which the two planes are joined, then the return will be back along the incoming beam. If the incoming beam is at an angle to that seam, then the return will not be retroreflected.

A special case of a 90-degree dihedral geometry is one in which two surfaces, at least one of which is not flat, meet at a right angle. A simple case of this, shown in the figure, is a "top hat" geometry. A common occurrence of this type of dihedral on an aircraft is at the joint between a wing/tail airfoil and the fuselage or the joint between a vertical and horizontal stabilizer. In the general case of the top-hat configuration, one can see that if the radar beam includes rays that would intercept the axis of the cylinder, then those rays will be retroreflected, resulting in an enhanced radar signature.

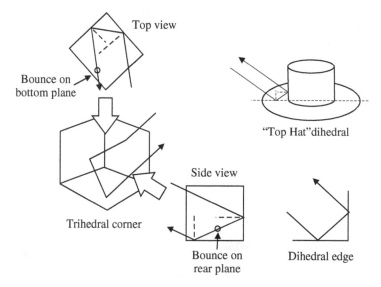

Figure 11.6 90-degree dihedral and trihedral geometries

11.5 Signature Reduction

The second rule is to avoid curved surfaces. The reason for this is that the radar reflection from a curved surface is spread out over a wide angle, increasing the probability that some of it will be reflected back to the radar receiver. This is not a desirable characteristic, as it makes it hard to accomplish the third objective, which is to concentrate all of the large radar signatures in directions that can only be observed from a few discrete directions located off to the sides of the aircraft.

As a simple example of applying these rules, an aircraft fuselage shaped like a pyramid, with the point forward and sharp edges where the flat surfaces of the pyramid meet, would only reflect energy back to the radar if the radar were in one of four directions, along the normal to each of the four sides of the pyramid, which are the top, bottom, and sides of the pyramid.

If the wings were also flat and were tilted in the vertical plane so that they did not form a 90-degree dihedral angle where they joined the fuselage, and the tail consisted of two combined horizontal/vertical stabilizers tilted so that they did not form a 90-degree angle with any other surfaces, this configuration would have little radar signature except when viewed from a small number of aspects, none of which would be from the front or rear of the aircraft. If the aircraft were maneuvering, it would never present one of those aspects to any radar for an extended time. While the radar might occasionally be located at one of the aspects that had a large radar cross-section, that geometry would be fleeting, and it would be difficult to confirm a detection or to track the aircraft.

Of course, this simple geometry might have some serious aerodynamic issues, but many of its features can be seen in the shape of the F-117 Nighthawk stealth fighter fielded by the United States a number of years ago. The XQ-58 Valkyrie UCAV features stealth technology with a trapezoidal fuselage, a chined edge V-tail, and an S-shaped air intake.

A full treatment to reduce radar signatures would include both shape and at least selective use of RAM and/or RAP and would have to include internal carriage of any weapons that might be carried on the AV, as external weapons on rails introduce significant unwanted radar signature elements.

11.5.5 Emitted Signals

Emitted signals can reveal the location of an aircraft if the opponent has intercept and direction-finding capability, or if the signal itself can be intercepted and interpreted and contains location information. The latter issue is discussed in connection with UAV data links in Chapters 13 and 14. Here, we address interception and direction finding, which does not depend on being able to "read" information contained in the signal.

Given an opponent that has the required intercept and direction-finding equipment, the only ways to avoid detection and location are to cease emitting the signals or to use low-probability of intercept (LPI) transmission techniques. Spread-spectrum approaches, which are among the most common LPI techniques, are discussed in the Data-Link section of this book (Chapter 13).

UAVs that use satellite data links to communicate with their controllers (e.g., Northrop Grumman RQ-4 Global Hawk) can orient their transmitting antennas so that little signal is radiated downward, making them hard to detect on the ground and reducing the accuracy of any direction finding. This can be combined with LPI transmission waveforms.

11.5.6 Active Susceptibility Reduction Measures

Military UAVs may face with incoming enemy missiles that either need to be evaded or diverted. Most surface-to-air missiles will be closing to a target at about 1.2–1.4 m/s, at 20 kilometers altitude.

This means 15–25 seconds to reach the target UAV. If the UAV is equipped with a radar warning receiver, this threat could be sensed. If the UAV is not capable of evading the missile due to its maneuverability limits, the only way to survive is to divert the missile.

Survivability is defined as the capability of an AV to avoid or withstand a hostile environment, while susceptibility is referred to as the inability of an AV to avoid the threats in a hostile environment. Sections 11.5.1 through 11.5.5 discussed passive susceptibility reduction measures, such as visual, IR, and acoustic signature reduction. These signature reduction techniques may be the only way to increase the survivability of small AVs due to their limited size.

Although these traditional techniques can be employed for large AVs to partially reduce their signature, they are not sufficient. For larger AVs (such as Global Hawk), active susceptibility reduction measures such as flares, chaff, and other decoys may also be utilized. Chaff and other countermeasures primarily blind the defenders' detection for a short period of time. For instance, active measures will blanket the surrounding area with chaff so that radar operators could not distinguish the UAV from the false signals on their scopes. Chaff is the oldest active susceptibility reduction measure and dates back to World War II.

Military UAVs may carry flares, chaff, and other decoys as active susceptibility reduction payloads. These payloads are released when a hostile missile is coming at the UAV and will cause the missile to divert. A flare basically generates a stronger heat signature so that the heat-seeking missile is diverted. Moreover, these payloads are deployed in a hostile environment to deceive the enemy radars. A chaff creates a small cloud of metal (i.e., a wide blanket) to mislead a radar. The aircraft susceptibility reduction concept is one of the active research areas in developing modern UCAVs.

To satisfy the system survivability requirement, Global Hawk, Block 10, is equipped to employ active countermeasures against radar and IR-guided threats to the system, as identified in the STAR.

11.6 Autonomy for Weapon Payloads

11.6.1 Fundamental Concept

The general subject of autonomy was discussed in Chapter 9. As one might expect, there are some special issues related to autonomy when one of the things that a truly autonomous AV might do is to employ lethal force against an enemy target. There is a fundamental question about whether it is a good idea to allow a "robot" (here a UAV) to make the decision to kill humans under any circumstances. That question is contentious and outside the scope of this book. We limit ourselves to identifying the technical and practical issues related to how one might attempt to achieve that level of autonomy.

The present well-established state of the art for UAVs allows for autonomous flight based on waypoints or other general directions from the operator and could allow for autonomous takeoff and landing or recovery if that were desired. The technology for automatic target detection and recognition is actively being pursued, but is still not an established capability. At present, it is not likely that any unmanned system is allowed to make an autonomous decision to apply lethal force.

One of the perceived advantages of using semiactive laser guidance on the armed Predator is that the weapon is guided by a laser spot that is controlled by the payload operator. The operator can divert the weapon at any time up to a very few seconds prior to impact by moving the spot to a new location. This keeps the human operator "in the loop" until the missile no longer has enough time to maneuver away from the point to which it has been homing, which is seen as desirable in an asymmetric war waged against targets mixed in with a civilian population.

11.6 Autonomy for Weapon Payloads 271

It may be that Predator operators often use an image auto-tracker to make the laser track the selected target as the AV, and perhaps the target, move. Nonetheless, the decision about what to shoot at and when to shoot and the decision actually to launch a missile are made by the operator, who can change his or her mind almost up until impact.

The engagement process is time sensitive, but not so time sensitive that the requirement for the operator to make the decision to launch is likely to create missed opportunities, at least in the present applications. It is conceivable that if an operator were simultaneously controlling more than one AV, there could be some advantage in making the system operate in a "fire and forget" mode by not monitoring the progress of the engagement and letting the image auto-tracker complete the engagement. This is not likely to be done at present for two reasons: (1) in a "surgical strike" scenario with targets embedded in civilian areas, most people would consider it good policy to keep the operator in the loop as late as possible in case something happens that makes it desirable to move the missile away from its original aim-point and (2) image auto-trackers are widely viewed as unreliable and the operator may be needed to restore the track or refine the aim-point.

At present, therefore, the lack of autonomy in detecting and selecting targets is driven by lack of reliable automatic target recognition algorithms; the lack of autonomy after launch is not much of a hindrance; and the advantages of keeping a "man in the loop" outweigh its minor disadvantages.

This conclusion applies to the medium surface attack mission in an insurgency environment and for present and near-term automatic target detection and recognition technology. One might ask whether it also applies to missions that may be added in the future and how it might be changed if there was a reliable way to detect and select targets that applied to some future combat scenario.

The question about new missions cannot be answered without guessing what those missions might be. Within the limitations of the armed utility UAV arena, it is not clear to the authors what missions other than a medium surface attack might be added. All other obvious missions are likely to require a more fully combat-ready AV – what we have been calling a UCAV.

Because anyone working in the area of armed UAVs needs to be aware of underlying issues, we relax our restriction against considering issues unique to a true UCAV in order to provide a brief discussion of their possible missions. Considering UCAVs, we add such missions as Suppression of Enemy Air Defense (SEAD), standoff interception of enemy aircraft using long-range air-to-air missiles (as in protecting a fleet against air attack), tactical and strategic bombing, and air-to-air combat in a "dogfight" situation.

It appears that all except the air-to-air dogfight mission could be performed under the constraint that targeting and the decision to launch or shoot have to be made by a human operator. This is not to say that it might not be possible to achieve efficiencies and higher rates of engagement if more of the process were automated. Rather, it is limited only to the conclusion that it should be possible to carry out the missions for surface attack, standoff air defense, or various kinds of bombing with a human in the loop, given a secure data link, even if that data link has significant delay and latency.

Air-to-air dogfights, however, are a notoriously split-second kind of activity. In 2002, an Iraqi MiG-25 (manned fighter) intercepted a Predator A. Both fired missiles at each other; MiG-25 evaded but Predator was shot down.

One could imagine that even if the UCAV were being operated by a pilot (and weapons officer, if applicable) in an immersion type of flight simulator provided with wide-angle video in near real time, there would be penalties for even a fraction of a second in latency in the video and similar delays in transmitting commands to the AV. If data-link bandwidth restrictions or latency resulting from relays via satellite or just due to transmission times from some distant parts of the world were to add up to a second or two, there might be a serious reduction in the ability to defeat a manned aircraft in a dogfight. The authors do not know the results of any studies that may have been performed in this area.

In answer, then, to the first question posed above, this assessment suggests that autonomous decisions to employ lethal force probably are not essential for most UCAV applications, but that there is at least one possible application that might not be possible without that level of autonomy.

With regard to the second question, related to what might happen in different kinds of wars, a conventional, symmetrical war between two nation-states with advanced militaries could lead to a different kind of battlefield than exists in the asymmetric conflicts presently underway.

If there were well-defined frontlines and relatively well-defined combat zones, one might be more willing to allow some sort of automatic target detection and recognition to be used to select targets. Given precision navigation, the autonomous process could be limited to particular areas on the ground in which, for instance, all or most moving vehicles are expected to be associated with the opposing military.

11.6.2 Rules of Engagement

Radar can detect moving vehicles, and perhaps distinguish between tracked and wheeled vehicles, so one might be able to allow a radar to cue an imaging sensor that would lock on to the moving vehicle and then allow it to be attacked. This would require rules of engagement that assume that anyone moving around in the selected area was an enemy combatant. It would raise questions about how to identify ambulances and other vehicles that should not be attacked.

The risk of attacking friendly vehicles that happen to be in a location at which only enemy vehicles are expected already is a serious one and is being addressed by various types of identification, friend or foe (IFF) systems. These systems typically involve a coded interrogation signal and a coded response that tells the interrogator that the vehicle or person being interrogated is a friend. The rule of engagement then might be to assume that anything or anyone that does not respond properly is a legitimate target.

This could work for friendly vehicles, but to apply it to noncombatant vehicles, such as ambulances, that are operated by the enemy, they would have to be equipped with IFF transponders compatible with "our" side's IFF systems and given the codes with which to respond. This might be viewed as equivalent to the marking of ambulances with highly visible red crosses or red crescents, which is how the protection of those vehicles is supposed to work when a human observer is making the decision whether or not to engage. However, it would raise issues of sharing technology and coding information that may well be very sensitive.

There presumably could be an effort to develop painted markings that could be "read" by the sensors on the UAVs, but this certainly would present difficulties in the dirty environment of a battlefield and also with the universal tendency of soldiers in combat to tie tents, sleeping bags, rolled tarpaulins, and numerous other types of equipment and stores to every available surface of any vehicle that they occupy.

IFF already is common for combat aircraft. Combined with the fact that civilian aircraft are unlikely to be present in a volume of airspace in which dogfights are occurring, this means that IFF might provide a solution for the air-to-air combat problem. If all aircraft in some volume of airspace were detected and interrogated by radars on the UCAVs or, perhaps, on the ground or on either manned or unmanned over-watch aircraft, the resulting three-dimensional map of all aircraft, each with a track and each labeled as friend of foe, could be distributed to all the friendly UCAVs and could, in principle, provide a basis for complete autonomy. The likelihood of noncombatant aircraft in a volume of airspace – in which a dogfight is going on – is probably low enough to be

Questions 273

neglected, although there may be medical evacuation helicopters present and that is a risk area that would need to be assessed. This approach also would almost certainly amount to giving some automated system the responsibility for telling a force of UCAVs which aircraft they should attack and which they should protect.

Based on these very general arguments, we conclude that in a conventional war there might be ways to solve the problem of sorting out possible targets into friendly and unknown but assumed unfriendly. If that were considered sufficient to allow an autonomous attack, then "complete" autonomy, probably within some geometrically constrained area on the ground or volume of airspace, might be feasible for UCAVs.

As discussed in connection with the general concept of autonomous UAV operations, the authors are of the opinion that the artificial intelligence capability required to make truly autonomous military decisions, as opposed to flying from one specified point to another, perhaps with some "smart" routing decisions, is still in the future.

The type of autonomy suggested above as being possible in surface attack would, at present, be based on something as simple as "shoot everything that moves" with some qualifiers such as "if it does not respond as a friend to an IFF interrogation," "if it is in some delimited area on the ground," "if it is classified as a tracked vehicle," or other conditions of a similar nature. This may at some point become "smart" enough to be acceptable, but it would not be likely to pass the "Turing" test that asked whether it would look to an outside observer as if a human operator were selecting the targets to be engaged.

Many will say that there are some very basic issues that are not technical in nature that need to be considered in deciding on whether or not to build a capability for autonomous decisions about applying lethal force. The technical community certainly has an important role in debating these issues, but they cannot be settled by technical arguments and analyses.

Questions

1) What do RF, dB, dBsm, IFF, RCS, UCAV, RPG, RAM, RAP, SEAD, FOV, BIT, HERO, and LPI stand for?
2) What does Hellfire stand for?
3) Briefly describe RCS.
4) List types of UAV signatures that might be reduced.
5) Name three classes of unmanned "aircraft" that may deliver some lethal warhead to a target.
6) Provide three groups of examples for lethal warheads.
7) Define: (a) ammunition, (b) missile, (c) machine gun.
8) What design issues related to carriage and delivery of weapons are discussed in this chapter? List topics.
9) Name one light and expendable military UAV.
10) What is the weapon payload capacity of AeroVironment Predator UAV?
11) Why did the US Air Force choose a US Army missile for integration on the Predator/Reaper and not any Air Force air-to-ground missile?
12) Briefly discuss the possibility to tradeoff fuel for weapon payload.
13) Briefly discuss structural issues for designing the hard points to support external weapon payloads.
14) What forces are applied to hard points that carry weapon payloads?
15) Using an illustration, explain the concept of a rotary launcher (include the weapons bay doors).

274   *11 Weapon Payloads*

16) What types of information may be transmitted into a weapon over the umbilical connection?
17) What is the reference level variation in air pressure for calculating the UAV noise level in dB?
18) What is the required noise level for airplanes – with a takeoff weight of 1,200 lb – to be certificated under FAR Part 36?
19) Compare the noise level of a twin turboprop transport (4–6 passenger) with that of a light single-reciprocating-engine aircraft.
20) Compare the noise level of a light twinjet corporate aircraft with that of a twin turboprop aircraft.
21) What is the perceived sound level for a UAV at 25,000 ft altitude with (a) a light single piston engine, (b) twin turboprop engines, and (c) twin jet engines?
22) What is the best way to make a UAV hard to see from the ground?
23) What is the best color for painting an aircraft to make it difficult to be seen if the flight operation is only during night?
24) What is "acquisition basket" for a UAV weapon sensor?
25) Briefly describe the constraints of the UAS in selection of which existing weapons are suitable for adaptation to UAV delivery.
26) Briefly discuss the electromagnetic interference issues for an AV that is armed with a missile, when it is operating in the vicinity of radar systems of a naval vessel.
27) Discuss the launch constraints of a new UAV when it is armed with legacy existing weapons.
28) What does a glancing blow refer to?
29) A UAV with a twin turbofan engine is flying at 20,000 ft. This UAV is generating a sound at 80 dBA level when flying at sea level. What is the perceived sound level in dBA?
30) Write four techniques to structurally mounting a weapon payload to the UAV hardpoint structure.
31) What are two fairly simple ways to reduce the noise from the blade tips?
32) Briefly compare visual and IR signatures.
33) Compare the main sources of aircraft noise during: (a) takeoff and (b) landing.
34) What is the resolution of most IR imaging systems (in pixels) within the height and width of the FOV?
35) Briefly discuss the techniques to reduce aircraft infrared signatures.
36) What is the goal of a silent aircraft concept?
37) Which primary features of an aircraft contribute to its visual signatures?
38) What is the "temperature" of the sky on a clear night when it is being looked at?
39) What is the speed of light?
40) How is a radar signature generated by an AV?
41) At what wavelength range is the radio wave transmission excellent?
42) In the transmission of radio waves, which one translates into a smaller antenna size, shorter or longer wavelength?
43) For the radar wave reflections, discuss the relationship between the radar wavelength to the dimensions of the object off which the wave is being reflected.
44) The radar signature is very sensitive to two features of an object. What are the features?
45) Define radar cross-section.
46) Briefly compare radar-absorbing materials and radar-absorbing paint.
47) What are the most important basic principles of shaping to reduce radar cross-sections?
48) With the use of a figure, explain how a "top hat" geometry is reflecting an incoming radar signal.
49) Why should curved surfaces be avoided to reduce RCS?
50) What are the three parameters impacting the RCS?

51) When shaping rules are applied to a UAV to reduce RCS, what does the aircraft fuselage look like? Then provide an example aircraft.
52) Briefly discuss techniques and rules to reduce RCS of UAVs.
53) Briefly define survivability and susceptibility.
54) Provide two examples for active susceptibility reduction measures.
55) What is the oldest active susceptibility reduction measure?
56) Open discussion: Is it a good idea to allow a "robot" (here a UAV) to make the decision to kill humans under any circumstances?
57) Briefly discuss whether keeping the "man (i.e., UAV payload operator) in the loop" is a good idea in employing a weapon payload.
58) Provide a few example missions that may be conducted by UCAVs.
59) Provide a few example missions that may be conducted by armed UAVs.
60) Briefly discuss how friend or foe (IFF) systems are operating.
61) What is the authors' opinion on the topic of allowing/possibility of a UAV to make truly autonomous military decisions? What do you think?
62) Briefly discuss rules of engagement in an automatic target detection and recognition by a UAV.
63) When and where was the first UAV military strike in history?
64) Briefly discuss how the weapon payload location is determined.
65) A large UAV with a piston engine is causing variations in air pressure with the amount of 350 microbar at sea level. Determine the noise level stated in dBA.
66) A small UAV with a turbofan engine is causing variations in air pressure with the amount of 14,000 microbar at 3,000 ft. Determine the noise level stated in dBA.
67) A UAV is transmitting an electromagnetic wave with a wavelength of 450 m. What is the frequency of this wave?
68) A UAV is transmitting an electromagnetic wave with a wavelength of 0.02 m. Is this wave within the RF band?
69) A fixed-wing UAV has an RCS of 2.6 m^2. What is the RCS in dBsm?
70) A fixed-wing UAV has an RCS of 7.3 m^2. What is the RCS in dBsm?
71) Name two UAVs that are a type of loitering munition.
72) What new changes are supposed to be applied to Army's Shadow RQ-7B unmanned aircraft by Textron Systems in 2021?
73) Name a UAV with stealth technology. Describe its features.

12

Other Payloads

12.1 Overview

There are a great many possible payloads for UAVs in addition to the imaging sensors and weapon-related payloads discussed in the preceding chapters (Chapters 10 and 11 respectively). Any list that could be provided would certainly be incomplete and would be out-of-date by the time that it could be published.

The most that can be attempted is to discuss a few of the most likely payloads, with a concentration on types that may place special requirements on the AV or data link or other portions of the overall UAS.

These discussions are not intended to be introductions to the design of any of these other payloads. Rather, they are intended to provide a very basic introduction to some of the technologies involved and as examples of how the various ways in which a UAV/UAS may be used affect the overall system design. In this chapter, the following payloads will be presented: (1) radar, (2) electronic warfare, (3) chemical detection, (4) nuclear radiation sensors, (5) meteorological/scientific sensors, (6) pseudo-satellites, (7) robotic arm, and (8) package and cargo.

12.2 Radar

12.2.1 General Radar Considerations

Radar – an electromagnetic sensor – is a target detection device that uses electromagnetic waves to measure the range, angle, and velocity of an object. A UAV radar is employed for the detection and tracking moving ground vehicles as well as flying aircraft within its range. The signal of a radar will be reflected from all objects (including surroundings) in the FOV, but the radar rejects signals from fixed or slow-moving unwanted targets. Four main elements of a radar are: (1) transmitter, (2) receiver, (3) signal processor, and (4) antenna.

To measure the distance to a target, a pulse (radio wave) is emitted/transmitted toward the target. Some of the radio waves will be intercepted by the targets. The intercepted radio waves that hit the target are reflected back (see Figure 12.1) in many different directions. When part of the reflected signal is received by radar, the time delay (Δt) is measured. Since the speed of the radio

---

Introduction to UAV Systems, Fifth Edition. Paul Gerin Fahlstrom, Thomas James Gleason, and Mohammad H. Sadraey.
© 2022 John Wiley & Sons, Inc. Published 2022 by John Wiley & Sons, Inc.
Companion website: www.wiley.com/go/fahlstrom/uavsystems5e

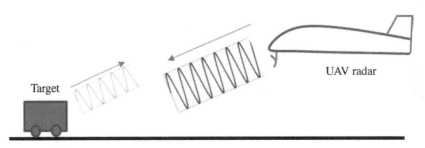

Figure 12.1 Emitted and reflected signals

wave is known (i.e., $V = 300{,}000$ km/s), the range is equal to velocity times the time delay divided by two ($R = V\,\Delta t/2$). To determine the velocity of a moving target, this time delay as well as the Doppler equation [19] is employed. This equation is based on the fact that the reflected wave has a different frequency (f_r) than that of the emitted wave (f_e).

Unlike many electro-optical sensors (e.g., a camera), a radar can operate independently of weather and lighting conditions. All-weather reconnaissance is possible using radar because electromagnetic radiation at radar frequencies (typically 9 to 35 GHz for a UAV) are less absorbed by moisture than at optical frequencies (visible through far-infrared) and can "see" through clouds and fog. A radar system provides its own source of energy and hence does not depend on reflected light or heat emitted from the target.

Radar sensors inherently have the capability to measure range to the target, based on round-trip time of travel of the radar signal. For pulsed radars, this measurement is made by timing the arrival of the reflected pulse relative to the transmitted pulse. For continuous-wave (CW) radars, a modulation superimposed on the continuous-wave signal is used to determine the round-trip time for the signal.

A major advantage of a radar sensor is that, as an active system, it can use Doppler processing to distinguish moving targets from a stationary background. Radar energy reflected from a moving surface has its frequency shifted by an amount that is proportional to the velocity component of the reflecting surface that lies along the direction of propagation of the radar beam. The Doppler shift (or Doppler effect) is defined as the change in frequency of a wave for an observer moving relative to its source.

If the return signal is combined with an unshifted signal in the receiver, "Doppler" signals are generated at difference frequencies corresponding to the Doppler shifts of the target returns. The receiver can ignore unshifted returns, thus separating returns from moving targets from returns from stationary background and clutter.

When the radar transmitter is moving, as is almost always the case for a radar system on a UAV, there is a relative velocity between the transmitter and the ground. Without compensation, the radar would detect fixed objects on the ground as moving targets. This difficulty can be overcome by using a "clutter reference" approach. The radar assumes that most of its returns are from stationary clutter objects on the ground and uses the Doppler shift in these returns to define a shifted frequency as the "zero velocity" point from which it measures Doppler shifts. It then is possible to detect returns from any target that is moving relative to its surroundings.

The component of relative ground velocity along the radar beam varies as a function of the angle between the radar beam and the air-vehicle velocity, so a new clutter reference is taken for each individual radar return. The detailed implementation of the clutter reference system varies depending on the size of the radar beam relative to the expected target size, the waveform of the radar, and other system-specific design characteristics.

12.2 Radar

The Doppler signal may be due to overall motion of the target, or due to motion of part of the target. For instance, the top of the tread loop of a tracked vehicle moves at a different velocity than the hull. The relationship between the two Doppler signals can be used to recognize the return from a moving tracked vehicle. Similar effects can be seen for helicopters (from the main and tail rotors and rotor hubs), rotating antennas, and other parts of a target that move relative to the body of the target.

Most radars do not have sufficient resolution to provide an image of a vehicle-sized target. They may be able to provide a low-resolution image of a ship or building, and sometimes can provide an image of the ground with sufficient resolution to display roads, buildings, tree lines, lakes, and hills. If such images are desired, they must be synthesized within the radar processor, since the sensor itself does not directly detect an image. Rather, the sensor provides a map of radar return intensity, with or without additional processing (such as Doppler frequency) versus angle and range.

The processor can use this information to generate a pseudo-image for display to an operator. Small targets, such as vehicles, may appear as bright points in such an image, or can be represented by icons that provide information about some of their characteristics that are known to the radar, but not directly related to their "image," such as their state of motion relative to the background, or identification based on internal Doppler signatures (e.g., a moving tracked vehicle).

In addition to Doppler processing, a radar can be designed to determine polarization changes in the reflected signal relative to the transmitted signal. This information can provide additional discrimination between targets and clutter and between different types of targets.

Radar sensors can use beams that are larger than the angular dimensions of the targets to be detected, particularly if Doppler processing is used and the targets of primary interest are moving. However, the performance of most radar systems is limited by the ability of the radar to separate targets from clutter, and this ability can be enhanced by keeping the radar beam from being too much larger than the target (i.e., using a "fill factor" near 1, where the fill factor is the ratio of the target area, projected perpendicular to the beam, to the beam cross-sectional area). Therefore, it is usually desirable to use a small beam, particularly for typical UAV applications, that attempt to detect vehicles and other small targets on the ground.

The minimum beam size (angular width) of a radar is governed by the same diffraction effects that apply to resolution of optical sensors (see Chapter 10). Thus, the beam width is proportional to the ratio of the radar wavelength (λ) to the antenna diameter (D). Since UAV antennas are restricted in size, it is generally desirable to use short wavelengths. However, even at 95 GHz, which is the highest frequency (shortest wavelength) for which off-the-shelf radar components presently are available, the wavelength is still about 3.2 mm. Thus, λ/D for a 30-cm antenna is only 1/94, compared to a value of λ/D of 1/100,000, which is common for optical and infrared sensors. The result is that the beam width for radars is measured in tens of milliradian (mrad), compared to the resolution of optical sensors, which is of the order of tens of microradian (μrad).

The desire for a small beam width when using a small antenna favors a short wavelength. However, attenuation by water vapor in the atmosphere increases significantly for frequencies above about 12 GHz. This may be acceptable for a short-range radar to be used on a UAV, but must be kept in mind when performing system tradeoffs.

There are many different types of radar systems, distinguished by frequency, waveform, and processing approaches. Selection of the appropriate type of system depends on the mission to be performed. In the context of UAV applications, there are additional constraints related to size, weight, antenna configurations and size, and cost (since the radar must be as expendable as the air vehicle itself). The details of radar sensor design are beyond the scope of this book, but one special type of radar sensor that has been used on UAVs and has very good resolution is described in the following section.

280   *12 Other Payloads*

12.2.2 Synthetic Aperture Radar

A new type of radar to provide a higher spatial resolution than conventional beam-scanning radars is the Synthetic Aperture Radar (SAR). In this radar, a sequence of acquisitions from a short antenna are combined to simulate an extremely large antenna (or aperture).

A synthetic aperture radar (SAR) takes advantage of the fact that radar frequencies, although very high, still are low enough to allow the radar data processor to operate on the raw signal at the carrier frequency. This allows the radar to perform what is known as "coherent detection" in which the phase of a return signal is compared with the phase of the transmitted signal. This means the distance that the signal has traveled in its round trip can be measured down to a fraction of the wavelength of the signal.

A SAR transmits a signal more or less perpendicular to the direction of motion of the AV and then receives the returns over a period of time during which the AV moves some significant distance. This effectively increases the aperture of the receiver by the distance traveled during the interval for which coherent data is available. Without getting into any detail about how all this is accomplished, the result is that a SAR can have enough angular resolution to generate "images" that show individual trees and vehicles and even people at significant ranges from the radar.

These images are the output of a relatively complicated computational process whose input is the time-resolved relative phases and amplitudes of the transmitted and received signal, as well as the velocity of the AV and, of course, a large number of parameters that depend on the details of the radar and the frequency at which it is operating. The "image" is produced in a strip that runs along one side of the AV flight path and a SAR is also sometimes referred to as a side-looking airborne radar (SLAR).

As is later discussed in connection with data links, the raw data rate for a SAR is so high that it will overwhelm most data links if an attempt is made to transmit it all to the ground as it is acquired. This can be addressed either by doing the processing onboard the AV, so that only the final "image" has to be transmitted, or by taking advantage of the massive digital storage now available to record the raw data for transmission at some rate less than real time. In the latter case, despite huge storage capabilities, it probably will be necessary to limit the maximum data collection period and downlink the data before taking anymore new data.

Northrop Grumman RQ-4 Global Hawk, General Atomics MQ-1 Predator, and General Atomics MQ-9 Reaper are all equipped with SAR as a Ground Moving Target Indicator for conducting reconnaissance missions.

12.3 Electronic Warfare

Electronic warfare (EW) is military action involving the use of electromagnetic energy to determine, exploit, reduce, or prevent hostile use of the electromagnetic spectrum and action that retains friendly use of the electromagnetic spectrum.

Electronic warfare payloads are used to detect, exploit, and prevent or reduce hostile use of the electromagnetic spectrum. The conduct of EW can be organized into the following three categories:

1) Electronic support measures (ESM) that involve intercepting and locating hostile signals and analyzing them for future operations. Intelligence gathering related to intercepted signals is known as "signal intelligence" (SIGINT). If the signals are radar signals the procedure is called "electronic intelligence" (ELINT) and COMINT for communication signals. The most common ESM payload used with current UAV systems is the radio direction finder. Basic direction

finding (DF) equipment consists of an antenna and signal processor that sense the direction or bearing of received radio or radar signals. A simple commercial type of scanner sold to listen to police or other emergency radio signals to find the received radio signal combined with a directional antenna could result in an effective UAV ESM payload.

2) Electronic countermeasures (ECM) are self-protecting actions taken to prevent the hostile use of the electromagnetic spectrum. Basic defensive ECM strategies are: (a) radar interference, (b) target modifications, and (c) changing the electrical properties of air. Two ECM interference techniques are: (a) jamming and (b) deception of homing missiles.

ECM often takes the form of jamming. Communication and radar jammers are relatively inexpensive and easy to incorporate in UAVs. Jamming is the deliberate radiation of electromagnetic energy to compete with an enemy's incoming receiver signals. All the energy of a jammer can be concentrated on the receiver frequency or the power can be spread across a band of frequencies. The former is called spot jamming and the later barrage jamming. There is a great deal of potential for the use of UAVs, integrated with other systems, to provide jamming.

3) Electronic counter-countermeasures (ECCM) are actions taken to prevent hostile forces from conducting ECM against friendly forces. UAVs may require ECCM techniques to protect their payloads and data links. In 2021, Boeing equipped F-15s with an immune satellite navigation system – built by Elbit Systems – to ensure uninterrupted GPS operation, which provides full jamming immunity for multiple satellite channels and can handle multiple interfering signals.

12.4 Chemical Detection

The purpose of chemical detection payloads is to detect the presence of chemicals in the air, or sometimes on the ground or surface of water. This may apply to military or terrorist situations in which the chemicals have been deliberately spread in an attempt to cause mass casualties or to civilian situations in which the chemicals are pollutants, leaks, spills, or products of fires or volcanoes. For the military and terrorist scenarios, the mission of the UAS would be to provide warning to allow troops to deploy protective equipment so as to prevent or reduce casualties and contamination or to allow the civilian population either to stay inside or to evacuate an area that is threatened by the chemicals. For the civilian scenario, the mission may be routine sampling and surveillance or might be similar to the military mission of warning the population in the case of a serious release of harmful chemicals.

There are two basic types of chemical sensors, point and remote. Point sensors require that the detecting device be in contact with the agent. These sensors require the AV to fly through the contaminated volume or drop the sensor into the site to be examined, so that the sensor is in contact with the agent and subsequently can transmit the information to a monitoring station directly or by a relay contained on the AV. The detection technologies available for contact sensors include wet chemistry, mass spectrometers, and ion mobility spectrometers.

Remote sensors do not have to be in direct contact with the chemical agent that they detect. They detect and identify the chemical agent by using the absorption or scattering of electromagnetic radiation passing through the chemical mass. Laser radars and FLIRs with filters can be used for remote chemical detection.

A UAV may be ideal for contact sensing, since it allows the sensor to be flown through the harmful agent without exposing any personnel. This is one of the classic justifications for use of an unmanned vehicle. However, unless the UAV is expendable after only one flight, it must be remembered that it will be recovered and must be handled by the ground crew. This requires that it be easy to decontaminate, which places restrictions on structures, seals, finishes, and materials that are not likely to be met by a UAV unless included in the design from the beginning.

In June 2018, the US Department of the Interior deployed UAVs to support the remote sensing data acquisition requirements for the monitoring of the Kīlauea Volcano eruptions in Hawaii. A team of sUAV operators monitored volcanic activity using thermal video imagery and on-board gas sensors.

12.5 Nuclear Radiation Sensors

Nuclear radiation sensors can perform two types of missions:

1) Detection of radioactive leaks or of fallout suspended in the atmosphere, to provide data for prediction and warning similar to that provided by a chemical-agent sensor.
2) Detection of radiation signatures of weapons in storage or of weapon production facilities, for location of nuclear delivery systems or monitoring of treaty compliance.

In the first role, considerations are similar to those that apply to chemical detection, including the requirement for ease of decontamination if the UAV is to be recovered.

Searching for nuclear delivery systems may require low and slow flights over unfriendly territory. Detection of low-level signatures is enhanced both by low altitude and long integration times for the weak signals to be detected. The relatively high survivability of a UAV, combined with its expendability, may make it a good choice for such missions.

Even if there is permission to overfly some country for treaty verification, a UAV may be considered less obtrusive than a manned surveillance platform. This UAV-based instrumentation system for radiation measurements may be used to assess potential radiation risks. A significant advantage of this technology is for the help given to people to enable them to avoid being exposed to radiation. They are employed to determine the dose equivalent rates and gamma spectra in a single measurement.

In 2011, there was a nuclear reactor disaster at Japan's Fukushima nuclear power plant, which was primarily caused by a major earthquake and 15-metre tsunami. The power supply and cooling system of three reactors were all disabled. Interest in the levels of radiation all over the nation increased dramatically after this accident. Some international organizations conducted special monitoring operations to assess the state of radiation levels near the Fukushima power plant. A new technology using UAVs, developed by the International Atomic Energy Agency (IAEA) for use by the authorities of the Fukushima Prefecture in Japan, allowed for the remote monitoring of radiation measurements in areas where contamination was too high for people to enter.

12.6 Meteorological and Environmental Sensors

Meteorological (MET) and environmental data are critical in many civilian situations. The potential for very long time-on-station without operator fatigue opens up many possibilities for UAVs as monitors of developing storms or other long-term weather phenomena.

Meteorological and environmental information is also vital to the successful conduct of military operations. Barometric pressure, ambient air temperature, wind speed and direction, and relative humidity are essential for determining the performance of artillery and missile systems and predicting future weather conditions that impact ground and/or air operations and tactics.

12.7 Pseudo-Satellites

Figure 12.2 NASA Global Hawk (Source: NASA Photo / Tom Miller / Public domain)

In either case, placing the sensor *in situ* (at the approximate point of interest) results in the most accurate observation. This is easily accomplished with a UAV. Simple, light and inexpensive "MET" sensors have been developed for UAVs, which can be attached to almost any air vehicle and when used in conjunction with UAV airspeed, altitude, and navigation data can provide a very accurate picture of the environmental conditions under which the various weapon systems must operate.

Meteorological sensors are used to measure climate and weather. For some missions, there is even a need to have the features of clouds, lightning, rain, snow, and even hurricanes. Environmental sensors may be utilized to measure pollution and greenhouse gasses in order to study and control the environment. These sensors may also be employed to study marine life in lakes, rivers, and oceans.

Environmental sensors are capable of providing various types of information: location, position, the individual's movements, and contextual elements. The application of these sensors (humidity sensors, air quality sensors, and HPM particle sensors) is endless.

NASA's Armstrong Flight Research Center operates two Northrop Grumman Global Hawk UAV (Figure 12.2) for high-altitude, long-duration missions. The two yellow-and-black pods under the wings of NASA Global Hawk house atmospheric measurement probes. Thirteen instruments were installed on the autonomously operated aircraft for the 2014 Airborne Tropical Tropopause Experiment over the western Pacific Ocean.

Between 2010 and 2017 the aircraft served NASA's Science Mission Directorate, the National Oceanic and Atmospheric Administration (NOAA), and the Department of Energy in performing Earth observation research. The Global Hawk aircraft proved itself to be a valuable asset for high-altitude hurricane and severe storm research performed over the Atlantic and Pacific oceans.

Another example is the Nordic Unmanned Company, which employs VTOL UAVs in several activities including emission monitoring, logistics and robotization, defense and security, and inspections. Moreover, Boeing Insitu concluded a series of tests of the biological combat assessment system on board ScanEagle.

12.7 Pseudo-Satellites

In recent years, there has been increasing interest in the concept of UAVs that fly at very high altitudes and have very long endurance, usually powered by electric motors and using solar cells to keep batteries charged indefinitely. These UAVs could loiter over a point on the ground to provide a platform with many of the characteristics of a satellite in stationary orbit, but at a small fraction of the cost.

A UAV designed for this application must have a very high maximum altitude and high maximum lift-to-drag ratio, $(L/D)_{max}$, at its operation altitude in order to minimize the power required to maintain that altitude. It needs to be able to maintain itself over a point on the ground despite whatever winds it encounters, so must have an airspeed capability comparable to those winds. However, it probably could vary its altitude to select favorable winds.

As a general rule, it is probably desirable for a UAV that is going to loiter for long periods at high altitudes to operate above the normal ceiling for commercial and military aircraft in order to minimize airspace management issues and possible conflicts.

It must be able to carry whatever payload is needed to perform its mission and also must be able to provide the prime power needed by the payload. Some of the missions that have been considered are:

- Forrest/brush fire monitoring
- Weather monitoring
- Communications relay
- Large-area surveillance

The details of any of these payloads will depend on the particular mission to be performed.

In the forest and brush fire monitoring case, the payload might consist primarily of visible and thermal imaging sensors. Combining the position and attitude of the UAV with the LOS angles of the imaging systems, the location on the ground of each bright hot spot in the images could be determined.

Weather monitoring could involve any of the sensors used in weather satellites as well as direct measurements of winds and other atmospheric information at the operating altitude.

In a communications-relay application, a UAV operating as a pseudo-satellite could provide a relatively inexpensive, wide coverage, line-of-sight communications node that could function somewhere between a super cell phone tower and a real satellite.

The line-of-sight distance to and from the surface would be much shorter than for a geostationary satellite. The altitude above sea level for a geostationary orbit is about 36,000 km (22,000 miles) while the likely altitude for a loitering UAV would be of the order of about 60,000 ft, only about 18 km (11.3 miles). For one-way losses that are proportional to R^2, this results in a factor of about 4,000,000 less transmitter power required to produce the same signal strength at the end of the one-way path.

That is one reason why a constellation of non-geostationary satellites is often used for applications that involve up- or downlinks using nondirectional antennas such as satellite telephones or TV broadcasting. However, even a low earth orbit is at 160–2,000 km, leading to R^2 ratios relative to a high-altitude UAV of around 80 to as high as 12,000. Even the lower end of this range is significant when designing a transmitter or receiver. On the other hand, the area covered by a single pseudo-satellite would be smaller than for the real satellite by a factor similar to the R^2 ratio.

Large-area surveillance applications of pseudo-satellites would have a similar relationship to the commercial and military imaging satellites presently in use. The pseudo-satellites would be much less expensive and would offer some advantages in resolution due to their lower altitude, but would have less coverage area when over any particular point on the ground. A major difference would be that they would be operating in the airspace of the country over which they were flying, even if they were at very high altitudes. Therefore, they would presumably be subject to over-flight restrictions, unlike satellites that operate outside of the atmosphere.

For UAVs functioning as a pseudo-satellite, there are interesting differences in the types of risk present in the overall system that lead to significant differences in the system-level tradeoff of cost versus risk of component or subsystem failure.

12.7 Pseudo-Satellites

Real satellites are at considerable risk during launch and then at lower risk, from an "aerodynamic" standpoint, once in orbit. UAVs acting as pseudo-satellites might be most at-risk during takeoff and climb out, as are most aircraft, but the level of risk would be lower than for a satellite launch. On the other hand, the risk of platform failure once "on station" might be higher for an aircraft than for a satellite as an aircraft has many more flight-critical moving parts and subsystems than a satellite.

The main risk to a real satellite after launch is failure of some mission-critical item, and failure of anything essential to the overall mission can be catastrophic. This includes all the mission-critical elements of the payload as well as of the satellite itself, in and on which the payload is mounted. Even if the satellite continues to "fly" perfectly, if the payload ceases to function it becomes a total loss.

For a UAV, however, if it can still fly and be controlled, it can be landed and whatever has failed can be replaced or fixed and it can then take off again and resume its mission. Some UAVs and light aircraft are designed to include a parachute capable of allowing the aircraft to achieve a non-catastrophic return to earth even in the extreme cases of having a wing break off.

Regardless of the ability, in principle, of a UAV to remain aloft indefinitely using solar cells to recharge batteries, it is likely that the batteries and the many moving parts will require maintenance and/or replacement of items that wear over time on some regular schedule. Satellites also wear out and have design lifetimes of the order of 10 years or so due to anticipated failures and expenditure of the fuel needed for the thrusters that maintain the satellite in its assigned orbit and location.

Therefore, the requirement to land a UAV pseudo-satellite periodically for maintenance need not be viewed as a disadvantage relative to a real satellite, while the ability to land and repair component and subsystem failures is a significant advantage.

The UAV designer can consider the use of commercial-grade components and subsystems for all but the flight-critical subsystems. Even for the flight-critical subsystems, the expectation of regular maintenance cycles can allow a more relaxed tradeoff of cost versus redundancy and reliability.

As an example, if a high-bandwidth data link that is required to support the mission of, say, rebroadcasting "satellite" TV were to be backed up by a very reliable, but also very limited capability data link that was adequate to land the UAV back at its base for repairs, that could be quite adequate for a UAV, but would not be an option for a satellite.

The tradeoffs between satellites in space and UAVs being used in a pseudo-satellite role would depend on such factors as:

- The consequences of a single UAV being out of service for some period of time or the cost of having a replacement ready to launch at once (and the time that it would take to reach its station at high altitude).
- The acceptability of a possible crash or parachute landing in the areas where impact might occur.
- The added life-cycle costs of performing periodic maintenance on the UAV and its payload, compared to the added cost of designing for very high reliability and redundancy in a satellite and the need to replace the satellite after the end of its useful lifetime in space.
- The ability to upgrade the UAV payload at any scheduled maintenance versus the very high cost, or complete impracticality, of making any repair or upgrade to the payload of anything in orbit.
- The advantages or disadvantages of lower altitude for a particular application.
- The issue of overflight in national airspace, which is avoided for satellites.
- The payload capability of a long-endurance UAV, which is likely for some time to be less than what can be put into orbit on a large booster. This tradeoff would be influenced by the second-order effects of lower altitude (lower transmitter power requirements, for instance), possible lower redundancy, and, perhaps, of using more than one UAV to replace one satellite.

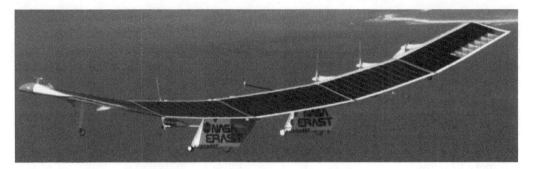

Figure 12.3 NASA Pathfinder (Source: NASA / Nick Galante / Public domain)

However, it is easy to see how the ability to recover and repair a UAV could have a great impact on system and subsystem tradeoffs in the UAV design, could lead to significantly lower cost, and could be the key advantage of the UAV over a satellite inserted into orbit.

As part of an evolutionary series of solar – and fuel-cell – system-powered UAVs (Environmental Research Aircraft and Sensor Technology program), NASA developed a few HALE UAVs (e.g., Helios, Pathfinder (Figure 12.3), and Centurion). The mission was to perform atmospheric research tasks as well as serve as communications platforms (i.e., pseudo-satellites). The Helios prototype reached an altitude of 96,863 feet, but on June 26, 2003, it broke up and fell into the Pacific Ocean. Due to the unsuccessful results of this project, it was formally terminated. The NASA investigation report identified structural design issues (lack of sufficient strength and stiffness in facing atmospheric disturbances) as the root cause of the accident.

12.8 Robotic Arm

One of the interesting capabilities of UAVs is to pick up objects and perform various tasks via a robotic arm. A robotic arm (or hand) is an electro-mechanical programmable arm with similar functions to a human arm. It can have various configurations and various numbers of degrees-of-freedom. It may have various numbers of segments, joints, and mechanisms that resemble the shoulder, elbow, and wrist. Demand for assistive robots is increasing and some types may be employed in UAVs.

Robotic arms allow operators to work independently and perform tasks that are impossible without them. A robotic arm can be directed to perform a variety of security, research, observation, maintenance, environmental, critical, and dangerous tasks. Examples of UAV robotic arm functions are: (1) taking a sample from a target pool, (2) removing an object, (3) disarming an explosive item, (4) fastening or unfastening a screw/bolt, (5) gas/arc welding a pipe, and (6) fixing a failed system.

The UAV arm could have a fully closed-loop control system or be remotely controlled by an operator from GCS. The feedback enables the robot to move in a controlled manner and conduct its task precisely. The links of such an arm are connected by joints allowing either rotational motion or linear displacement.

The position and orientation of the robotic arm may be controlled by a number of electric stepper motors and mechanical gears, and can be measured by a number of measurements devices and motion sensors such as potentiometers and gyroscopes. Moreover, force sensors may be placed at the gripper for detecting the amount of pressure applied on an object.

12.9 Package and Cargo

A growing commercial application of UAVs is carrying and delivery of package/cargo/parcel/mail/goods. Quadcopter delivery capable of para-dropping a package or cargo near a desired location has already been operational. More and more fire and police departments from around the world are adopting, integrating, and deploying UAV platforms into their day-to-day workflows. In June 2021, drone-based postal delivery has been tested in Slovenia.

The advantages of using UAVs include faster delivery, reducing pollution, optimizing logistics routes, and the delivery of essential goods to remote areas such as mountain settlements.

Amazon Prime Air (Figure 12.4) – a future delivery system – is designed to safely and efficiently deliver packages to customers in 30 minutes or less using UAVs. In August 2020, Amazon received approval from the Federal Aviation Administration (FAA) to operate its fleet of Prime Air delivery drones. Amazon joins UPS and Alphabet-owned Wing, who previously won FAA approval for their drone delivery operations.

As an example, a Hexacopter UAV, the Aurelia X6 Standard – Made by Aurelia Aerospace – is a platform for a variety of payloads weighing up to 5 kg. The UAV, which is powered by Pixhawk & Ardupilot, has a range of 5 km and an endurance of about 30 minutes. The Wingcopter 198, an all-electric VTOL fixed-wing UAV becomes [3] the first UAV (in 2021) to deliver up to three packages to multiple locations during one flight. Moreover, Flytrex has received approval from the FAA to expand backyard drone delivery (to operate above people) to homes in North Carolina.

A UAV can handle such payloads in a very similar method to that of other types of payloads, except the delivery technique. A level of autonomy is required for identifying the location and height of the delivery target area, as well as the release/drop operation.

There are a number of design issues related to carriage and delivery of packages. A complete discussion is presented in Chapter 11 (Section 11.4) for the weapon payload. In this section, a few new items are discussed.

A multicopter UAV (including quadcopters), which is capable of vertical takeoff and landing, is a better candidate than a fixed-wing UAV for carrying and delivery of packages. The UAV should be capable of taking off with a useful load of package. There should be some flexibilities in the size, shape, and weight of the payload. Thus, the UAV is a great platform for a wide variety of payloads.

Carrying and dropping payload requires provisions for mounting the payload on a platform under the fuselage of the AV. The mechanism for holding and releasing the payload needs to be robustly designed. During payload delivery operation, there should be no contact/interaction between the payload and the landing gear. The hold/drop mechanism should be controlled by a

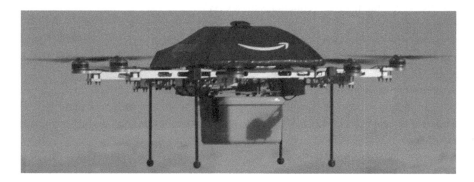

Figure 12.4 Amazon Prime Air UAV (Source: Amazon, Inc.)

feedback (i.e., closed loop) control system. Thus, when the payload is dropped, a dedicated sensor confirms its delivery.

It is recommended that the payload be placed such that its cg is matched with the UAV cg. This ensures that the UAV cg will not be changed after the payload delivery.

12.10 Urban Air Mobility

Urban Air Mobility (UAM) is a future generation technology and will be an exciting and revolutionary application of UAVs. It envisions a safe and efficient aviation transportation system that will use highly automated aircraft. The program will operate and transport passengers or cargo – similar to the driverless car – at lower altitudes within urban and suburban areas.

The FAA expects to develop a new vertiport standard in the coming years. In 2020, FAA in collaboration with NASA developed and published the Advanced Air Mobility (AAM) concept of operation. This document [34] describes the envisioned operational environment that supports the expected growth of flight operations in and around urban areas. To implement such a vision, levels of autonomy have to be increased and operational rules – across a range of environments including major metropolitan areas and the surrounding suburbs – have to be developed.

For the near future, the UAM will employ manned aircraft [35]; however, in the long term, the goal is to utilize autonomous UAVs. For this objective, the human will be the main UAV payload (i.e., drone taxi, passenger-carrying air taxis, and passenger drone). There is currently a market for a drone taxi, but the actual service is predicted not to happen everywhere before 2030. Ehang 184 (China), Vahana (Airbus), and Volocopter (USA) are currently developing drone taxies.

In order for this technology to happen, a number of challenges has to be addressed and a number of core technology enablers has to be developed: (1) manned–unmanned aircraft teaming, (2) flight safety, (3) framework for operation, (4) airspace management software, (5) infrastructure development, (6) avionics, (7) vertical takeoff and landing (VTOL), (8) detect and avoid (DAA) system, (9) autonomy, (10) intelligence, and (11) integration of drones into the airspace.

Questions

1) What do SAR, SLAR, NOAA, FLIR, SIGINT, UAM, DAA, and EW stand for?
2) What does Hellfire stand for?
3) List the payloads discussed in this chapter.
4) What are typical radar wave frequencies for a UAV?
5) Define Doppler shift.
6) Briefly describe the function and main elements of a radar.
7) Provide one advantage and one disadvantage of a radar compared with an electro-optical sensor.
8) Define "fill factor".
9) Compare a typical beam width for radars and optical sensors.
10) What parameters need to be considered in the selection of an appropriate radar for a UAV?
11) Compare a typical ratio of the radar wavelength to the antenna/lens diameter for radars and optical sensors.
12) Name three UAVs equipped with SAR.

13) How is the range to a target measured by a radar?
14) Define electronic warfare.
15) List three categories for conducting electronic warfare.
16) What are electronic support measures?
17) What are electronic counter-countermeasures?
18) What is the function of an electronic counter-countermeasures system?
19) Briefly discuss electronic countermeasures.
20) Define jamming.
21) Define spot jamming.
22) Briefly describe two basic types of chemical sensors.
23) What are two types of missions that nuclear radiation sensors can perform?
24) What is a significant advantage of nuclear radiation sensors used by UAVs?
25) Name three environmental sensors.
26) Name three environmental items of information.
27) Discuss why meteorological and environmental data is important.
28) How many instruments were installed on the NASA Global Hawk for the 2014 Airborne Tropical Tropopause Experiment?
29) What was the NASA Global Hawk's mission between 2010 and 2017?
30) What do meteorological sensors measure?
31) What are possible missions of operating UAVs as pseudo-satellites?
32) Discuss the benefit of using a UAV as a pseudo-satellite (in a communications-relay application) in terms of communication transmitter power.
33) Provide one restriction for using UAVs as pseudo-satellites – compared with real satellites – in large-area surveillance applications.
34) Discuss the differences in the types of risk present in the overall system for a UAV functioning as a pseudo-satellite.
35) Is "the requirement to land a pseudo-satellite UAV periodically for maintenance" viewed as a disadvantage or advantage? Why?
36) Compare the line-of-sight distance to and from the surface for a geostationary satellite and a pseudo-satellite UAV.
37) Briefly discuss the factors that impact the tradeoffs between satellites in space and UAVs being used in a pseudo-satellite role.
38) What is the key advantage of the UAV in a pseudo-satellite role over a satellite?
39) Briefly discuss the NASA Environmental Research Aircraft and Sensor Technology program.
40) What was the root cause of break-up of Helios UAV during the last flight?
41) Name three UAVs developed by NASA as part of an evolutionary series of solar – and fuel-cell – system-powered UAVs.
42) Define a robotic arm.
43) List at least five functions of a robotic arm.
44) Discuss how the position and orientation of the robotic arm may be controlled.
45) What is Urban Air Mobility?
46) What is the main UAV payload for the Urban Air Mobility program?
47) What companies have won FAA approval for their drone delivery operations?
48) What is the recommendation for the location of package payload?
49) Which type of UAV is more suited for delivery of packages: (a) multicopter or (2) fixed-wing?

50) Discuss the desired features for a package hold/delivery mechanism.
51) What is the maximum payload mass that an Aurelia X6 Standard UAV can handle?
52) Name three drone taxis that are currently under development.
53) Which UAV became the first UAV to deliver up to three packages to multiple locations during one flight?

Part V

Data Links

This part of the book introduces the functions and characteristics of UAV data links (i.e., communications subsystem); identifies the primary performance, complexity, and cost drivers for such data links; and discusses the options available to a UAV system designer for achieving required system performance within the constraints of various levels of capability for the data-link subsystem. Emphasis throughout is on generic characteristics and the interaction of these characteristics with overall UAV system performance, rather than on the details of data-link design.

The intent of this part of the book is to assist the reader in understanding how to structure system tradeoffs and/or plan test bed and technology efforts associated with UAV systems, with the objective of balancing and integrating the data link with all other aspects of the system, particularly including sensor design, onboard and ground processing, and human factors.

A data link may use either: (1) wireless – a radio-frequency (RF) transmission – or (2) wired (e.g., a fiber-optic cable). An RF data link has the advantage of allowing the AV to fly free of any physical tether to the control station. It also avoids the cost of a fiber-optic cable that usually will not be recoverable at the end of the flight.

In computing, communications, and data link, *bandwidth* refers to the transmission capacity and is the maximum rate of data transfer in a unit of time across a given path. Typically, bandwidth is expressed in bits per second (bps). For instance, for streaming an HD quality movie, the rate is about 4 Megabits per second (Mbps). A fiber-optic cable has the advantages of having an extremely high bandwidth and of being secure and impossible to jam.

However, there are serious mechanical issues associated with maintaining a physical connection between the ground station and the AV as the AV flies. Any attempt to allow the UAV to maneuver or orbit over a location to turn around and come back to the ground station quickly can raise issues with the cable trailing behind it.

Most UAV systems will use an RF data link. The exceptions are likely to either be very short-range observation systems, such as a rotary-wing UAV launched from and tethered to a ship to provide an elevated vantage point for radar and electro-optic sensors, or short-range lethal systems that are fiber-optic-guided weapon systems rather than recoverable UAVs. Currently, the US Air Force uses a concept called "Remote-Split Operations," where the satellite data link is located in a different location and is connected to the GCS through fiber optic cabling.

Introduction to UAV Systems, Fifth Edition. Paul Gerin Fahlstrom, Thomas James Gleason, and Mohammad H. Sadraey.
© 2022 John Wiley & Sons, Inc. Published 2022 by John Wiley & Sons, Inc.
Companion website: www.wiley.com/go/fahlstrom/uavsystems5e

292 | Part V Data Links

The functions of a data link are the same regardless of how it is implemented, but we concentrate in this textbook on issues that apply to RF data links.

This part is divided into four chapters: (1) Chapter 13, Data-Link Functions and Attributes, (2) Chapter 14, Data-Link Margin, (3) Chapter 15, Data-Link Reduction, and (4) Chapter 16. Data-Link Tradeoffs.

13

Data Link Functions and Attributes

13.1 Overview

This chapter provides a general description of the functions and attributes of the data-link subsystem of a UAS and describes how the attributes of the data link interact with the mission and design of the UAV.

The data link provides a communications link between the UAV and its ground station, and is a critical part of the complete UAS. It is very important for the designers of a UAS to realize that the characteristics of the data link must be taken into account in the design of the total system, with numerous tradeoffs between the mission, control, and design of the AV and the design of the data link. If the UAS designer assumes that the data link is a simple, near-instantaneous pipeline for data and commands, there are likely to be unpleasant surprises and system failures when the system must deal with the limitations of real data links. On the other hand, if the system, including the data link, is designed as a whole, adjusting AV and control concepts and designs in tradeoffs with data-link cost and complexity, it is possible to achieve total system success while accommodating the fundamental limitations of data links.

Currently, the US Air Force uses a concept called "Remote-Split Operations" where the satellite datalink is located in a different location and is connected to the GCS through fiber optic cabling. This allows Predators to be launched and recovered by a small "Launch and Recovery Element" and then handed off to a "Mission Control Element" for the rest of the flight.

13.2 Background

The highest level of difficulty and complexity associated with data links results from the special needs of military UAV data links in such areas as resistance to deliberate jamming and deception. However, even the most routine civilian application must avoid unintentional interference from the vast number of RF systems that are constantly emitting in any developed and inhabited area, so the difference between the military and civilian requirements are not basic. This treatment addresses the full military requirements so that the reader will be aware of what is really only the worst case of a generally difficult environment in which UAV data links must operate.

Introduction to UAV Systems, Fifth Edition. Paul Gerin Fahlstrom, Thomas James Gleason, and Mohammad H. Sadraey.
© 2022 John Wiley & Sons, Inc. Published 2022 by John Wiley & Sons, Inc.
Companion website: www.wiley.com/go/fahlstrom/uavsystems5e

This chapter draws many of its specific examples from US Army experience with the Aquila RPV and its data link, the MICNS. MICNS was a sophisticated, anti-jam (AJ), digital data link designed to provide two-way data communication and a position-measurement capability to command the AV, transmit sensor data to the ground, and assist in AV navigation by providing precision position fixes relative to the ground-station location.

It was designed to operate in a severe jamming environment. Delays in the MICNS program resulted in early testing of the Aquila AV using high-bandwidth commercial data links of the sort used by mobile television operations to link their video back to the studio. When MICNS became available and was integrated into the Aquila UAS, a number of serious problems were discovered.

Perhaps the primary lesson that can be drawn from the MICNS/Aquila development and testing history is that integration of a data link with a system of the complexity of a UAV is far from trivial. Unless the data link is a simple, real-time, high-bandwidth communications channel that can be treated much like a hard wire, its characteristics are likely to have significant impacts on system performance. If these impacts are taken into account in the rest of the system design, essential system performance can be preserved.

If the system is designed assuming essentially unlimited data-link capability, there is a good chance that major redesigns will be required when a real, limited data link is installed. No data link – that operates beyond the line of sight and is interference-resistant – is likely to be simple, real-time, and high-bandwidth. Therefore, the constraints and characteristics of the objective data link must be considered during the initial system design.

Many of the key data-link issues are related to the time delay that the data link introduces into any control process that closes a loop between the air vehicle and the controller on the ground. In the Aquila era, these delays were most likely to be due to bandwidth restrictions and AJ processing times. More recently, it has become common to control UAVs from great distances via communications satellites. An example is the practice of conducting combat missions in Southwest Asia from ground control stations in the western United States. All of the issues discussed in this chapter with regard to image and command latency using the MICNS AJ data link apply equally to the case in which the time delay is introduced by sheer distance combined with an accumulation of small delays as the signal is relayed from point to point.

The design tradeoff between the data link and the rest of the UAV system should occur early in the overall system design process. This allows a partitioning of the burden between the data link, processing in the air and on the ground, mission requirements, and operator training.

As with most technologies, there are natural levels of data-link capability that are separated by jumps in both cost and complexity. Cost-effective system definition requires that these levels and the jumps between them be recognized so that an informed decision can be made as to whether the next jump in cost is justified by the increment in capability that it provides.

The interaction between the data link and the rest of the UAV system is complex and multifaceted. The critical characteristics that are responsible for most of the complexity in the interaction are bandwidth restrictions and time delays, whether they are associated with AJ capability, or distance, or relays, or limitations on a general-purpose communications network that is being used by the UAS. We will start with a general description of data-link functions and attributes and how they interact. With this background, we then will assess the tradeoffs associated with AJ capability and establish the likely limits of data-rate capacity for AJ and "jam-resistant" data links under various conditions. Finally, we will consider the impact of data-link restrictions, from whatever source, on RPV mission performance, and consider how a total system approach to UAV design may allow these impacts to be reduced.

13.3 Data-Link Functions

The basic functions of a UAV data link are illustrated in Figure 13.1. They are as follows:

- An uplink (or command link) with a bandwidth of a few kbps that allows the ground station to control the AV and its payload. The uplink must be available whenever the ground control unit wants to transmit a command, but may be silent during periods when the AV is carrying out a previous command (e.g., flying from one point to another under autopilot control).
- A downlink that provides two channels (which may be integrated into a single-data stream): (1) a UAV status (or telemetry) channel that transmits to the ground control station such information as the present AV airspeed, altitude, engine RPM, etc. and (2) a payload status channel that transmits information such as pointing angles. The UAV status channel requires only a small bandwidth, similar to the command link. The second channel transmits sensor data to the ground. This channel requires a bandwidth sufficient to deal with the amount of data produced by the sensors, typically anywhere from 300 kbps to 10 Mbps. Normally the downlink operates continuously, but there may be provisions for temporary onboard recording of data for delayed transmission.
- The data link may also be used to measure the range and azimuth to the AV from the ground station (antenna), which will assist in navigating the AV and to increase the overall accuracy of target locations measured by the sensors on the AV.

The data link typically consists of several major subsystems. The portion of the data link that is located on the AV includes an air data terminal (ADT) and antennas. The ADT includes the RF receiver and transmitter, whatever modems are required to modulate/demodulate the signal, and, sometimes, processors to compress the data to be transmitted to fit within the bandwidth limitation of the downlink. The antennas may be omnidirectional or may have some gain and require pointing.

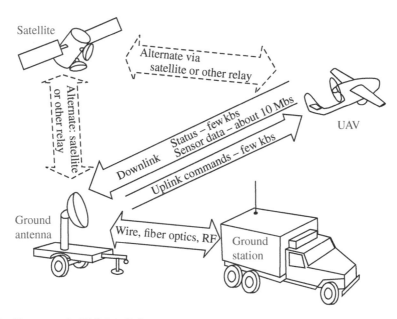

Figure 13.1 Elements of a UAS data link

The equipment on the ground is called a ground data terminal (GDT) and consists of one or more antennas, an RF receiver and transmitter, modems, and processors to reconstruct the sensor data if it has been compressed before transmission. (The question of whether data compression and reconstruction should be internal or external to the data link is discussed later.) The GDT may be packaged in several pieces, often including an antenna vehicle that can be placed some distance from the UAS GCS, a local data link from the ground antenna to the GCS, and processors and interfaces within the GCS.

Moreover, in both AV and GCS, an amplifier is required to increase the power of the signal. Thus, the main subsystems of data link in both AV and GCS are: (1) transmitter, (2) receiver, (3) antennas, (4) modems, (5) processor, and (6) power amplifier. A transmitter consists of an oscillator that generates an oscillating radio wave, a modulator that impresses an information signal on the wave, and an amplifier that increases the power of the signal.

The functional elements of the data link do not change in a fundamental way if the data streams to and from the GDT to the ADT are "transmitted" via some combination of high-bandwidth hard wires or fiber optics, uplinked to a satellite that connects to a second or third satellite, and then finally linked to the ADT from space. There still is an uplink channel and a downlink channel, as viewed from the perspective of the UAS. The required functions of these channels are as shown in the figure and their capabilities are set by the weakest link in the chain.

Three important parameters of a data link are: (1) signal frequency, (2) data link margin, and (3) data rate. The signal frequency is discussed in Section 13.6, the data link margin is explored in Chapter 14, and the data rate will be presented in Chapter 15.

13.4 Desirable Data-Link Attributes

If a UAV data link had only to operate under highly-controlled conditions on a particular test range, it probably would be adequate to use simple telemetry receivers and transmitters. Interference from other emitters on the range might be encountered, but could be controlled by careful selection of operating frequencies and, if necessary, control of the other emitters.

However, experience has shown that such a simple data link is not adequate to ensure reliable operation if the UAV system is moved from the test range on which all frequency conflicts have been resolved to another test range, let alone to a realistic battlefield or "urban" electromagnetic environment. (We will use *"urban"* as shorthand for the electromagnetic environment in a highly developed, at least moderately densely inhabited area, which, today, implies a lot of possible conflicting and interfering signals.)

At an absolute minimum, a UAV data link must be robust enough to operate anywhere that the user might need to test, train, or operate in the absence of deliberate jamming. This requires that the link operates on frequencies that are available for assignment at all such locations and that it is able to resist disruption by inadvertent interference from other RF emitters that are likely to be present.

On a battlefield, the UAV system may face a variety of Electronic Warfare (EW) threats, including direction-finding used to target artillery on the ground station, anti-radiation munitions (ARMs) that home in on the emissions from the GDT, interception and exploitation, deception, and both inadvertent and deliberate jamming of the data link. It is highly desirable that the data link provides as much protection against these threats as reasonably can be afforded.

13.4 Desirable Data-Link Attributes 297

There are nine desirable attributes for a UAV data link related to mutual interference and EW:

1) Worldwide availability of frequency allocation/assignment: operate on frequencies that are available for test and training operations at all locations of interest to the user in peacetime as well as being available during wartime.
2) Resistance to unintentional interference: 0perate successfully despite the intermittent presence of in-band signals from other RF systems.
3) Low probability of intercept: difficult to intercept and measure azimuths at the ranges and locations available for enemy direction-finding systems.
4) Security: unintelligible if intercepted, due to signal encoding.
5) Resistance to deception: reject attempts by an enemy to send commands to the AV or deceptive information to the GDT.
6) Anti-ARM: difficult to engage with an ARM and/or minimize damage to the ground station if engaged by an ARM.
7) Anti-jam: operate successfully despite deliberate attempts to jam the up- and/or downlinks.
8) Digital data links: digital data and modulation is a natural choice for a UAV data link.
9) Signal strength: the signal should be strong at the receiver location. High-power transmitters may have additional constraints with respect to radiation safety, generation of X-rays, and protection from high voltages.

The relative priorities of these desirable attributes depend on the mission and scenarios for a particular UAV. In general, the priorities will be different for the uplink than for the downlink. The general considerations that affect these priorities are discussed in the following sections.

13.4.1 Worldwide Availability

This issue is the most important desirable attribute of data link for general-purpose civilian or military systems. Special-purpose systems might be designed for use only in specific locations. In principle, they might use frequency bands available only at that location (at the cost of a potential redesign if it later were desired to use the system somewhere else). Even in such a special case, however, one should not forget that the system probably will have to be usable at one or more test site, which may have frequency restrictions different from the eventual operational site.

For general-purpose systems, the most restrictive area with regard to frequency availability presently is probably Europe. As development accelerates worldwide, other areas may become similarly restrictive and may have a different set of rules. Even today, some "non-developmental item" (NDI), that is, off the shelf, data links sold outside of Europe use frequency bands that may not be available for peacetime use in Europe.

If a data link uses frequencies not available worldwide, it may have to be designed with alternate frequencies depending on where it is to be used. For a civilian system, this is a simple necessity. For a military system, it can be argued that the frequencies could be used in wartime despite peacetime restrictions, but a military user has a test and training requirement that makes it essential that the data link be usable in peacetime. UAV operator skills are likely to need constant refreshing. It may be difficult to find places where a UAV can be flown in some areas, but training can be conducted with manned aircraft carrying the UAV payload and data link. This training should use the operational data link so that its characteristics are embedded in the training.

If an NDI data link without worldwide availability is accepted for initial procurement, replacement with a later version that is available for use everywhere should be a preplanned feature of the program. In this situation, the cost of scrapping the initial data link should be considered an unavoidable cost, which might be justified if the NDI data link allowed earlier fielding and could be replaced before too many systems have been delivered.

13.4.2 Resistance to Unintentional Interference

The ability to operate without significant risk of mission failure due to unintentional interference is a second essential requirement for a UAV data link. The electromagnetic environment used to specify and test this capability should be the worst case thought to be possible for the system in question. For military systems, this should include joint operations and a realistic mix of emitters to include those that might be encountered on test ranges, in training areas, and in the operational area. Simple telemetry links have been shown by experience not to meet this requirement with regard to test ranges and training areas, let alone with regard to the intense electromagnetic environment that would exist on a modern battlefield.

In addition to avoiding frequency conflicts, resistance to unintentional interference can be enhanced by use of error-detection codes, acknowledgement, and retransmission protocols, and many of the same techniques that are used to provide resistance to jamming. Electromagnetic interference (EMI) has been discussed in Section 11.4 of Chapter 11.

13.4.3 Low Probability of Intercept (LPI)

LPI is highly desirable for a military uplink, since the ground station is likely to have to remain stationary for long periods of time while it has AVs in the air, making it a sitting target for artillery or homing missiles if it is located. The survivability of the ground station can be improved by locating it well to the rear (in some cases) or by allowing AVs to be handed off from one ground station to another, so that the ground stations can move more often.

Furthermore, the emitting antenna for the uplink can be remote from the rest of the ground station, and all parts of the ground station can be provided with ballistic protection (at the cost of larger and heavier vehicles). However, it is highly desirable to reduce the vulnerability of the ground stations at its source by reducing the probability that the enemy will be able to get a good location by direction finding.

LPI is less important for the downlink. However, if the mission is covert, whether that might be surveillance and/or attacks against a terrorist meeting place or similar operations related to law enforcement, there would be a potential benefit from preventing the people who might be targets of the aerial surveillance from knowing that an AV is overhead and transmitting. Depending on the frequencies used, it might not be too difficult to acquire some sort of scanning receiver that could provide that type of warning.

LPI can be provided by frequency spreading, frequency agility, power management, and low duty cycles. At higher frequencies, the uplink signal may also be masked from ground-based direction finders by lack of a clear line of sight to the GDT antenna. Within the "low cost" constraint, LPI may have to be considered as a "nice to have" attribute that is present as a bonus because of characteristics that are primarily driven by anti-ARM and AJ requirements.

13.4.4 Security

For many of the tactical functions considered for UAVs, when the first modern systems were fielded in the 1980s, there probably would be little benefit to an enemy who could listen in on

13.4 Desirable Data-Link Attributes 299

either the uplink or the downlink, unless he could also break in with deception based on information gleaned from the intercept.

In recent years, however, some of the primary applications of UAVs have become missions in which maintaining operational covertness is critical. As mentioned earlier, terrorists or criminals can use knowledge that they are being watched from above to alter their behavior or take cover from attacks launched from the UAV. Under these circumstances, security, which requires encryption of both the uplink and downlink data streams, becomes an important requirement.

13.4.5 Resistance to Deception

Deception of the uplink would allow an opponent to take control of the AV and either crash, redirect, or recover it. This is worse than jamming, since it leads to loss of the AV and payload, while jamming typically only denies the performance of a particular mission. Furthermore, the opponent could attack many AVs in sequence with a single deception system if he could cause them to crash, while jamming ties up assets for long periods of time since the AV can continue its mission if the jammer moves on to another AV. Deception of the uplink only requires getting the AV to accept one catastrophic command (e.g., stop engine, switch data-link frequency, deploy parachute, change altitude to lower than terrain, etc.).

Deception on the downlink is more difficult, since the operators are likely to recognize it. Deception related to the sensor data on the downlink would require believable false sensor data, which would be very difficult to provide. Deception of the status downlink might cause a mission to abort or even to crash. For instance, a steadily ascending altitude reading might lead the operator to try setting a lower altitude, leading to a crash. However, this would require more sophistication than issuing a single bad command to the AV.

Resistance to deception can be provided by authentication codes and by some of the techniques that provide resistance to jamming, such as spread-spectrum transmission using secure codes. Some protection of the uplink seems prudent, particularly if the intent is to deploy a family of tactical UAVs that use a common data link and some common command codes, which might result from use of a common ground station.

Resistance to deception might be implemented external to the data link, since authentication codes could be generated in system software and checked by the AV computer without any direct participation by the data link (other than transmitting the message that includes the authentication).

13.4.6 Anti-ARM

ARMs are not an issue for the downlink, since the AV is not an appropriate target for such weapons. It is desirable to make the ground station a difficult target for anti-radiation munitions (ARMs) since it is stationary, radiates signals toward the enemy, and is a reasonably high-value asset. Considerable protection against ARMs can be provided by using a remote transmit antenna and a low duty cycle for the uplink.

Ideally, the uplink should not transmit unless there is a command to send up to the AV, which would allow it to remain silent for long periods of time. This is partly a system issue, since the whole system should be designed to make minimum use of the uplink, but it is also a data-link issue, since some data links may be designed to emit signals regularly even if no new commands are awaiting transmission.

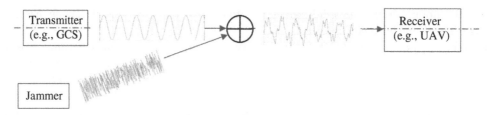

Figure 13.2 Jamming via sending noise to a data link signal

Additional protection from ARMs can be provided by LPI, frequency agility, and spread-spectrum techniques, which are desirable from other standpoints as well. If the ARM threat were judged to be severe enough, various active approaches, such as decoys, are possible to add to the protection of the ground station.

13.4.7 Anti-Jam

Similar to the case for GPS, data link and satellite communications are vulnerable to noise jamming. Jamming is the deliberate blocking or interference with a wireless communication. There are various jamming techniques; one way is to transmit (i.e., add) radio signals (e.g., noise) that disrupt communications (Figure 13.2) by decreasing the signal-to-noise ratio.

The ability of a data link to operate effectively in the presence of deliberate efforts to jam it is "Anti-Jam," or "AJ," or "jam-resistant" capability. Sometimes "Anti-Jam" is equated with full protection against a worst-case jamming threat and "jam-resistant" is used to describe some lesser degree of protection against jamming. As used here, jam resistance is a subset of AJ.

It is useful to introduce the concept of an AJ margin without, at this point, defining it mathematically. The AJ margin of a data link is a measure of the amount of jammer power that the link can tolerate before its operation degrades below an acceptable level, normally determined by the specified maximum acceptable error rate for the link. The AJ margin is usually stated in dB.

In the particular case of the AJ margin, the margin being described in dB is the actual signal-to-noise ratio available to the system in the absence of jamming divided by the minimum signal-to-noise ratio required for successful system operation. Thus, an AJ margin of 30 dB means that the jammer must reduce the signal-to-noise ratio at the receiver by a factor of greater than 1,000 (i.e., $10 \log (1,000) = 30$) in order to interfere with the successful operation of the system.

In discussing AJ margins expressed in dB, it is important to keep in mind that a factor of 2 in dB is not a factor of 2 in jammer power. Thus, reducing an AJ margin of 40 dB by a factor of 2, to 20 dB, would reduce the required jammer power by a factor of about 100. In other words, if the jammer power is increased by a factor of 100, the data link AJ margin has to be increased by a factor of 2 in dB unit. The difference between, say, a 10,000-W jammer and a 100-W jammer is much more significant than might be assumed when dealing with a simple factor of 2. The AJ margin is discussed at greater length in Chapter 14.

The overall priority of AJ capability depends on the threat that the UAV is expected to face and the degree to which the UAV mission can tolerate jamming. The data link is unlikely to be jammed everywhere all of the time. In one limit of mission tolerance, it may be possible to record sensor data onboard the UAV while performing preprogrammed mission profiles and send the data down when a hole in the jamming is found or even bring the data home in memory (i.e., hard disk). In this limit, it might not be necessary to use the uplink at all until nearly at the recovery point, so uplink AJ might be of little importance. For some UAV applications this might be an acceptable degraded mode of operation, even if not planned as the primary mode. In such a case, it might be possible to use a data link with little AJ capability.

The other limit is represented by a mission similar to that of Predator, where many key functions can be performed only in real time. The most obvious examples are acquisition, location, and attack of moving targets or surveillance of a border crossing area. A replay of a recording even a few minutes old would be quite useless in most cases. For these missions, the level of AJ capability, combined with the jamming threat, determines whether or not the enemy can deny mission effectiveness to the UAV system.

In many cases jamming of the downlink is more harmful to the mission than jamming the uplink. Many missions can be performed using preprogrammed flight profiles and sensor search patterns. If the uplink is jammed the operator is denied the flexibility of looking again at an item of interest from a different angle, but he or she can record data on the ground and replay it, if another look is required at what has already been seen in real time.

The AV can be programmed to return to the vicinity of the ground station, where the uplink is unlikely to be jammed, for recovery at the end of the mission. Therefore, most missions are more tolerant of uplink jamming than downlink jamming, which denies real-time data.

13.4.8 Digital Data Links

A data link may transmit either digital or analog (e.g., voltage) data. If it transmits digital data, it may use either digital or analog modulation of the carrier. Many simple telemetry links use analog modulation, at least for their video channel. Most AJ data links use digital modulation to transmit digital data.

Modulation is the process of converting data into radio waves, and varying one or more properties of a periodic waveform (i.e., the carrier signal), with a separate signal, called the modulation signal, that typically contains information to be transmitted. Therefore, the modulation is the process of adding the desired information (data) to a radio wave. This is one of the functions of a transmitter. Primary techniques for modulation are: (1) amplitude modulation, (2) frequency modulation, and (3) phase modulation. A modem (short for modulator–demodulator) is a device that converts data from a digital format into one suitable for a radio wave.

Modern UAV systems are equipped with a digital processor for control. Moreover, autopilot functions in the GCS and AV, and the sensors data onboard the AV are also almost certain to be digital, at least in its final stages. Digital data formats are essential for most, if not all, approaches to error-detection, tolerance to intermittent interference through redundant transmission, encryption, and authentication codes. For all of these reasons, digital data and modulation is a natural choice for a UAV data link. This treatment assumes that the data link is digital unless explicitly stated otherwise.

13.4.9 Signal Strength

The signal should be strong enough at the receiver location, such that the receiver has a quality and clear data. When a UAV is close to the GCS, the datalink signal strength is very strong. However, as the UAV flies away from the GCS, the datalink signal strength decreases rapidly with distance. There is a distance between UAV and GCS such that the signal is not strong enough to be picked up. This is referred to as the range of a data link. Radio wave interference reduces the range of the datalink.

Electromagnetic waves are absorbed by clouds, humidity, rain, moisture, smoke, dust, and fog. Thus, as a radio wave is travelling through the atmosphere, it gets weaker and weaker.

In the mid- and far-IR, atmospheric attenuation mechanisms include scattering and absorption, mainly due to water. The attenuation mechanisms are different for vapor, rain, or fog. Water vapor primarily absorbs the IR energy, while rain primarily scatters energy, and fogs both absorb and scatter. More discussion is presented in Chapter 10, section 3.

In order to carry a strong data link and a higher data rate, a higher power/energy (at the transmitter station) to propagate the signal is required. Consequently, a higher frequency and a higher data rate incur more cost for transmission. High-power transmitters may have additional constraints with respect to radiation safety, generation of X-rays, and protection from high voltages.

13.5 System Interface Issues

There are several major areas in which interface issues are likely to arise with regard to UAV data links. They include: (1) mechanical and electrical, (2) data-rate restrictions, (3) control-loop delays, and (4) interoperability, interchangeability, and commonality. In the next four sections, these data link interface issues are discussed.

13.5.1 Mechanical and Electrical

General mechanical and electrical interfaces are difficult to discuss in a generic manner and are outside the scope of this treatment. Clearly, weight and power restrictions on the AV may be a significant constraint on ADT design. Ground antenna size and pointing requirements may have an impact on the configuration of the ground station. These factors are more likely to be system drivers for AJ data links than for non-AJ data links, although ground antenna size and pointing can be driven by navigation requirements, even in the absence of AJ requirements.

The antennas on the AV can be an issue if the link uses a relatively high frequency and uses steerable, medium-gain antennas to achieve either longer range (for the same transmitter power) or AJ margin. Steerable antennas typically must project from the body of the AV and may be vulnerable to damage during recovery. No single antenna location can provide full coverage for all AV maneuvers (e.g., left and right bank turns), so at least two antennas (typically dorsal and ventral) are required. More may be required to fill holes in coverage or if the receive and transmit antennas are separate. MICNS used three transmit and two receive antennas on Aquila.

The electrical interface to a data link includes more than just power and data-in and data-out. Typically, the data link should inform the rest of the system when it is operating, whether the data in its output buffers are good (i.e., new data that pass the error-detection checks), and other status information that may be needed by the operators, such as fading signal strength or increasing error rates that may indicate an eminent loss of link. In addition, built-in test capability, with appropriate interfaces to the rest of the system, is highly desirable.

13.5.2 Data-Rate Restrictions

Restrictions on the available data rate on the sensor downlink may be the area with the greatest impact on the rest of the UAV system. Many sensors are capable of producing data at much higher rates than can be transmitted by any reasonable data link.

For instance, a high-resolution TV camera or forward-looking infrared (FLIR) sensor will produce a data rate of order 75 megabits per second. If such sensors are operating at the standard 30 frames per second (fps), they can produce about 74 million (i.e., Mega) bits per second (Mbps) of

13.5 System Interface Issues   303

raw data (640 by 480 pixels at 8 bits per pixel at 30 fps, $640 \times 480 \times 8 \times 30 = 73.7 \times 10^6$). No data link that meets UAV size, weight, and cost constraints is likely to have enough capacity to transmit this raw data rate.

It is necessary to limit the maximum data collection period and downlink the data before taking on any more new data. There are a variety of ways that are transparent or nearly transparent to the operators to reduce the transmitted data rate without loss of information. However, as discussed later, if AJ capability or even "jam-resistance" is required, it will probably be necessary to reduce the transmitted data rate to a point where there will be an impact on the mission.

The key issue here is the nature of that impact. If the overall system design is built around an unrestricted data rate, then it is very likely that the impact of reducing the data rate will be to degrade the performance of the mission. On the other hand, if the system design, including operator procedures and mission planning, is established with the restricted data rate in mind, it is entirely possible that the mission can be performed without degradation.

The transmitted data rate can mainly be reduced by two methods: (1) compression or (2) truncation. Data compression processes consist of converting the data to a more efficient representation that allows the original data to be reconstructed on the ground. Ideally, when data are compressed and then reconstructed no information is lost. In practice, there is often a small loss of information due to imperfections or approximations in the process. On the other hand, data truncation involves discarding some of the data in order to transmit the remainder. A typical example would be to throw away every other frame of TV video to reduce the data rate by a factor of two. Some information is lost in this process, but the loss may not be perceptible to the operator, for whom a new frame every 1/15 of a second may provide all of the information that he needs.

A more severe form of truncation is to throw away the borders of each frame of video, reducing the effective field of view of the sensor by a factor of two in each dimension. This reduces the data rate by a factor of four, but it also reduces the area that can be observed on the ground. In the second case, useful information may be lost in the truncation process.

A combination of compression and truncation may be required to stay within the data-rate limits of the downlink. The choice of compression and truncation techniques should be made as a part of a total system-engineering effort that considers the characteristics of the sensor and how the data will be used to perform the mission, as well as the characteristics of the data link.

Data compression and truncation requirements are driven by bandwidth restrictions of the data link, which, in turn, are driven by AJ considerations. Chapter 14 explores the likely bandwidth implications of AJ requirements. Chapter 15 then discusses possible data compression and truncation approaches for accommodating the resulting data rates.

13.5.3 Control-Loop Delays

Some UAV functions require human-in-the-loop closed-loop control from the ground. Manual pointing of a sensor at a target and initiating auto-tracking of that target is one example. Another is flying the AV into a recovery net or landing it on a runway under manual control. For both cases, a camera/sensor will be employed for providing feedbacks to the GCS operator (i.e., the controller in Figure 5.7).

The control loops that carry out these functions involve two-way transmission over the data link. If the data link uses data compression or truncation, message blocking, time multiplexing of the up- and downlinks on a single frequency, or any block processing of data before or after transmission, there will be delays in transmission of the commands and feedback data in the control loop. Delays in control loops are generally detrimental, and the UAV system design must take these delays into account if it is to avoid serious or even catastrophic problems.

As an example, consider a particular data link that might provide only one time slot per second for transmission of commands on the uplink, and have a capability to downlink only one frame of video per second.

If the UAV is landed by an operator who aims a TV camera at a predetermined descent angle and manually flies the AV toward a runway threshold based on the TV picture, there might be a total loop delay of 2 seconds or more: (1) one second to send up a command, (2) one second waiting to see a frame of video that reflects any resulting change in the AV flight path, (3) a fraction of a second for the operator to react, and (4) minor delays within the electronics of the operator console. This is a case where the data rate is too low and unacceptable in this data link.

Note that the delay would not be constant, since the delay waiting for the next uplink time slot would depend on the phasing between the operator input and the data-link time multiplexing. It is very likely that this delay would make it impossible to land the AV with any reliability, particularly in windy conditions.

Two solutions are possible. For recovery operations, the simplest solution is to add an auxiliary data link that has low power, no AJ capability, and a wide bandwidth. This link would be used only during the final approach to a net or runway. The other solution is to use a recovery mode that is not sensitive to data-link delays. Possibilities include an automatic landing system that closes the control loop onboard the AV (e.g., a system to track a beacon on the runway or net and use the tracking data to drive the autopilot for a landing) or a parachute or parafoil recovery. For a parachute, only a single command is required to deploy the parachute.

A second or so of uncertainty in the time of deployment would be of little consequence. A parafoil landing may be slow enough that 2- or 3-s delays in the control loop are acceptable, at least for recovery on the ground. However, this may not be true for recovery on a moving platform such as a ship underway.

For the case of sensor pointing, the option of using a short-range, back-up data link does not exist. The solution in this case is to design a control loop that can operate successfully with time delays of 2–3 s. Studies performed for the Aquila RPV program show that this is possible if the control loop operates in a mode that automatically compensates for the motion of the sensor field of view during the time delay between the video that the operator is observing and the arrival at the AV of the operator's command for the field of view to move [36].

Techniques similar to those described in Reference [36] were successfully applied to the Aquila/MICNS system. However, it should be noted that these techniques require a good inertial reference on the AV and high-resolution resolvers on the sensor pointing system so that pointing commands can be computed and executed relative to a reference that is stable over times of the order of the maximum delay to be accommodated.

Of course, in either case it would be possible to solve the problem of transmission delays by requiring a higher data rate from the data link. However, this solution may have an impact in terms of AJ capability or data-link complexity and cost that is less desirable than the impact of either an auxiliary data link for recovery or a more sophisticated control system design. Also, if satellite transmissions are present in the data-link chain, the delays may be fundamental and would not go away with a higher bandwidth. A balanced system design will consider all solutions in terms of total system impact to avoid asking too much or too little of any of the subsystems.

A major source of control-loop delays is data-rate reduction in the downlink, leading to effects such as the video frame-rate reduction described in the example earlier. The effects of frame-rate reduction are discussed in detail in Chapter 15 as part of the discussion of data compression and truncation. However, significant delays can be caused by other factors in data-link and UAV system

13.5 System Interface Issues 305

design. As mentioned earlier, there may be significant delays, which may not be completely predictable, associated with locating the ground station and pilot halfway around the world from the AV.

A similar, but more easily avoided, delay could be caused by a data link that waits for a multi-command message block to be filled before transmitting that block to the AV. In this case, the command that fills the block gets transmitted almost immediately, but the first command in the next block has to wait until enough additional commands have been accumulated to fill the block. Even the data link might transmit an incomplete block after some maximum wait-time or might assume that commands will be issued by the GCS at a rate that ensures that the wait to fill a block will never be unacceptable.

Additional delays can be caused by waiting for a complete block of compressed sensor data to be received before beginning to reconstruct the block and by the time taken to do the reconstruction. Also, the cumulative delay through the loop due to the computer in each subsystem waiting for the computer in the next subsystem to be ready to accept a data transfer can be significant for some applications.

The effect of control-loop delays on a system not designed to accommodate them can be quite serious. Unless provisions are made for these delays, the substitution of an AJ data link for a simple telemetry link or operating a UAV thousands of miles away via satellite instead of nearby with a direct data link are likely to require significant redesign of system software and may require changes to hardware.

13.5.4   Interoperability, Interchangeability, and Commonality

Interoperability and interchangeability are not the same thing. In the context of UAV data links, interoperability would mean that an ADT from one data link could communicate with a GDT from another, and vice versa. This is very unlikely unless the two data links are actually identical, built to the same design. The only level at which interoperability is likely to be achievable is for simple telemetry links using independent simplex channels (i.e., separate up and down channels on different frequencies and operating independently). Any more sophisticated link involves details of modulations, timing, synchronization, etc., that would be very difficult to specify adequately to ensure that the systems would work together.

In particular, different AJ techniques, such as direct spread spectrum and frequency hopping, are fundamentally not interoperable, even when they use the same portion of the spectrum. Therefore, the only practical way to have interoperable data links for different UAV systems would almost certainly be to have a common data link, although the actual hardware could be manufactured by competitive sources to a common design.

Interchangeability is a lesser requirement. This would only require that two different data links could be substituted for each other in one or more UAV systems. Both ADTs and GDTs would have to be changed to allow operation. Interchangeability would allow use of a low-cost, non-AJ data link for training and permissive environments, and a higher-cost, AJ data link for intense EW environments. The cost tradeoff would be between buying smaller numbers, each of the high- and low-capability data links, and supporting both links in the field, versus buying a larger number of the high-capability links and supporting only one link in the field.

Interchangeability would require common mechanical and electrical specifications (form and fit) and common characteristics as seen by the operators and the rest of the UAV systems (function). If, as is likely, the AJ data link had limited bandwidth, delays, and effects on the video or other data transmitted, then the system would have to be designed to accommodate those characteristics and the operators would have to be trained to work with them. One possible approach would be to require the non-AJ data link to have a mode that emulated the AJ data link.

Rather than building the emulation into the data link, this could be achieved in an interface or "smart buffer" within the GCS, using the high-bandwidth, non-AJ data link like a hard wire to connect the AV with a simulation of the AJ data link built into the interface. The interface would present commands to the AV with the timings and formats that would be present with the AJ data link and process the sensor and status data downlinked from the AV in such a way that the GCS would see the same effects with regard to timing and data compression and reconstruction as would be present with the other data link. While this concept appears possible in principle, the complexity of actually implementing it should not be underestimated.

In any environment in which more than one type or model of UAV is expected to be in use, there is likely to be interest in a common data link to lower acquisition and support costs and ensure interoperability. A common data link might serve two or more UAV systems and might be intended to have applications across service or agency lines.

It is interesting to know that the MICNS was originally a common data link to meet the requirements of the Stand-Off Target Acquisition System (SOTAS), Aquila, and the Air Force Precision Location and Strike System (PLSS). SOTAS and PLSS were terminated, but the MICNS design tested with Aquila still had the provisions to meet their requirements.

A common data link would have a single set of data-link hardware, possibly with optional modules for different applications. The common data link could, by definition, be interoperable and interchangeable between systems since it would hardly qualify as common unless all versions used the same RF sections and modems, at least through conversion of the signals from or to digital data in an input or output buffer. This would not preclude different antennas for different applications and, possibly, different transmitter powers for different range capabilities, but it would ensure that any ADT could talk to any GDT if the associated air and ground systems provided appropriate commands and inputs.

A major innovation that began in the latter part of the twentieth century has been the introduction of standardized networks, such as the Internet, that allow communications between distant systems based on a common protocol. In UASs, communications between AVs and ground control stations, or even between multiple air vehicles, may use such networks instead of direct, one-to-one data links.

Data interoperability then consists of providing the appropriate communications links and protocols for the AV and ground station to connect to the network and designing the message formats to transmit the required information up and down via the network. This approach potentially leads to a situation in which the data link is independent of the UAV system, but may also result in significant and unpredictable latencies in the transmission process.

The potential advantages of a common data link are reduced acquisition and support costs and interoperability. The disadvantage is whatever penalty must be paid in terms of burdening the common data link with the worst requirements of all of the systems in which it is to be used. This burden can manifest itself in two ways:

1) If the resulting data link becomes complex and expensive enough, the potential cost savings may be cancelled out.
2) Some capabilities that might be possible and affordable in a data link that was optimized for a single application might not be possible in a common data link that meets multiple requirements.

Issues related to a common data link are further discussed in Chapters 14 and 15 with regard to AJ capability, data rates, and data-link range. At this point, it is worth noting that, if AJ capability is required, then the best solution for a short-range data link may be basically incompatible with the best solution for a long-range data link, suggesting that a single common data link should not try to cover both range categories.

13.6 Antennas

Both transmitter and receiver are equipped with an antenna (in both UAV and GCS). Four types of antennas are in common use with UAV systems: (1) omnidirectional, (2) Yagi arrays, (3) parabolic reflectors or dishes, and (4) lenses. The first three types are more common on the ground (i.e., GCS), while the fourth (a lens antenna) is one of the few types of directive antennas that are suitable for use on an AV. The first one is most suitable for RC model planes, small UAVs, and their GCSs. The third one is needed for MALE and HALE UAVs for satellite communications. An important performance feature of an antenna is the gain (which will be defined in Chapter 14).

13.6.1 Omnidirectional Antenna

An omnidirectional antenna – which commonly consists of a conductive element such as straight metal wire or rod – is the simplest and most widely used class of antenna in communications. This monopole antenna radiates equal signal power in all directions perpendicular to the antenna axis. Omnidirectional antennas – oriented vertically – radiate equally in all horizontal directions. Directional antennas provide a powerful means of achieving signal selectivity, when various signal sources observed by the receiver are separated spatially.

The signal power is decreasing with increasing angle to that axis, where declining to zero on the axis. The power radiated drops off with the elevation angle, so little radio energy is aimed into the sky (for a GCS) or down toward the earth (for a UAV). The five-eighth wave monopole, with a length of 5/8 of a wavelength, is recommended, since, at this length, the antenna radiates maximum power in horizontal directions.

The omnidirectional antenna for GCS of a Switchblade UAV is illustrated in Figure 2.5. The ANAFI Ai UAV (see Figure 20.2) has four omnidirectional antennas with a reflector (with a gain of 2.5 dBi/antenna). The UAV automatically determines the best antenna pair depending on its orientation and position in relation to the GCS's position.

13.6.2 Parabolic Reflectors

A parabolic reflector or dish can be used to design a relatively inexpensive high-gain antenna that provides a beam width of a few degrees. A paraboloid is a mathematical surface with cylindrical symmetry whose cross-section is a parabola (bowl-shaped). It focuses parallel rays to a point called the focus, as shown in Figure 13.3. If a source of radiation is placed at the focus of the paraboloid then the energy will be reflected from the paraboloid into a beam whose beam width, in the ideal case, is determined by the diffraction limit associated with the diameter of the reflector. Similarly, incoming parallel or received radiation is focused and sent to a receiver in the receiving mode of a system utilizing a parabolic reflector antenna.

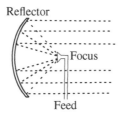

Figure 13.3 Parabolic reflector antenna for a transmitter or receiver

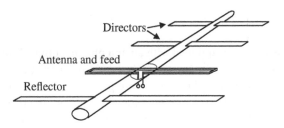

Figure 13.4 Yagi-Uda antenna

Large parabolic reflector antennas can have *gains as high as 30–40 dB* when used at short wavelengths.

13.6.3 Array/Directional Antennas

For lower frequency applications (e.g., VHF) and over-the-horizon transmission, arrays of two or more parallel horizontal dipoles, arranged to give highly directive radiation patterns, are used with UAV systems. These antennas are also referred to as directional antennas or beam antennas. An arrangement of a combination of driven (excited) dipoles, reflector, and director rods is the Yagi-Uda antenna shown in Figure 13.4.

The Yagi-Uda uses a driven dipole and as many as 30 parasitic dipoles, parallel and closely spaced. One of the elements is called the reflector and is slightly longer than and located behind the driven element. The remaining parasitic elements are called directors, and are shorter than the driven dipole and located in front of it. The spacing between the elements is approximately 0.1–0.4 λ.

Yagi-Uda antennas have good directivity and gains as high as 20 dB. They are commonly seen as household TV antennas. Due to the size and high drag configuration, this antenna is not employed in UAVs, while they are utilized in GCSs.

13.6.4 Lens Antennas

Lens antennas operate on the same principles as optical lens and for UAV applications operate at frequencies higher than 10 GHz. A point source of microwave energy spreads out spherically and just as an optical lens collimates a source of light, electromagnetic radio waves can be collimated by a dielectric or metal-plate lens placed in front of the point source of radiating electromagnetic energy. This is illustrated in Figure 13.5.

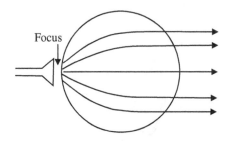

Figure 13.5 Lunberg lens antenna

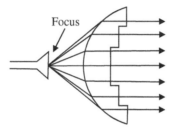

Figure 13.6 Zoned lens antenna

A particular kind of lens antenna that is formed from a solid sphere made from a dielectric material whose refractive index varies from its center to the surface is called a Lunburg lens. This antenna will collimate rays from the opposite side of the sphere from the feed point and has application with the AV.

The material is usually *polystyrene or Lucite* (an *acrylic* plastic resin). The waves passing through the lens slow down. The waves that pass through the center of the lens are delayed more than those that have a shorter path within the dielectric material. This collimates an outgoing beam, as shown in the figure, or focuses an incoming beam. If the lens is correctly shaped, the energy emanating spherically is focused into a beam and therefore has gain.

A dielectric lens tends to be heavy and in addition absorb a significant amount of power because the center must be several wavelengths thick. Sections of the lens can be removed by cutouts or steps if the phase shifts associated with each cutout are multiples of the wavelengths so as not to affect the wave front. These are called zoned lenses and an example of a zoned-lens antenna is shown in Figure 13.6.

HALE UAVs such as Global Hawk and MALE hunter UAVs such as Predator and Reaper all have dish antenna – with direction pointing capabilities – for satellite communications.

The Global Hawk UAV is equipped with a number of antennas: (1) 1.2 m K_u-band SatCom antenna, (2) UHF antenna, (3) LOS antenna, (4) SAR combined antenna, and (5) dual-band common data link antenna. The top front surface of the fuselage houses the 48 in Ku-band wideband satellite communications antenna.

NASA's Ikhana Predator B is equipped with five antennas: (1) one parabolic with a K_u band for a SatCom (inside top part of the fuselage nose), (2) one for Inmarsat (on top of the mid fuselage), (3) one C-band for line of sight (underneath fuselage), and (4) two ARC-210 radio antennas.

The Predator and Reaper primary satellite link (at the GCS) consists of a 6.1-m (20-ft) satellite dish antenna and associated support equipment. The satellite link provides communications between the ground station and the aircraft when it is beyond line-of-sight and is a link to networks that disseminate secondary intelligence.

13.7 Data Link Frequency

An important step in the design of the data link is the selection of frequency. All radio waves are transmitted with frequencies between about 30 Hz and 300 GHz. However, to prevent interference between different users of the radio spectrum, transmitters are strictly regulated by national radio laws, and are restricted by governments to specific frequencies and power levels. The regulations prevent someone from setting up a transmitter that blocks radio signals and cellphones from working.

Table 13.1 Commonly used frequency bands in communications and data links

No.	Band	Frequency	Applications and Examples
1	HF	3–30 MHz	Over-the-Horizon radar *surveillance*
2	VHF	30-300 MHz	• Very long-range *surveillance* • *RC plane, Model airplane*
3	UHF	300–1,000 MHz	• Very long-range *surveillance* • ScanEagle
4	L	1–2 GHz	• Long range *surveillance* • Cell phone signal – 3G • GPS satellites broadcast • *En route traffic control*
5	S	2–4 GHz	• Moderate range *surveillance* • *Terminal traffic control* • *Long-range weather* • *WiFi*
6	C	4–8 GHz	• *Long-range tracking* • *Airborne weather detection* • MQ-1B Predator • WiFi
7	X	8–12 GHz	• Military application
8	K_u	12–18 GHz	• High-resolution mapping • Satellite altimetry • MQ-1B Predator
9	K, K_a	18–40 GHz	• Cell phone signal – 5G • Very high-resolution mapping • Airport surveillance

Table 13.1 shows the commonly used frequency bands [37] for communication systems and data links. The communications for MALE and HALE UAVs usually use satellite links and BVLOS mode. The most common frequency bands of this type of data link are K_u, K, S, L, C, and X bands.

A certain number of frequencies in the 72 MHz VHF band have been designated by law for use with RC planes. Most current RC planes are using the frequency of 72 MHz, but the trend is to move toward 2.4 GHz. The L band (1–2 GHz) seems to become the preferred route for flight of UAVs.

WiFi frequency bands are 2.4 GHz and 5 GHz. The wireless router uses these *frequencies* to transmit the Internet to any *Wi-Fi*-connected device like smartphones and laptops. The signal with a frequency of 5 GHz will transmit more data per second, but to a shorter range. In contrast, the signal with a frequency of 2.4 GHz will transmit less data per second, but to a longer range. Thus, if you are using a UAV inside a lab using Wi-Fi, use 5 GHz for video transmission from UAV to GCS, while 2.4 GHz for sending a command from GCS to UAV.

GPS satellites transmit signals on two frequencies: 1.575 GHz and 1.227 GHz. Cell phone companies are using UHF and L bands for 4G communications, but are employing very high frequencies for 5G communications. For instance, Verizon is using 28 GHz, T-Mobile is using 28 GHz and 39 GHz, and AT&T is utilizing 39 GHz for 5G communications.

A certain number of radio frequencies have been designated by federal law for use by aeromodellers and with aircraft being controlled with an FM RC system. They fall into the 72 MHz band and each separate frequency has been given a unique channel number. Note that remote controlled cars and boats use frequencies in the 27 and 75 MHz bands.

The ScanEagle (see Figure 2.2) has a 900 MHz (i.e., UHF band) data link and a 2.4 GHz S-band downlink for video transmission. Data link frequency in MQ-1B Predator is employing Ku-band for LOS and C-band for BVLOS.

The Global Hawk (see Figure 1.4) is equipped with multiple antennas for various transmitting and receiving functions. The Garmin GSX 70 radar used in a NASA Global Hawk is employing an output frequency range of 9.3 to 9.5 GHz (X-Band). The mission control center has data up- and down-links to the Global Hawk vehicle directly and via the Ku satellite and the UHF satellite systems. The launch and recovery station has up- and down-data communications links to the Global Hawk vehicle and to the UHF communications satellite.

The Predator air vehicle and its payloads/sensors are controlled from the GCS via a C-band for LOS missions and a K_u-band satellite data link for beyond-line-of-sight operations.

ANAFI Ai became the first UAV (in 2021) to use 4G technology as the main data link between the UAV and the operator. ANAFI Ai integrates a 4G module in its data link, which allows the transmission of a video signal (with a data rate of 12 Mbps) with a very low latency (300 ms) without range limit and everywhere in the world. Users will no longer experience transmission limitations, which enables precise control at any distance.

In order to make sure that the UAV has the worldwide availability of frequency, the selection of an operating frequency requires a compromise, while the governing law must be followed. A lower frequency offers a better and more reliable propagation, while having reduced data-rate ability. In contrast, a higher frequency is capable of carrying high data rates, but requires higher power/energy to propagate the signal. Thus, a higher frequency incurs more cost for transmission.

In selection of frequency of a data link, not only must the national and international laws be observed but also the UAS mission requirements, such as AJ, and security should be considered. Issues related to data link frequency are further discussed in Chapter 15.

Questions

1) What do RF, ADT, GDT, AJ, EW, Mbps, GHz, ARM, RF, MICNS, SOTAS, PLSS, HERO, and EMI stand for?
2) Provide two advantages of a wireless over a wired data link.
3) Provide one advantage of a data link using fiber-optic cable over a wireless data link.
4) What are the main differences between uplink and downlink in UAV communications?
5) Write data link typical major subsystems.
6) Briefly compare air data terminal with ground data terminal.
7) What are three important parameters of a data link?
8) What are basic functions of the UAV data link?
9) What is the definition of bandwidth in communications?
10) Provide at least five Electronic Warfare (EW) threats that a UAV system may face on a battlefield.
11) What are seven desirable attributes for a UAV data link related to mutual interference and EW?
12) What is the typical feature of an "urban" electromagnetic environment?
13) What is the main function of a modem?

14) Write the main subsystems of a data link.
15) Write the main elements of a transmitter.
16) Why are frequencies of communications strictly regulated by governments?
17) In selecting the frequency of communications, which frequency requires more power/energy/ cost to propagate the signal? Low frequency or high frequency?
18) What factors impact on the selection of frequency for a data link?
19) Briefly compare the effect of water vapor, rain, and fog in attenuating electromagnetic waves.
20) Define the range of a data link.
21) How can data link resistance to unintentional interference be enhanced?
22) Provide cases where unintentional interference might occur for a data link.
23) Why is a low probability of intercept highly desirable for a military uplink?
24) Provide three techniques for improving the survivability of the ground station.
25) Write four techniques to provide a reduction in the probability of intercept for a data link.
26) Discuss how security of a data link is provided.
27) Between deception and jamming, which one is worse? Why?
28) Provide examples of a catastrophic command transmitted via deception to an AV.
29) Which one is more difficult and requires more sophistication: (a) deception on the downlink or (b) deception of the uplink? Why?
30) Write two techniques to provide resistance of a data link to deception.
31) How can protection of the ground station against ARMs be provided? Provide five techniques.
32) If an ARM threat was judged to be severe, what technique may be employed to add to the protection of the ground station?
33) Explain the difference between jam-resistant and anti-jam.
34) Define the anti-jam margin of a data link.
35) A data-link AJ margin is desired to be 20 dB. How much must a jammer reduce the signal-to-noise ratio at the receiver in order to interfere with the successful operation of the data-link system?
36) Briefly discuss the data link overall priority of an AJ capability.
37) Provide examples of Predator UAV key functions that can be performed only in real time.
38) Most UAV missions are more tolerant of uplink jamming than downlink jamming. True or false?
39) In many cases, jamming of the downlink is less harmful to the mission than jamming the uplink. True or false?
40) What is modulation?
41) What are the primary techniques for modulation?
42) What is a modem?
43) Is UAV sensors data digital or analog in its final stage?
44) Provide one example of analog data.
45) Write four data link interface topics that are discussed in this chapter.
46) Why are at least two antennas (typically dorsal and ventral) required for a UAV?
47) How many antennas were used on Aquila MICNS?
48) What is the range of frequencies for the K_u band in a data link?
49) A UAV camera is operating at 20 frames per second. How much raw data (in bits) is produced in each second?
50) What are two main methods used to reduce the transmitted data rate?
51) Describe the data compression process.
52) Describe the data truncation process.

Questions 313

53) A UAV video camera is producing 40 frames per second. In a data compression process, every other frame of video is thrown away. What is the data rate that an operator is receiving in the GCS (in frames per second)?

54) In a data truncation process, the borders of each frame of video are thrown away. How much data rate is reduced (in frames per second)?

55) Provide two examples for UAV functions that require human-in-the-loop closed-loop control from the ground.

56) Discuss the general impact of control-loop delays for a UAV mission when a human operator is in the loop.

57) A UAS data rate is too low. Its data link can provide only one time slot per second for transmission of commands on the uplink and have a capability to downlink only one frame of video per second. A net recovery system using a human in the loop is not effective. Provide a solution for this recovery issue.

58) Write at least three sources of control-loop delays.

59) Write one major source of control-loop delays.

60) Define interoperability. Are different AJ techniques interoperable?

61) Define interchangeability. What are the requirements?

62) The MICNS was originally a common data link to meet the requirements of two systems. What are they?

63) Briefly describe the characteristics of a common data link.

64) Discuss the likelihood, requirements, and issues of using the Internet for the data link of a UAS.

65) Write advantages and disadvantages of a common data link.

66) What frequencies do GPS satellites transmit signals?

67) What frequencies are employed by wireless routers to transmit on the Internet to any *WiFi*-connected device like smartphones and laptops?

68) What frequency is employed by these cellphone companies: (a) Verizon 5G, (b) AT&T 3G, and (c) T-Mobile 5G?

69) What frequency and band are designated by law for use with RC planes?

70) What frequency bands are employed in the data link of MQ-1B Predator?

71) What frequency and band are used in the data link of ScanEagle?

72) What frequency and band are employed in the downlink for video transmission of the data link of ScanEagle?

73) Name types of antennas that are in common use with UAV systems.

74) Briefly describe features of an omnidirectional antenna.

75) Briefly describe features of a parabolic reflector.

76) Briefly describe features of a Yagi-Uda antenna.

77) Why is a length of 5/8 of a wavelength for a monoplane antenna recommended?

78) How many antennas are there in ANAFI Ai UAV? What is the type?

79) What are other two names for an array antenna?

80) Briefly describe features of a lens antenna.

81) What is the type of antenna for satellite communications in: (a) Global Hawk, (b) Predator, and (c) Reaper?

82) What is the diameter of a satellite data-link dish antenna at the GCS for the Predator and Reaper UAVs?

83) What is the typical rate of data produced by a high-resolution video from a TV or forward-looking infrared (FLIR) sensor (in bps)?

84) What frequency bands are employed by Global Hawk data link?

85) What frequency bands are employed by Predator data link?
86) Which UAV became the first to use 4G technology as the main data link between the UAV and the operator?
87) List antennas of a NASA Ikhana Predator B.
88) What frequency band is designated by federal law for use by aeromodellers?
89) What frequency bands are used by remote controlled cars and boats?

14

Data-Link Margin

14.1 Overview

In Chapter 13, data-link functions and attributes, system interface issues, antennas, and frequency of data link are presented. This chapter describes the basics of how to determine how much margin a data link has against all of the things that can attenuate or interfere with its signal. This margin and how it can be increased are among the central driving factors in the selection and design of a data link for a UAS.

The data link margin – measured in dB – is the difference between the minimum expected signal power received at the receiver and the received signal power at which the receiver will stop working. The latter is referred to as the receiver's sensitivity. The signal strength is decreased along the path – from transmitter to receiver – due to a number of factors including atmospheric variables such as humidity. The data link margin is a provision to compensate the signal power/strength lost in travel between a transmitter and a receiver.

Consider a data link with a transmitter that transmits a signal with a power of 40 dB. On the other end, the expected signal power received at the receiver should be at least 32 dB. The margin for such data link should be at least 8 dB (i.e., 40 – 32 = 8).

The margin is discussed in the context of an electronic-warfare (EW) environment because that is the situation that places the greatest requirements on the data-link margin. All of the same principles apply to more benign environments in which there is no deliberate attempt being made to jam the data link, but in which the link must still deal with natural signal losses and unintentional interference. It is a common practice to include two terms in the total data-link margin, one for the margin required in a non-jamming environment and one to the AJ margin. In the design of the link, the total margin is the goal, and the breakout is only for the purpose of providing visibility to the extra margin allocated to dealing with deliberate jamming.

The discussion of AJ capability here is primarily placed in the context of a point-to-point data link, although the sections related to processing gain are equally applicable to network connections.

A general discussion of AJ capability in a network environment is beyond the scope of this text. Since "the network" will not, in general, be part of the UAV system, the designer of the UAV may place requirements on the network, and must understand the tradeoffs for a particular network, but will not, in general, directly carry out the design tradeoffs for the network.

Introduction to UAV Systems, Fifth Edition. Paul Gerin Fahlstrom, Thomas James Gleason, and Mohammad H. Sadraey.
© 2022 John Wiley & Sons, Inc. Published 2022 by John Wiley & Sons, Inc.
Companion website: www.wiley.com/go/fahlstrom/uavsystems5e

14.2 Sources of Data-Link Margin

There are three possible sources for margin in a data link: (1) transmitter power, (2) antenna gain, and (3) processing gain. The features of these three sources are provided in this section.

14.2.1 Transmitter Power

Transmitter power is a straightforward way to add a signal margin and may be adequate for the most benign situations. However, in the presence of severe interference or deliberate jamming it is the least useful approach to defeating the jamming, at least for a UAV downlink. In this case, one would be trying to achieve AJ performance by simply beating the jammer in a power contest. A downlink transmitter on a UAV is almost certain to lose such a contest. Even for the uplink with a ground-based transmitter, it is not very profitable to attempt to produce clean, modulated power at levels that are competitive with a jammer.

14.2.2 Antenna Gain

One way to achieve the benefits of a higher-power transmitter without actually radiating any more power is to concentrate the radiation in the direction of the receiver. An everyday example of this process is provided by a flashlight. If a flashlight bulb were simply connected to a battery with no reflector or lens, it would radiate equally in all directions (an "isotropic" radiation pattern). Such an arrangement would not provide any useful amount of illumination at ranges of more than a few feet. However, if a reflector and/or lens are added to the system, much of the light can be concentrated in a narrow beam. This beam may produce a brightly lit spot at distances of tens or hundreds of yards. In the radio-frequency (RF) world, this type of concentration is called "antenna gain," since it is accomplished by using an antenna that concentrates the radiation in a preferred direction.

14.2.2.1 Definition of Antenna Gain

Antenna gain is the ability of the antenna to radiate more or less in any direction compared to a theoretical antenna. If an antenna could be made as a perfect sphere, it would radiate equally in all directions. When no direction is specified, the antenna gain refers to the peak value of the gain, the gain in the direction of the antenna's main lobe.

An ideal isotropic antenna radiates energy uniformly in all directions, so that received power is equal anywhere on the surface of an imaginary sphere surrounding the isotropic point source. Radiating uniformly in all directions is not always desirable and sometimes it is necessary to concentrate the radiated energy in a particular direction. If the energy from our isotropic source could be directed to just half of a sphere without any power going into the opposite half, the "half-isotropic" source would generate twice the radiation on the surface of the hemisphere. This concept is called the directive gain of the antenna.

The ratio of the power density radiated in a particular direction, at a given distance, to the power density that would be radiated at the same distance by an isotropic radiating source at the same total power is the Directive Gain, or just "gain" of an antenna.

14.2 Sources of Data-Link Margin 317

Gain applies both in transmission and reception. In a transmitter antenna, the gain represents effectiveness of the antenna to convert input power into radio waves. In a receiver antenna, the gain represents how well the antenna converts arriving radio waves into electric power. Thus, if an antenna with gain is used to receive a signal, the effective signal strength at the input of the receiver is increased by the gain of the antenna.

In UAV applications, it is often desirable to provide significant gain at the ground antenna (GCS) in order to allow the use of lower-power transmitters on the AV. This also increases the resistance of the system to jamming.

It also may be desirable to provide gain in the antennas on the AV, but high gain generally requires a larger antenna and a directive antenna must be pointed toward the point with which communications are being maintained. Both the size and the pointing requirements for high-gain antennas are easier to meet on the ground than on the AV, although this distinction can disappear for large AVs.

Antenna gain is defined as the ratio of the intensity of the radiation emitted in the preferred direction to the intensity that would be present in that direction if the transmitter had an isotropic pattern (i.e., radiated equally in all directions). Since it is a dimensionless ratio, it is conveniently expressed in dB. If the antenna has a radiation pattern that varies with angle and has a peak in some direction, which commonly is the case, then the peak antenna gain is the gain measured in the direction of the peak. An approximate estimate of peak antenna gain is provided by the following equation, where θ and ϕ are the full beam widths at the half-power point in the vertical and horizontal directions (in radians), respectively:

$$G_{dB} = 10 \log\left(\frac{4\pi}{\theta\phi}\right) \tag{14.1}$$

Using the fact that the antenna beam widths are proportional to the dimensions of the antenna, height (h) and width (w) and the wavelength (λ) of the transmitted signal, this expression (still as an approximation) can be rewritten as

$$G_{dB} = 10 \log\left(\frac{hw}{\lambda^2}\right) \tag{14.2}$$

Example 14.1

An antenna has a beam width of 35 degrees wide × 8 degrees high. Determine the peak antenna gain in dB.

Solution:

$$G_{dB} = 10 \times \log\left(\frac{4\pi}{\left(\frac{35}{57.3}\right)\left(\frac{8}{57.3}\right)}\right) = 21.7 \text{ dB} \qquad \text{from (14.1)}$$

The division by 57.3 is used to convert degrees to radian.

Example 14.2

A GCS is operating at 15 GHz. Its transmitter antenna has a height and a width of 30 cm by 20 cm respectively. Determine the peak antenna gain in dB.

Solution:
First, we need to calculate the wavelength. The speed of the radio wave is 300,000 km/s.

$$f = \frac{c}{\lambda} \Rightarrow \lambda = \frac{c}{f} = \frac{300{,}000{,}000}{15 \times 10^6} = 0.02 \text{ m} = 2 \text{ cm} \quad \text{from (11.2)}$$

so

$$G_{dB} = 10 \log\left(\frac{hw}{\lambda^2}\right) = 10 \times \log\left(\frac{0.3 \times 0.2}{0.02^2}\right) = 21.7 \text{ dB} \quad \text{from (14.2)}$$

Thus, the peak antenna gain is 21.7 dB.

Figure 14.1 shows how we define the beam width (BW) in a single dimension (height or width) or single plane (top or side). Various definitions are possible, but we use the common definition of the full angular width of the beam measured at the "half-power" or "3-dB" points (i.e., $10 \times \log(0.5) = -3$).

Most antennas show a pattern of "lobes" or maxima of radiation. In a directive antenna, the largest lobe, in the desired direction of propagation (i.e., 0°), is called the "main lobe." The figure shows the "main lobe" of the beam as an oval plotted on a polar grid with power per unit area as the radial coordinate and the angle off axis as the angular coordinate.

14.2.2.2 Applications of Antenna Gain for Data Links

Antenna gain can provide a data-link margin in two different ways. At the transmitter, antenna gain concentrates the signal power in a beam that is directed at the receiver. This produces an effective radiated power (ERP) that is equal to the actual transmitter power times the antenna gain. An antenna small enough to be carried on a UAV can easily have a gain of 10 dB, leading to an effective

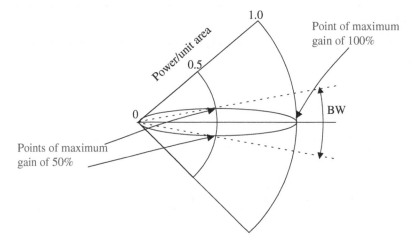

Figure 14.1 Definition of beam width – top or side view

14.2 Sources of Data-Link Margin

multiplication of the transmitter power by a factor of 10. If the transmitter is on the ground, an antenna gain of 30 dB or more is practical in the shorter wavelength (e.g., microwave bands).

In order to benefit from this gain, the transmitter antenna must be aimed at the receiver, which requires that the antenna be steerable (e.g., Global Hawk satellite communication antenna) and the system keep track of the direction to the receiver antenna and provide pointing commands. The gain of the transmitter and receiver antennas directly increases the power delivered to the receiver just as if the transmitter power were increased by the same factor. When dealing with atmospheric and range losses, this increase can easily multiply the data-link effective range significantly. For unintentional interference, the benefits also can be significant as the transmitter that is causing the interference, even if it has an antenna with gain, will not, in general, be pointed at the UAS receiver with which it is interfering.

Unfortunately, when there is deliberate jamming, the jammer can also use an antenna with gain and attempt to point it at the UAS receiver that it is trying to jam. It may not be able to use a very narrow beam, since it may not know exactly where the receiver is located, forcing it to use a wide enough beam to cover all possible locations. However, a beam 50 degrees wide × 10 degrees high has a gain of about 19 dB, so a jammer might easily have as much antenna gain at the transmitter as a UAV downlink.

Any gain available from the UAV transmitter antenna is of value and makes its contribution to the total margin of the link. However, even for data links operating at 15 GHz, most UAVs are not large enough to carry an antenna with enough gain to provide a significant advantage over a jammer through transmitter antenna gain alone.

Gain at the receiver antenna contributes to the AJ margin by discriminating between the signal and jammer energy based on the directions from which the energy arrives at the antenna. Figure 14.2 illustrates this mechanism. If the receiver antenna is pointed at the AV transmitter that it wishes to receive, then the data-link signal experiences the full gain of the main beam of the antenna (G_S in the figure). If an interfering or jamming signal arrives from a direction that is outside of the main beam, it experiences only the gain in a side lobe of the antenna (G_J). The effect of this is to enhance the signal over the jammer by a factor of G_S/G_J, which depends on the exact angles of arrival of the jammer energy and the structure of the side lobes of the antenna.

It is important to note that this enhancement depends on the source of interference being outside the main beam of the receiver antenna. If it is within the main beam there will still be some difference in gain, unless the jammer is directly in line with the data link transmitter, but the difference in gain will be much smaller than if the jammer were in a side lobe, and becomes negligible as the angle between the two sources goes to zero.

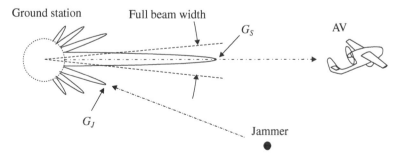

Figure 14.2 Geometry for antenna gain

For data links, it is common to claim a contribution to the data link margin from receiver antenna gain that ignores all of the structure in the side lobes of the antenna and assumes that the jammer is outside of the main lobe (and sometimes the first or second side lobes), defined by a specified beam width, as shown in Figure 14.2. The contribution to the margin is then stated as the ratio of the peak gain in the main lobe to the gain in the highest side lobe that is outside of the specified beam width.

The discrimination between the signal and interference or jammer can be enhanced by active elements at the antenna that suppress signals in the side lobes and/or provide steerable antenna nulls (to null out the jammer) that can be placed in the direction of the interference. These techniques can lower the gain in the side lobes and narrow the effective beam width for a particular level of discrimination.

Depending on the carrier frequency of the data link and how large and expensive an antenna is acceptable, the effective discrimination between the signal and jammer at a ground antenna can be as high as 45–50 dB. At the upper end of this range, there may be problems with leakage of the jammer energy into the main lobe due to multipath propagation and small antenna imperfections. These effects place a limit on the practical level of discrimination.

Airborne antennas on UAVs usually cannot be big enough to have much gain. Even at 15 GHz, the carrier wavelength is 2 cm. An 8-cm diameter antenna would be fairly large for a steerable antenna on typical small UAVs. It would have a diameter of only 4 wavelengths, resulting in a theoretical peak gain of about 21 dB. Leakage of off-axis signals into the main beam due to reflection off surfaces of the UAV can also be a serious problem for the airborne system. The airborne system can use steerable nulls to improve its discrimination, at least for a high carrier frequency, at the expense of a significant increase in cost.

At lower carrier frequencies, it is much harder to get high antenna gains. At 5 GHz, the carrier wavelength is 6 cm and a 10-cm antenna would be only 1.7 wavelengths in diameter, suggesting antenna gains of 14 dB or less. Even a ground antenna would have to be three times as large at 5 GHz as at 15 GHz to have the same gain.

To achieve a significant AJ margin through angular discrimination at the receiving antenna requires that the data link operate in a line-of-sight mode with the receiving antenna pointed at the transmitter. If one of the terminals is on the ground, this puts a limit on the range of the data link that depends on the altitude of the airborne terminal.

In reality, the Earth is curved and prevents LOS propagation to a very long distance. A circle with the radius of R_E is representing the Earth (Figure 14.3). Using Pythagorean equation for the triangle shown, one can write:

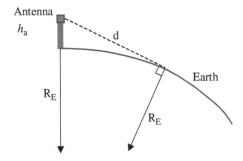

Figure 14.3 Antenna height and Earth radius

14.2 Sources of Data-Link Margin

$$d^2 + R_E^2 = (R_E + h_a)^2 \tag{14.3}$$

where h_a is the antenna height and d is the distance between the antenna and the horizon. Since the radius of Earth (about 6,378 km) is much longer than the antenna height, this can be simplified as

$$d = \sqrt{2 R_E h_a} \tag{14.4}$$

This equation represents the relation between height of a terrestrial transmitter (e.g., radar) and maximum distance (i.e., line-of-sight range) to a ground receiver (LOS). However, the atmosphere is refracting the radio waves. The common correction [38] for refraction is to consider the Earth to have a radius that is 4/3 of its actual radius. This leads to the following modified equation:

$$d = \sqrt{2 \times 4/3 \times R_E h_a} \tag{14.5}$$

Figure 14.4 shows the minimum altitude at which the AV is above the radar horizon as a function of horizontal range. This curve is for a smooth Earth. Masking by elevated terrain will usually result in higher minimum altitudes at the same range for an UAV operating over land. As an example, with an antenna height of 500 m, the line-of-sight range is about 80 km.

When using optical sensors, including FLIRs, the inherent range limitations of the sensors and the requirement to stay below any cloud cover will force the UAV to operate at altitudes of 1,000 m or less much of the time. The sensor range could be increased to allow operation at higher altitudes, but weather conditions are not under the control of the system designer. For instance, median daytime cloud ceilings in Europe in the winter (October through April) are less than 1,200 m [39].

Therefore, the maximum range for a direct ground-to-air, line-of-sight link will be about 124 km when using optical sensors in a European climate. Unless the ground terminal is favorably located on ground as high as any between the terminal and the AV, it is likely that the practical limit on the range of a direct link will be significantly less than 100 km due to terrain masking of the line of sight at long ranges.

A relay can be used to extend the range of the data link beyond those for which direct, line-of-sight propagation is possible from the ground to the UAV. When using a relay, there are really two different data links that might be jammed: the link between the ground and the relay and the link

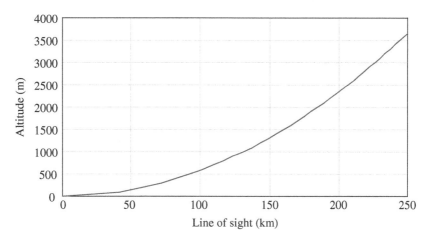

Figure 14.4 Line-of-sight range versus altitude

between the relay and the UAV. The relay can be located at high altitude where a direct line of sight to the ground is available, so the link between the ground and the relay can benefit fully from a high-gain antenna on the ground. However, the antenna gain available between the relay and the UAV will be limited by the antenna size that can be carried on the relay aircraft. If the relay is carried on a small UAV, the direct antenna gain will be limited to 15–20 dB. Some additional effective gain could be provided through steerable nulls.

Lower gain at the receiving antenna not only reduces the AJ margin when the jammer is not in the main beam, but it also increases the probability that the jammer will be within the main beam, which eliminates the contribution of receiver antenna gain to the AJ margin. This effect is discussed later under the topic of jamming geometry.

When a satellite link is used, there usually will be no issue in the links between the satellite and the ground station or the links between satellites, both of which will normally use relatively high-gain antennas. The remaining issue is between the last satellite in the chain and the AV. Larger AVs can carry antennas with at least moderate gain and directional pointing (as in Global Hawk, Predator, and Reaper), particularly as they would generally need to cover only the upper hemisphere, assuming that an occasional short dropout due to AV maneuvering is tolerable.

The sources of interference and jamming will tend to be on the ground, so they will have little or no access to an upward pointing directional antenna on the AV. Under these circumstances, the advantage of antenna gain is to provide an increase in signal strength, particularly in the "downlink" from the AV to the satellite, which will, in this case, be pointed upward geometrically. The additional signal strength supports a higher signal bandwidth. In conclusion, the antenna gain is a source of data link margin; as the antenna gain is increased, less data link margin is required.

14.2.3 Processing Gain

In the context of an AJ margin, processing gain refers to enhancement of the signal relative to the jammer that results from forcing the jammer to spread its power out over a bandwidth that is greater than the information bandwidth of the signal communicated by the data link. There can be a corresponding advantage against unintentional interference if the interfering source is operating at a single frequency, which is typical of the ordinary kinds of radiating RF sources that may interfere with a data link.

Processing gain is accomplished by encoding the data-link information in some way that increases its bandwidth before transmission and then decoding it at the receiver to recover the original bandwidth. The jammer, unable to duplicate the coding, must jam the bandwidth of the artificially broadened transmitted signal, which prevents it from concentrating its power within the true bandwidth of the original data-link information. A non-jammer interfering source will only interfere with the portion of the signal that overlaps with its narrow operating wavelength band. Because this type of processing gain is particularly important when dealing with jammers and its design usually is aimed at defeating jamming, we will discuss it in terms of AJ margin and effectiveness.

14.2.3.1 Direct Spread-Spectrum Transmission

Figure 14.5 illustrates one form of processing gain: direct spread-spectrum transmission. In this case, a pseudo-noise modulation is added to the signal to create a transmitted signal that has a larger bandwidth and lower power per unit frequency interval than the original signal. A jammer must jam over the entire bandwidth of the spread transmission. If the jammer is more powerful than the data-link transmitter it can create a signal-to-jammer ratio (S/J) less than one over the spread bandwidth. However, the data-link receiver knows the form of the pseudo-noise modulation added at the

14.2 Sources of Data-Link Margin

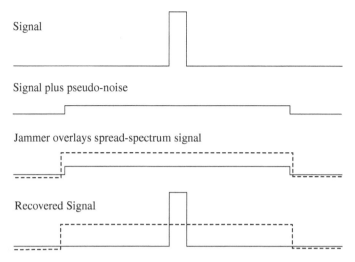

Figure 14.5 Direct spread-spectrum processing gain

transmitter and can subtract it from the received signal, which re-creates the original signal within its original bandwidth.

The receiver can then reject all jammer energy that is outside of the original signal bandwidth. Since the jammer energy was spread out over the transmitted bandwidth, the jammer power within the original signal bandwidth is reduced by a factor equal to the ratio of the original bandwidth to the transmitted bandwidth, compared to an unspread transmission and jammer, and the S/J in the receiver after recovering the original signal may be greater than one, which is the desired effect.

Processing gain is defined as

$$PG_{dB} = 10 \log \left(\frac{B_{Tr}}{B_{Info}} \right) \tag{14.6}$$

where PG is the processing gain, B_{Tr} is the bandwidth of the transmitted signal, and B_{Info} is the bandwidth of the information contained in the transmitted signal.

Direct spread-spectrum transmission, in which the instantaneous RF signal is spread over a wide bandwidth, as illustrated in Figure 14.5, is one way to provide processing gain. It has the advantage of making the transmitted signal look much like noise and making it difficult to intercept or on which to perform direction finding. It has the disadvantage that the modulation rates required to produce a signal that is broad compared to downlink information bandwidths are very large and the entire RF system must be able to accommodate the resulting bandwidth.

For instance, if the downlink signal has a 1-Mbps information bandwidth and 20 dB of processing gain is required, then the pseudo-noise modulator must operate at 100 Mbps (since $10 \times \log (1/100) = -20$ dB) and the instantaneous bandwidth of the RF system must also be 100 Mbps. Fast modulators (and demodulators at the receiver) and wide instantaneous bandwidths are likely to increase the cost of the data link.

14.2.3.2 Frequency Hopping

Another way to produce processing gain is frequency hopping. In this case, the signal transmitted at any instant is a normal, unspread signal. However, the carrier frequency for the transmission changes over time in a series of pseudo-random hops. This is illustrated in Figure 14.6. If a jammer

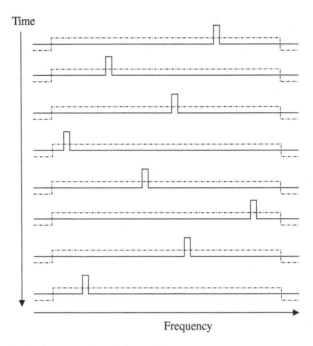

Figure 14.6 Schematic of a frequency-hopping waveform

does not know the pattern of the hops and cannot follow the pattern in real time, then it must jam the entire band over which the transmitter hops.

The receiver, of course, knows the pattern and tunes itself so that it always is receiving the signal with a matched bandwidth set at the correct carrier frequency. The result is the same as for a direct spread-spectrum signal: the jammer power must be spread over a wide bandwidth and the receiver can reject all jammer energy that is not within the bandwidth of the transmitted signal. Processing gain is again equal to the ratio of the transmitted bandwidth to the information bandwidth, where the transmitted bandwidth is the bandwidth over which the system hops.

There are two classes of frequency hoppers. Slow hoppers change frequencies at rates that are relatively slow compared to the rates at which electronic systems can process data and switch operating modes. A slow hopper might switch frequencies at rates of 1–100 hops per second. A jammer capable of detecting the new frequency and tuning to that frequency within a few milliseconds could jam most of the information carried by a slow-hopping data link.

Fast frequency hoppers hop at rates that are more comparable to the maximum rates at which information can be acquired, processed, and reacted to by an interception and jamming system. For instance, a fast hopper might hop at a rate of 10 kHz (i.e., 10,000 hops per second) or more, so that a frequency-following jammer would have only a few microseconds to detect the hop, determine the new frequency, and tune the jammer if it wanted to jam most of the dwell time of the data link on each frequency.

The signal from a frequency hopper at any given instant is concentrated in a normal bandwidth. However, an intercept receiver may still have to use a wide-band front end since it does not know where to tune to find the signal at any given time. It may not be able to make effective use of occasional brief bursts of signal if it sits on a single frequency and waits for the hopper to hit that

14.2 Sources of Data-Link Margin 325

frequency. Clearly, the faster the hop the more difficult it is to intercept and to direction find against a frequency hopper.

Slow frequency hoppers are easy in principle to intercept, locate through direction finding, and jam. Dwell times of a significant fraction of a second allow plenty of time to find and follow to the new frequency. In practice, the problem may be complicated if there are many emitters operating in the same band as the frequency hopper, particularly if some of the other emitters are also frequency hoppers.

The enemy EW system must then "fingerprint" the data-link signal and sort it out from the other emitters. For a UAV downlink, however, the "fingerprint" could simply be that the transmitter is located in the air many kilometers to the enemy's side of the front lines. A judgment of the vulnerability of a slow frequency hopper will revolve around details of scenario and assumptions about enemy intents, since the technology to jam such a link is clearly feasible.

14.2.3.3 Comparing Two Techniques

Frequency-hopping data links may be less expensive, for equivalent processing gain, than direct spread-spectrum links because the switching rate to frequency hop is much lower than the modulation rate for direct spreading and the instantaneous RF bandwidth required is much lower. It may also be possible to achieve larger processing gains with frequency hoppers since an RF system is likely to be able to tune (hop) over a wider frequency band than it can reasonably cover with an instantaneous spread-spectrum emission. Active antenna enhancement, such as side-lobe canceling is also likely to be easier, at least for relatively slow hop rates.

The disadvantages of frequency hopping relative to direct spread spectrum seem mainly to be that it is more susceptible to interception, location, and jamming (with frequency followers). For fast enough hop rates, the difference in susceptibility becomes smaller, but some of the cost and complexity advantages also become smaller. Different, but equally competent organizations have chosen each approach for UAV data links based on detailed tradeoffs, illustrating the fact that either may be the optimum choice for a particular application.

Frequency allocation and assignment are issues sometimes raised with regard to spread-spectrum data links. Direct-spread links may require tens of MHz of instantaneous bandwidth, while frequency hoppers may be designed to hop over entire frequency bands, such as the UHF band, or to use as much as a GHz of one of the higher-frequency bands. However, the very nature of spread-spectrum transmission minimizes the probability of interference from or to other systems operating in the same band.

A direct-spread system appears as a slight increase in background noise to a nonspread-spectrum receiver that happens to be operating within its transmitting band, while the spread-spectrum receiver averages any nonspread signals that it sees over its spread bandwidth when it subtracts the pseudo-noise modulation, converting those signals to noise at a low level.

Frequency hoppers have an inherent resistance to unintentional interference since with each hop they move away from any frequency on which such interference is present, making the interference intermittent worse. Moreover, they are usually designed to allow the hopping pattern to be programmed to avoid any frequencies where such interference is expected, preventing the data link from either experiencing or causing any interference. For training purposes, a frequency hopper can be programmed either not to hop at all or to hop over a very restricted frequency band without having any essential effect on the operation of the link.

Frequency allocation and assignment have been available both in the United States and Europe for specific direct-spread and frequency-hopping systems using bandwidths or hopping ranges typical of an AJ UAV data link. There do not appear to be any fundamental barriers to either type of link in peace or war, and nor is it clear that either offers any advantage in this respect. The nature of the frequency management process is such that each specific link design must be considered as a separate case. The only general rule that seems to apply is that the frequency band to be used must be available for basic UAV data-link applications before one can even consider whether it can be used in a spread-spectrum mode.

14.2.3.4 Scrambling and Redundant Transmission

A special case of direct spread-spectrum transmission that might be used in combination with frequency hopping is the use of scrambled, redundant transmissions with error-detection codes. This is illustrated in Figure 14.7 for the case of a frame of video information. After digitizing the video data, the data link scrambles the frame so that contiguous portions of the picture (e.g., successive lines of the video) would be transmitted at widely separated times within the frame of transmitted data. In addition, the data link adds extra bits to each small block of data within the frame that allow the receiver to check for errors in the received signal.

Finally, the data link transmits each block of data twice, at different times within the frame. This approach has two effects. First, the bandwidth of the transmitted signal is increased by the addition of the error-detection coding and the redundant transmission, which can be decoded at the receiver to produce processing gain. Second, the effect of intermittent jamming, by a swept jammer for instance, is reduced because: (1) a brief interval of jamming would not affect contiguous portions of the picture, but rather it would spread isolated blemishes over the entire picture, and (2) redundant transmission may allow recovery of all of the information jammed during one interval from information transmitted during other intervals.

Scrambling and redundancy might typically increase the transmitted bandwidth by a factor of 2 or 3 at most, producing a processing gain of 3–5 dB. This gain adds to any gain due to pseudo-noise modulation or frequency hopping. These techniques are particularly useful with frequency hoppers that hop several times within one frame of data. Scrambling and redundant transmission can reduce or prevent loss of data if the hopper happens to fall on top of an interfering emitter for one or more hops during the frame.

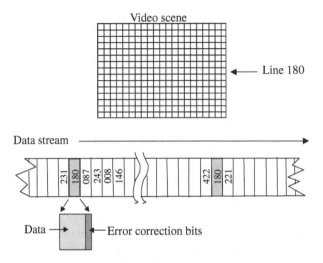

Figure 14.7 Scrambling, redundancy, and addition of error-detection

14.3 Anti-Jam Margin

It is recommended to include one term in the total data-link margin for a deliberate jamming environment. The Global Hawk is usually encrypting its communicated data to protect the UAV against cyber-attacks. By using the satellite communication and the encrypted data link, Global Hawk UAVs can safely fly anywhere in the world.

In this section, after defining the anti-jam margin, the jamming geometry for up- and downlinks for multiple jammers is presented. Then, data link implications of AJ capability and anti-jam uplinks are discussed.

14.3.1 Definition of Anti-Jam Margin

The common mathematical definition of the AJ margin is

$$AJM_{dB} = PG_{dB} + FadeM_{dB} \tag{14.7}$$

PG_{dB} is the processing gain as defined in Equation 14.3 and $FadeM_{dB}$ is the "Fade Margin," which is defined as the ratio of the available signal-to-noise ratio (S/N) of the system under normal conditions to the required S/N. Any well-designed data link will have some fade margin and a jammer must overcome that margin before it can degrade performance of the system. However, before the jammer can contribute effective noise to the system its effective power is reduced by whatever processing gain is present in the system.

Note that the definition of the AJ margin does not explicitly include the gain of the receiving antenna. The gain of this antenna, if present, contributes to the fade margin by increasing the signal relative to the noise. All other things being held constant, increasing the antenna gain by some number of dB will increase the fade margin, and thus the AJ margin by the same number of dB. Therefore, the definition of the AJ margin implicitly assumes that the jammer is outside the main beam and does not benefit from the gain of the receiver antenna.

Similarly, increases in transmitter power or gain in the transmitter antenna enter the AJ margin via the fade margin. The effect of active side-lobe canceling or steerable nulls at the receiver antenna is not directly accounted for in the simple definition above. They could be added to the processing gain, although they are qualitatively different from the spread-spectrum processing gain, or they could be represented by an additional term in the equation. The basic fade margin of a data link is discussed further in Section 14.5, Data-Link Signal-to-Noise Budgets.

It should be recognized that any simple, single number for the AJ margin is only an approximation: useful for gross comparisons and general discussions, but not likely to be precisely correct in all scenarios. If a precise estimate of the AJ performance is required, the actual signal and jammer power budgets should be worked out for a particular situation and the final S/N determined for that case.

It is also important to emphasize that any data link will have some AJ margin, since any well-designed link will have some fade margin. However, it is not the first 10 or 20 dB that is hard to come by. The hard goal to achieve is the last 10 or 12 dB that is required for a severe jamming environment. As mentioned earlier, it is also important to remember that the difference between, say, 50 dB and 40 dB of the AJ margin is a 10-times decrease in jammer power [50 – 40 = 10 log(100,000) – 10 log (10,000) = 10. Or, 100,000/10,000 = 10], not a 20% decrease as it would be on a logarithmic scale. In other words, if the AJ margin of a data link is increased from 40 dB to 50 dB, the power of a jammer must be increased 10 times to have the same effect.

14.3.2 Jammer Geometry

Whether or not a data link is jammed at a particular instant depends on the characteristics of the data link and jammer and on geometrical relationships between the data link and jammer beams at that instant. Typically, the data link will be jammed for certain AV locations, relative to the jammer positions, and not jammed for other locations. As with the AJ margin, a precise determination of where the link will be jammed required a system- and scenario-specific calculation of power budgets for the link and the jammer. However, it is possible to make some general comments about jammer (or jamming) geometry that apply qualitatively to typical mini-UAV data links.

If the data link uses omnidirectional (or very low gain) receiver antennas, the geometry for which jamming occurs depends only on the relative ranges from the jammer to the receiver and the data link transmitter to the receiver. There will be some value of S/J for which the link can just continue to function adequately. If S/J decreases below this value, the link will be jammed. If, for simplicity, atmospheric losses and terrain effects on propagation are neglected, then the data link signal and jammer strengths at the receiver are proportional to the inverse of the squares of the ranges between the data link transmitter and jammer to the data link receiver.

Figure 14.8 illustrates this case for a UAV uplink (i.e., at the UAV). The ranges of interest are the range from the link antenna on the ground to the UAV (R_L) and the range from the jammer to the UAV (R_J). Under the simplifying assumptions, S/J is inversely proportional to R_L^2 / R_J^2.

Let the value of the ratio R_L^2 / R_J^2 that produces a barely acceptable value of S/J be represented by the value "k." It turns out that the locus of points for which this ratio is constant is a circle whose center is displaced a distance $D = R_{LJ}/(k-1)$ behind the jammer along the line through the jammer and the link ground station, where R_{LJ} is the distance from the link ground station to the jammer. The radius of the circle is

$$\text{Radius} = \frac{R_{LJ}\sqrt{k}}{(k-1)} \tag{14.8}$$

The area within the circle is the region in which the jammer will defeat the uplink. If the value of k for a particular link and jammer is known, the geometry shown in Figure 14.8 could easily be calculated and plotted on a map for any scenario. Unfortunately, there is no simple way to calculate k without performing a complete signal-strength analysis for both the link and the jammer. The value of k is related to the AJ margin, and increases as that margin increases.

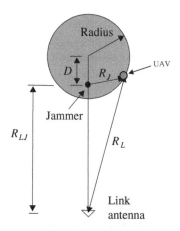

Figure 14.8 Uplink jamming for omnidirectional receive antenna

14.3 Anti-Jam Margin

The radius of the circle is proportional to $k^{-1/2}$, so the circle gets smaller as the AJ margin and k increase. A more exact solution would have to take into account the losses other than $1/R^2$ and would distort the circle into an oval region around the jammer in which the link would be jammed. However, the simple example in Figure 14.8 illustrates the qualitative characteristics of uplink jamming geometry for omnidirectional antennas.

Figure 14.9 illustrates a similarly simplified analysis of the geometry for jamming a downlink (i.e., at the GCS) that uses omnidirectional receiver antennas. In this case, the jammer range to the receiver (R_J) is fixed and is equal to the range from the jammer to the ground station (R_{LJ}). For a fixed S/J, requiring a fixed ratio of jammer range to link range, the link range (R_L) must also be fixed. Therefore, the link will be able to operate anywhere inside a circle centered on the ground station and will be jammed anywhere outside of this circle. If the downlink transmitter antenna has a high gain, the gain will contribute to making the circle larger, but will not change the shape of the region in which the link is jammed.

If the data link uses high-gain antennas on the ground to reject jammers that are not in line with the UAV, then the geometry for downlink jamming changes. Figure 14.10 illustrates this case

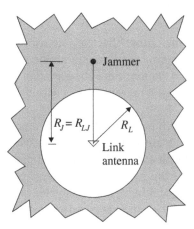

Figure 14.9 Downlink jammer geometry

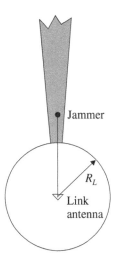

Figure 14.10 Geometry for a downlink with a high-gain antenna

qualitatively. If the jammer is in the main beam of the receiver antenna, the jamming situation is similar to that for omnidirectional antennas, where the link can operate as long as the UAV is within a circle centered on the ground station whose radius depends on the link and jammer parameters. However, if the UAV is far enough out of line with the jammer for the jammer to be outside the main beam of the antenna, then the link has an additional AJ margin equal to the gain of the main beam.

Figure 14.10 assumes that this additional gain is sufficient to allow the link to operate successfully any time that the jammer is not in the main beam. This results in a wedge-shaped region in which the downlink is jammed. The wedge is centered on the axis from the ground station to the jammer and has an angular width equal to the width of the main beam of the ground-station antenna. The point of the wedge is truncated at the perimeter of the circle around the ground station within which the link can operate without the assistance of the additional AJ discrimination provided by the antenna gain.

The geometry in Figure 14.10 is simplified by the assumption that the gain of the ground antenna is described by a step function with a sharp boundary at some angle off the axis. A real antenna has a more complicated gain pattern, which would be reflected in the shape of the "wedge." Also, the simplified analysis assumes a clear RF line of sight from the jammer to the ground antenna.

It also assumes that the antenna cannot effectively discriminate between the UAV signal and the jammer on the basis of elevation. The last assumption is correct for small UAV scenarios unless the AV is near the ground station, since the elevation angle to the AV is quite low and signals from the ground under the AV are very narrow within the main beam or can easily enter it via a multipath.

If there are several jammers, each jammer will establish a protected area in one of the shapes shown in Figures 14.8 to 14.10. Figure 14.11 summarizes a typical jamming scenario for a UAV having a high-gain ground antenna for the downlink and essentially omnidirectional receiver antennas for the uplink. Each of the two jammers targeted on the downlink creates a wedge-shaped area of effective jamming, while one jammer targeted on the uplink creates an oval area of effective uplink jamming. If enough jammers are used against the downlink, the total area of jamming created by a number of wedges can become great enough to have a significant impact on mission effectiveness. The wedges go out to the maximum range of the link.

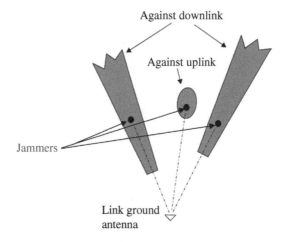

Figure 14.11 Jamming geometry for up- and downlinks for multiple jammers

14.3 Anti-Jam Margin

It is clear that very narrow antenna beams are an important asset to a UAV data link that depends on antenna gain for an AJ margin. While narrow beams and high gain go together, there may be a point at which side-lobe canceling or steerable nulls are more valuable than higher gain, since the angular discrimination of the antenna, rather than gain alone, is the key to minimizing the area over which the link will be jammed.

A particular case worth discussing is a low-gain antenna such as a Yagi used for non-line-of-sight reception. Such an antenna might have 10 or 15 dB of gain with a beam width of 50 degrees. Such an antenna adds to the fade margin of the link and thus adds to the AJ margin using the common definition. However, the beam is so wide that the jammer would almost always be in the beam and would be enhanced by the same gain as the signal. Therefore, the low-gain antenna makes no real contribution to the ability of the data link to operate in the presence of jamming. This is an illustration of the need to do a system-specific assessment of AJ margin before drawing any conclusions about the robustness of the link.

14.3.3   System Implications of AJ Capability

A requirement for AJ capability in a UAV data link has a number of system implications. These can be summarized in three interrelated areas: (1) operating frequency, (2) range, and (3) data rate.

The AJ capability tends to favor higher operating frequencies. Shorter wavelengths (due to higher frequencies) allow more antenna gain for the same antenna dimensions and antenna gain is, in many ways, the easiest way to improve the AJ margin. However, higher frequencies tend to require line-of-sight propagation from the transmitter to the receiver, which limits the range of the data link unless relays are used. If relays are used, it may be difficult to use high-gain antennas on the air-to-air stage of the link.

A higher basic frequency allows greater processing gain for the same fractional bandwidth. For instance, for a 1-Mbps data rate and a 400 MHz operating frequency, a processing gain of 20 dB requires a spread-spectrum frequency bandwidth of 100 MHz, or a fractional frequency bandwidth of 25%. When operating frequency is around 15 GHz, the same processing gain and spread-spectrum frequency bandwidth requires only a 1.33% fractional bandwidth, which is likely to be easier to achieve and to make active antenna processing, such as side-lobe cancellers, easier to build.

Short-range data links can fully exploit both antenna gain and processing gain advantages at higher frequencies. Long-range data links may only be able to benefit from the processing gain. Therefore, the case for higher frequencies for AJ capability is stronger for a short-range link than for a long-range link.

The disadvantages of higher frequencies are that the components are more expensive and the overall technology is less mature. The restriction to line-of-sight propagation is also a disadvantage for a long-range data link.

The interaction between AJ capability and data rate is very strong for any data link that cannot depend on line-of-sight propagation and high-gain antennas for most of its AJ margin. A high-gain antenna can easily provide 30 dB of AJ margin for a downlink. To get the same 30 dB margin from processing gain would require a reduction in the transmitted data rate from, for example, 10 MHz down to 100 kHz, or an equivalent increase in a transmitted, spread-spectrum frequency bandwidth. Higher frequencies are needed to carry more bits of data (i.e., faster speed). For short-range, line-of-sight data links, it is possible to have a fairly high data rate at moderate frequency bands with fair jam resistance by combining large, high-gain antennas with moderate amounts of processing gain.

For a long-range data link, there are three choices:

1) Low-frequency, non-line-of-sight operation at low data rates. Non-line-of-sight operation forces relatively low frequencies, which, in turn, limit the transmitted bandwidth that is available for processing gain. This limits the transmitted data rate that can be handled while still providing an AJ margin.
2) High-frequency, line-of-sight operation with relays having low-gain antennas and moderate data rates. At the cost of limiting the available antenna size and gain, this option allows use of UAVs for the relay vehicles. Moderate data rates can be supported since the high-frequency operating band allows large instantaneous, spread-spectrum bandwidths.
3) High-frequency, line-of-sight operation with high-gain relay antennas and high data rates. By using large relay vehicles with large, high-gain, tracking antennas, this option provides the AJ advantages of both antenna gain and wide bandwidth, allowing relatively high data rates at a high AJ margin.

Two simple Examples, 14.3 and 14.4, illustrate the gross features of these tradeoffs. Recall that processing gain is equal to the AJ margin minus antenna gain, and transmitted bandwidth is equal to processing gain times data rate.

Example 14.3

Consider a data link that must transmit a 10-MHz data rate bandwidth on its downlink and requires 40 dB of AJ margin over and above whatever routine signal margin is available. If the data link can accept a limited maximum range and operate in a line-of-sight mode, it can use a ground antenna with, say, 30 dB of gain and only needs 10 dB of processing gain. This would require a 100-MHz transmitted bandwidth and is consistent with any convenient frequency band down to as low as UHF (presuming that a frequency assignment was available). Table 14.1 demonstrates the summary of calculations.

Table 14.1 Summary of calculations for Example 14.3

Required data rate bandwidth	10 MHz
Required AJ margin	40 dB
Antenna gain	30 dB
Processing gain (= AJ margin – antenna gain)	10 dB
Transmitted bandwidth (= processing gain × data rate)	100 MHz

---

Example 14.4

Consider the data link introduced in Example 14.3. If, on the other hand, the data link had to operate in a non-line-of-sight mode at a long range with omnidirectional antennas, it would need 40 dB of processing gain, resulting in a 100 GHz transmitted bandwidth, which is not possible at any frequency that can propagate non-line-of-sight. (In fact, it would not be possible at any frequency for which traditional RF technology applies.) Therefore, the requirements for the 10 MHz data

14.3 Anti-Jam Margin

rate, 40-dB AJ margin, and long range without a relay are mutually incompatible. Table 14.2 demonstrates the summary of calculations. Recall that the gain 40 in dB is equivalent to 10,000 in non-dB units.

Table 14.2 Summary of calculations for Example 14.4

Required data rate bandwidth	10 MHz
Required AJ margin	40 dB
Antenna gain	0 dB
Processing gain (= AJ margin − antenna gain)	40 dB
Transmitted bandwidth (= processing gain × data rate)	100 GHz

In fact, the only way to achieve a long range with a 40-dB AJ margin is to use high-gain relay antennas at high frequencies. 40 dB of processing gain is not available for a 10-MHz data rate bandwidth and the only other possible source of AJ margin (neglecting transmitter power, as has been done in all of the simplified examples) is antenna gain.

If one attempted to use low-gain relay antennas, the antenna beams would be so wide that the jammer would always be in the antenna beam and the apparent contribution from antenna gain would not provide any real discrimination against the jammer. Even if the data rate bandwidth were reduced to 1 MHz, 40 dB of AJ margin from processing gain alone would require an unreasonably high transmitted bandwidth of 10 GHz.

Figure 14.12 displays the relationship between data rate bandwidth, processing gain, and transmitted bandwidth. The dotted line indicates a maximum transmitted bandwidth of 1 GHz, although slightly higher values (2 or 3 GHz) might be possible in the 15-GHz band. From the figure, it is clear that a 40-dB processing gain is not available for data rates greater than about 100 kHz. At the

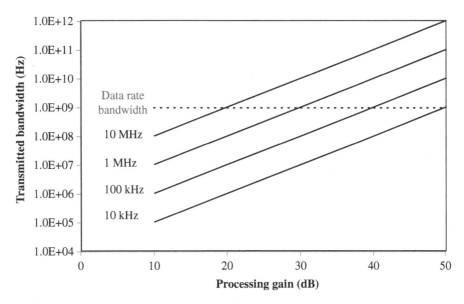

Figure 14.12 Transmitted bandwidth versus processing gain for several data rates

10 MHz data rate used in the examples above no more than 20 dB of processing gain is available in the highest frequency band likely to be used for a data link in the near future. Aside from entering into a power contest with the jammer, this leaves only high-gain antennas, and whatever vehicles are required to carry them, as an available option for AJ margins greater than 20 dB at data rates bandwidth as high as 10 MHz.

14.3.4 Anti-Jam Uplinks

Most of the discussion of AJ issues in this chapter focuses on the downlink. The situation for the uplink is significantly different in two ways:

1) The uplink receiver antenna, located on the UAV, probably cannot be large enough to have high gain, although it can have steerable nulls to suppress a limited number of jammer signals.
2) The data rate for the uplink is very low, so it can have large processing gain.

The total data rate required for an uplink is of the order of 1–2 kbps. A processing gain of 40 dB would require only 10–20 Mbps of transmitted bandwidth if the uplink signal does not have to be time multiplexed with the downlink. This would be the case if the uplink and downlink use two independent sets of transmitters and receivers, operating in a dual-simplex mode.

If, on the other hand, the data link operates in a duplex mode with a single receiver and transmitter at each end, then the uplink data is transmitted during short time slots taken away from the downlink. In this case, the uplink data will be transmitted at an instantaneous data rate similar to that used for the downlink and will have essentially the same processing gain as the downlink.

In a dual-simplex (i.e., duplex) system, the uplink would share the ground antenna with the downlink. If the antenna has high gain, then the uplink will benefit from a high ERP. However, a jammer operating against the uplink might use a high-gain tracking antenna aimed at the AV. This would leave only processing gain and active antenna processing, such as steerable nulls, for AJ protection of the uplink.

Therefore, it is fairly easy to provide high AJ margins for an uplink in a dual-simplex system where the link can provide high processing gain without having to share transmission time with the downlink. In a duplex system, uplink processing gain is similar to that for the downlink. If this does not provide adequate AJ margin, then the uplink must be provided with the equivalent of receiver antenna gain through active antenna processing.

14.4 Propagation

It is beyond the scope of this discussion to consider, in any detail, all of the factors that contribute to losses when an RF signal propagates through the atmosphere near the Earth. However, some familiarity with the basic factors that affect propagation is essential for an understanding of data-link design. The following sections describe the general characteristics of three of the basic issues in data-link signal propagation: (1) obstruction of the propagation path, (2) atmospheric absorption, and (3) precipitation losses.

14.4.1 Obstruction of the Propagation Path

Electromagnetic waves generally propagate in straight lines. However, the simple, straight-line mode of propagation can be modified by several effects. These include:

14.4 Propagation

1) Refraction by variations in atmospheric index of refraction (caused by variations in atmospheric density);
2) Diffraction caused by obstructions near, but not actually within, the atmosphere;
3) The nominal straight line between the transmitter and receiver; and
4) For long-enough wavelengths, complicated channeling and multiple propagation paths within a "waveguide" consisting of the layers of the atmosphere and the surface of the Earth.

The latter effect is what permits very long-range communications at relatively long wavelengths. It will not be addressed here in any detail, since it applies to wavelengths too long for most UAS data-link applications.

As a general rule, frequencies above a few GHz are considered to be useful only for "line of sight" communications. That is, they require a direct, unobstructed line of sight from the transmitter to the receiver. Slight amounts of refraction in the atmosphere allow beams at these frequencies to curve slightly over the horizon. If the Earth is considered to be a smooth sphere, the common correction for refraction is to consider the Earth to have a radius that is 4/3 of its actual radius. This moves the "radar horizon" out by a similar fraction, accounting for the fact that the beam is refracted slightly over the actual horizon. It is an appropriate factor for use at sea, where the surface of the sea reasonably approximates a smooth Earth.

However, for data-link operations over land, the limiting horizon is more likely to be determined by high ground under the data-link path than by the smooth-Earth model. In this case, diffraction effects require that the straight-line path clear the closest obstruction by a margin that depends on the wavelength of the signal. In particular, it has been found that the beam must clear any obstruction by a distance that allows about 60% of the first Fresnel zone to pass over the obstruction.

For a propagating electromagnetic beam, at distances from the antenna such that the antenna dimensions are negligible compared to the distance to the antenna, the Fresnel zones are defined as the circles (in a plane perpendicular to the beam) such that the path distance for energy passing from the transmitter to the receiver via the perimeter of the Fresnel zone, relative to energy that travels directly from the transmitter to the receiver, is a half-integral multiple of the wavelength of the signal. Figure 14.13 illustrates this geometry.

The locus of points that meet the requirement – path TBR minus path TAR is equal to $n \times \lambda/2$ – is a narrow ellipse with the transmitter and receiver at its foci. "Unobstructed" propagation requires that this ellipse be free of any substantial obstruction. For the special case where the nearest obstruction occurs at the midpoint of the path, it turns out that the radius of the first Fresnel zone is given approximately by $r = 0.5(\lambda R)^{1/2}$, where R is the distance from the transmitter to the receiver.

For $R = 50$ km, and requiring that the obstruction is at least $0.6r$ out of the direct line of sight, this means that the direct line of sight must clear the obstruction by about 0.05 m for visible light ($\lambda = 0.5$ μm), 15 m for $\lambda = 5$ cm, and 150 m for $\lambda = 5$ m. Most data links will operate with

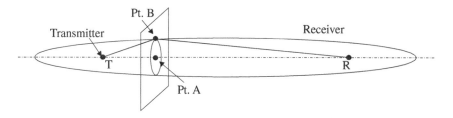

Figure 14.13 Fresnel zones of an electromagnetic beam

wavelengths in the cm or mm range, so a clearance of the order of 100 m or less will ensure truly unobstructed propagation in the line-of-sight beam.

Diffraction can be beneficial, since it can allow non-line-of-sight communications. In this case, the presence of obstacles (such as hills) can diffract energy out of the line-of-sight beam and into the valley behind the hill. This effect is most important at moderate frequencies (below 1 GHz). It explains why it is possible to receive a TV signal in a location that does not have a clear line of sight to the TV transmitter, albeit with reduced signal strength. As mentioned earlier, at frequencies below a few tens of MHz, other, more complicated, propagation modes become important. In this regime, the concept of obstruction of the line of sight no longer is meaningful.

14.4.2 Atmospheric Absorption

Various molecules in the atmosphere can absorb part of the energy in the signal. The primary sources of such absorption at the wavelengths of interest for data links are water vapor and oxygen molecules. At frequencies up to about 15 GHz, this absorption is very small (typically less than 3 dB at 100 km propagation distance). It should be noted that atmospheric absorption becomes more significant at higher frequencies. In particular, at frequencies of the 95 GHz to 120 GHz "window", atmospheric absorption can become the limiting factor on a data-link range. This absorption also prevents the use of frequencies outside of the windows for any but very short-range communications.

14.4.3 Precipitation Losses

Above frequencies of about 7–10 GHz, losses due to rain in the propagation path can become significant. Below about 7 GHz, losses even in heavy rain will be less than 1 dB at all ranges of interest for data links. The loss in rain depends on both the frequency of the signal and the elevation angle of the beam. Higher elevation angles cause the beam to "climb" above the rain at a shorter range, reducing the total losses. A typical UAV data link is likely to have a low elevation angle when used at long range, causing it to stay below the rain clouds for a large part of its path. Under these circumstances, the losses at 15 GHz in heavy rain (precipitation of 12.5 mm/h) can be as high as 100 dB for ranges of about 50 km. At 10 GHz, under the same conditions, the losses can be over 30 dB.

These losses are substantial and must be considered when designing a system that must operate in climates that include significant probabilities of heavy rain. Even in light rain (precipitation of 2.5 mm/h) the losses at 15 GHz can be of the order of 6 dB for ranges of the order of 50 km.

14.5 Data-Link Signal-to-Noise Budget

The data-link signal-to-noise (S/N) budget is an extremely useful conceptual framework for determining the fade margin of a data link as a function of the parameters of the link and of the environment in which it will operate. The S/N budget provides a tabular form for solving the range equations for the data link. By stating each "gain" or "loss" in dB, it reduces that solution to a process of adding up the gains and losses to find the net fade margin of the link.

The signal strength (S) at the output of the receiver antenna is given by

$$S = ERP_{\mathrm{T}} G_{\mathrm{R}} \left(\frac{\lambda}{4\pi R} \right)^2 \tag{14.9}$$

14.5 Data-Link Signal-to-Noise Budget 337

where ERP_T is the effective radiated power of the transmitter relative to an isotropic radiator, taking into account gain of the transmitter antenna and all losses in the transmitter and its antenna system, and G_R is the net gain of the receiver antenna, including the effects of losses in the antenna system.

Equation (14.9) neglects a number of losses that normally are relatively small for UAV data links. These include:

1) Losses due to atmospheric absorption, which are less than 3 dB over path lengths of up to 100 km at all frequencies below 15 GHz;
2) Losses due to polarization mismatches between the transmitter and receiver antennas, which are usually 1–2 dB at most for a well-designed system;
3) Losses due to precipitation in the propagation path, which may be significant at some wavelengths.

For a rough calculation, an additional loss of 3 dB will usually be sufficient to account for the effects of absorption and polarization. However, losses due to precipitation can be quite large at high frequencies. For instance, as mentioned earlier, at 15 GHz, 12.5 mm/h of rain can introduce a loss of nearly 50 dB over a 50-km path. This clearly must be taken into consideration if the data link is to be used under heavy rain conditions. Numbers for the loss in precipitation can be found in appropriate engineering handbooks.

The noise (N) in a data-link receiver is given by

$$N = kTBF \tag{14.10}$$

where k is the Boltzmann constant (1.3054×10^{-23} J/K), T is the temperature of the part of the receiver that contributes the limiting noise in K, B is the noise bandwidth of the receiver, and F is the "noise figure" of the receiver. By convention, most calculations are performed using $T = 290$ K, so that $kT = 4 \times 10^{-21}$ J $= 4 \times 10^{-21}$ W/s $= 4 \times 10^{-18}$ mW/Hz.

The S/N ratio is then given by

$$\frac{S}{N} = \frac{ERP_T G_R \left(\dfrac{\lambda}{4\pi R}\right)^2}{kTBF} \tag{14.11}$$

This equation can be written in logarithmic form as:

$$\log(S) - \log(N) = \log(ERP_T) + \log(G_R) - \log(kT) - \log(B) - \log(F)$$
$$- \log\left(\left[\frac{4\pi R}{\lambda}\right]^2\right) - \log(\text{precip. loss}) - \log(\text{misc. loss}) \tag{14.12}$$

The term involving λ is the "free space loss" and has been inverted, so that the logarithm is positive and the term is explicitly subtracted. In this format, the net excess of signal over noise, expressed as logarithms, is seen to be the sum of the "gain" terms minus the "loss" terms. Terms for the precipitation loss and miscellaneous losses due to absorption and polarization have been explicitly added to Equation 14.12.

If Equation 14.12 is expressed in dB by multiplying all of the logarithms by 10, the results can be arranged as in Table 14.3 to constitute the "data-link budget."

The fade margin is the amount of excess S/N available to deal with additional losses, such as geometrical fades, from which it takes its name. It also is the AJ margin available to the system before additional AJ steps, such as processing gain, are added. Good design practice normally would require a fade margin of at least 10 dB.

338   *14 Data-Link Margin*

Table 14.3   Format for a data-link budget

ERP_T	_____dB
Plus G_R	_____dB
minus kT	_____dB
minus B	_____dB
minus F	_____dB
minus $(\lambda/4\pi R)^2$	_____dB
minus precipitation loss	_____dB
minus misc. losses	_____dB
EQUALS AVAILABLE S/N	_____dB
minus required minimum S/N	_____dB
EQUALS fade margin	_____dB

Example 14.5

Consider a line-of-sight data link operating at 15 GHz with the following characteristics: air-vehicle transmitter power = 15 W, air-vehicle antenna gain = 12 dB, and losses in the transmitter system = 1 dB.

Furthermore, assume that the following characteristics describe the receiver system at the ground station: range from air vehicle = 30 km, bandwidth = 5 MHz, noise figure = 6 dB, antenna gain = 25 dB, maximum rain rate = 7.5 mm/h (medium), and minimum required S/N = 10 dB.

Complete the data-link budget for this data link. Does this data link operate successfully at the given conditions?

Solution:

First, we convert the 15 W power to 41.8 dBm (decibels above 1 mW), and then apply the antenna gain and internal losses to get a net ERP_T of

$$41.8\,\text{dBm} + 12\,\text{dB} - 1\,\text{dB} = 52.8\,\text{dBm}.$$

Table 14.4   Completed data-link budget

ERP_T	52.8 dB
plus G_R	+25.0 dB
minus kT	−(−174.0) dB
minus B	−67.0 dB
minus F	−6.0 dB
minus $(\lambda/4\pi R)^2$	−146.0 dB
minus precipitation loss	−15.0 dB
minus misc. losses	−3.0 dB
EQUALS AVAILABLE S/N	14.8 dB
minus required minimum S/N	10.0 dB
EQUALS fade margin	4.8 dB

Questions 339

The precipitation loss at 7.5 mm/h and 15 GHz, over a 30-km path, is approximately 15 dB. The free-space term for these parameters $10 \log(\lambda/4\pi r)^2$ is 146 dB, 10 log (5 MHz) is 67 dB, and 10 log (kT) at 290 K is -174 dBm/Hz (notice the negative value of this parameter, which means that we will add 174 dBm/Hz in the tabulated form of the S/N equation). If we now fill in Table 14.3 given above, we find the result shown in Table 14.4.

In other words, this data link should operate successfully at the range and under the conditions considered, but with very little margin for error in the calculation or for excess losses due to either the environment or malfunctions in the hardware. Only about 5 dB of additional losses would be required before the signal at the receiver would drop below the required S/N.

Questions

1) What do ERP, FLIR, GHz, AJ, S/J, and S/N stand for?
2) Define data link margin.
3) Name three possible sources for margin in a data link.
4) Name at least three sources that can attenuate or interfere with its data link signal.
5) Consider a data link with a transmitter that transmits a signal with a power of 45 dB. On the other end, the expected signal power received at the receiver should be at least 36 dB. What is the minimum margin for this data link?
6) Define antenna gain.
7) What does antenna gain represent in a transmitter antenna?
8) What does antenna gain represent in a receiver antenna?
9) Provide two benefits for a high gain antenna of the transmitter of a GCS.
10) Provide two requirements for a high gain antenna of the UAV.
11) Define the concept of a directive gain.
12) Define the beam width of an antenna.
13) What is effective radiated power?
14) What is a typical gain for a small antenna in a UAV?
15) Describe two ways that an antenna gain can provide a data-link margin.
16) Provide an example UAV that has a steerable antenna.
17) What is the benefit of a steerable antenna for a data link?
18) An antenna has a beam width of 30 degrees wide × 10 degrees high. Determine the antenna gain in dB.
19) By drawing a figure, discuss the interference of a jammer signal with the UAV receiver when it arrives from a direction that is outside of the main beam of the transmitter.
20) Discuss how the discrimination between the data link signal and interference or jammer can be enhanced.
21) What are typical values for the effective discrimination between a data link signal and a jammer at a ground antenna in dB?
22) What are the median daytime cloud ceilings in Europe in the winter?
23) An antenna height (i.e., UAV altitude) is 300 m. What is the line-of-sight range?
24) Provide one way to extend the range of the data link beyond those for which direct, line-of-sight propagation is possible from the ground to the UAV.
25) Discuss the possibility of "steerable antenna nulls" or "steerable nulls" to null out the jammer?
26) List three UAVs that carry antenna that have a directional pointing capability.

27) Discuss the advantage of UAV upward pointing directional antennas with regard to jamming.
28) Define processing gain in the context of the AJ margin.
29) How is the processing gain accomplished?
30) Using a figure, explain direct spread-spectrum transmission.
31) When a pseudo-noise modulation is added to the signal, what happens to its bandwidth and power per unit frequency interval?
32) Discuss the outcome when the jammer is more powerful than the data-link transmitter.
33) Define processing gain.
34) Write one advantage and one disadvantage of direct spread-spectrum transmission.
35) The downlink signal of a data link has a 2 MHz information bandwidth, and 30 dB of processing gain is required. What bandwidth must the pseudo-noise modulator operate at?
36) What is frequency hopping to produce processing gain?
37) Write two classes of frequency hoppers.
38) Compare fast and slow frequency hoppers from the view of a jammer.
39) Compare frequency-hopping data links and direct spread-spectrum data links in terms of cost.
40) Compare frequency-hopping data links and direct spread-spectrum data links in terms of interception.
41) Discuss features of scrambling and redundant transmission with error-detection codes.
42) Define fade margin.
43) Define anti-jam margin.
44) A jammer was designed for a data link with an AJ margin of 30 dB. The AJ margin of that data link is increased to 40 dB. How much the power of the jammer must be increased to have the same effect?
45) What is the relationship between the data link signal and jammer strengths at the receiver, with ranges between the data link transmitter and jammer to the data link receiver?
46) What is the relationship between S/J with ranges between the data link transmitter and jammer to the data link receiver?
47) By drawing a figure, show and discuss the region in which the jammer will defeat the uplink.
48) By drawing a figure, show and discuss the region in which the data link will defeat the jammer for a downlink signal.
49) Write at least five parameters that impact the jammer strength at the uplink receiver (i.e., at the UAV).
50) Write at least five parameters that impact the jammer strength at the downlink receiver (i.e., at the GCS).
51) Discuss ways to minimize the area over which the data link will be jammed.
52) Discuss the contribution of a low-gain antenna such as a Yagi to fade margin, the AJ margin, and to the ability of the data link to operate in the presence of jamming.
53) Write two disadvantages of higher frequencies in a data link.
54) With regards to frequency and line-of-sight, there are three choices for a data link in having a long range. What are the choices?
55) What is the relationship between processing gain, AJ margin, and antenna gain?
56) What is the relationship between transmitted bandwidth, processing gain, and data rate?
57) Consider a data link that must transmit a 20 MHz data rate bandwidth on its downlink and requires 30 dB of AJ margin over and above whatever routine signal margin is available. If the data link operates in a line-of-sight mode (i.e., has a limited range), and use a 25 dB gain ground antenna. Determine the required transmitted bandwidth (in Hz).

Questions 341

58) Briefly describe the differences between the situation of the downlink and of the uplink for the AJ.

59) Briefly discuss the processing time of the data in the uplink of a data link when it is operating in a duplex mode with a single receiver.

60) Briefly discuss the role of a duplex system for AJ protection of the uplink.

61) Write three of the basic topics in data-link signal propagation discussed in this chapter.

62) Write at least four effects that a straight-line mode of electromagnetic waves propagation can be modified.

63) What frequency range is considered to be useful only for "line of sight" communications?

64) Define the Fresnel zone.

65) Drawing a figure and discuss the requirement for un-obstruction propagation of radar radio waves.

66) An obstruction is located at a distance of 70 m of the direct line of sight for a transmitter. The transmitted radio wave has a wavelength of 10 cm. Is this propagation obstructed for a receiver located at a distance of 100 km?

67) Discuss how diffraction can be beneficial for propagation of electromagnetic waves (e.g., a TV signal) when obstacles (such as hills) are present.

68) What are two primary sources of atmospheric absorption for radio waves?

69) At which frequencies can atmospheric absorption become the limiting factor in a data-link range?

70) What is the typical radio wave propagation loss, at frequencies up to about 15 GHz at 100 km distance?

71) At which frequencies can losses due to rain in the radio wave propagation path become significant?

72) Which factors impact the radio wave propagation loss in rain?

73) For a UAV data link at low elevation angle when used at long range, briefly compare the radio wave propagation loss in heavy rain for frequencies of 10 GHz and 15 GHz.

74) What is the typical rain precipitation (in mm/h) for: (a) heavy rain and (b) light rain?

75) What parameters have considerable impact on the signal strength at the output of the receiver antenna?

76) What parameters have negligible impact on the signal strength at the output of the receiver antenna?

77) What is the typical radio wave (when the frequency is 15 GHz) propagation loss (in dB) due to heavy rain over a 50-km path?

78) According to Equation 14.10, what factors contribute to noise (N) in a data-link receiver?

79) Describe the signal-to-noise ratio (S/N) in a few words.

80) Provide two definitions for the fade margin.

81) What is the recommended fade margin (in dB) for a good design practice?

82) Consider a line-of-sight data link operating at 12 GHz with the following characteristics: air-vehicle transmitter power = 16 W, air-vehicle antenna gain = 10 dB, and losses in the transmitter system = 1.5 dB. Further, assume that the following characteristics describe the receiver system at the ground station: range from air vehicle = 50 km, bandwidth = 4 MHz, noise figure = 5 dB, antenna gain = 22 dB, maximum rain rate = 8 mm/h, and minimum required S/N = 10 dB. Complete the data-link budget for this data link. Does this data link operate successfully in the given conditions?

83) Consider a line-of-sight data link operating at 8 GHz with the following characteristics: air-vehicle transmitter power = 14 W, air-vehicle antenna gain = 13 dB, and losses in the

transmitter system = 1.7 dB. Further, assume that the following characteristics describe the receiver system at the ground station: range from air vehicle = 80 km, bandwidth = 2 MHz, noise figure = 3 dB, antenna gain = 24 dB, maximum rain rate = 6 mm/h, and minimum required S/N = 10 dB. Complete the data-link budget for this data link. Does this data link operate successfully in the given conditions?

84) What are fundamental factors that limit the total available bandwidth in wireless networks?

85) What are the functions performed by the receiver on an incoming signal from the antenna?

86) A GCS transmitter antenna with a width of 60 cm is located at a height of 30 m. The transmitted signal has a frequency of 2 GHz. Determine the peak antenna gain (in dB).

87) A UAV is flying at an altitude of 2,500 m. The GCS is at the distance of 200 km. Is there a line-of-sight between these two? Why?

15

Data-Rate Reduction

15.1 Overview

For any network or data link, one of the most valuable commodities is bandwidth or data rate, which is the rate at which data pass a point in the data transmission path. The typical unit is bit per second. For wireless networks, there are fundamental factors that limit the total bandwidth that can be available in any part of the electromagnetic spectrum. Of course, there is a limited total spectrum to be divided up between all the users that want to transmit information. These are important issues for a UAS data link, particularly for the downlink, which may have masses of data that would require a very large bandwidth to transmit in its raw form.

An uplink has a typical data rate of a few kbps to provide commands to UAV control surfaces and engines; and commands to payloads/sensors. In contrast, downlink has a high data rate (1–10 Mbps) for sensor data, and a low data-rate channel to transmit flight status information.

As discussed in the two preceding chapters, an AJ data link, or even a "jam-resistant" data link, for a UAV is likely to have a data rate that is significantly lower than the maximum raw data rate available from the sensors on the UAV.

The raw data rate from a high-resolution TV camera or FLIR sensor can be as high as 75 Mbps. However, as calculated in Example 14.3 in Chapter 14, it is estimated that the highest data rate likely to be practical for an AJ data link is about 10 Mbps. The result of this mismatch is that it is not possible to transmit the raw sensor data to the ground. The onboard processing unit must somehow reduce the data rate to a level that can be accommodated by the data link.

This chapter discusses the ways that this can be accomplished and introduces the tradeoffs that must be made between the data rate and the ability to perform functions that depend on the transmitted information.

15.2 Compression Versus Truncation

There are two ways to reduce the data rate: data compression and data truncation. Data compression processes the data into a more efficient form in such a way that all (or almost all) of the information contained in the data is preserved and the original data can be reconstructed on the ground if so desired. Ideally, no information is lost, whether or not the information is useful. In practice, information is lost due to imperfections in the compression and reconstruction

---

Introduction to UAV Systems, Fifth Edition. Paul Gerin Fahlstrom, Thomas James Gleason, and Mohammad H. Sadraey.
© 2022 John Wiley & Sons, Inc. Published 2022 by John Wiley & Sons, Inc.
Companion website: www.wiley.com/go/fahlstrom/uavsystems5e

processes. Data compression involves algorithms for eliminating redundancies in the raw data and then reinserting them on the ground if they are required to make the data intelligible to the operator.

A very simple example of data compression addresses data from an air-temperature sensor that gives a reading every second. If the temperature had not changed from the previous reading, data compression might consist of not transmitting the new (redundant) reading, while data reconstruction at the ground station would consist of holding and displaying the old reading until a new temperature was sensed and transmitted. This process could reduce the number of bits transmitted over a period of time by a large factor with no loss of information on the ground.

Data truncation throws away data to reduce the transmitted data rate. Information is lost in this process. However, if it is done intelligently, the information that is lost is not necessary for completing the mission so that the truncation process has little or no effect on mission performance. For example, video data is often acquired at a rate of 30 frames per second (fps), for reasons that are mostly cosmetic (to avoid flicker and jerkiness in the display). A human operator cannot make use of new information at a rate of 30 Hz, so discarding every other frame to reduce the data rate by a factor of two has little or no effect on operator performance, even though it certainly does discard some information.

If compression and truncation of unneeded data cannot reduce the data rate sufficiently, it may become necessary to discard data that would be useful on the ground as if it were transmitted. At this point, there is a potential for degrading the performance of the system. However, it may be possible to tolerate significant reduction in the transmitted information without affecting the performance of the mission. This is often under the control of the system designer and user, since different approaches to performing the mission can result in different partitions between what information is essential (e.g., aircraft height) and what is only nice to have (e.g., color of a target).

The key point is that data rate does not come free in a data link, particularly if the data link must provide significant AJ capability. In fact, data rates above about 1 Mbps may not be feasible in a "long-range, moderate-cost, jam-resistant" data link, depending on how some of the adjectives describing such a data link are translated into numerical specifications. Whether or not higher data rates are technically feasible, the data rate may be the only major parameter in the design tradeoff that can be varied in an attempt to maintain the goal of low or moderate cost, since range is linked to basic mission considerations and the jamming environment is under someone else's control.

15.3 Video Data

The most common high-data-rate information produced by UAV sensors is video from imaging sensors such as TV cameras or FLIRs. These data consist of a series of still pictures (frames), typically at a rate of 30 fps. Each frame consists of a large number of picture elements (pixels), each of which has a numerical value that corresponds to its brightness on a gray scale.

15.3.1 Gray Scale

In digital images, gray scale is a group of shades without any visible color. Gray scale means that the value of each pixel represents only the intensity information of the light. Such digital image typically displays only the darkest black to the brightest white. In another words, gray scale only

15.3 Video Data

contains brightness information, not color. The image contains only three colors: (1) black, (2) white, and (3) gray colors. However, gray color has multiple levels.

The value of each pixel is related to the number of bits of data used to represent the pixel. A typical raw video, after digitization, consists of 6 or 8 bits of gray-scale information per pixel. If the resolution of the picture is 640 pixels horizontally × 480 pixels vertically, there are 307,200 pixels. At 8 bits/pixel and 30 fps, this leads to a raw data rate of nearly 75 Mbps. If the video is in color, more bits are required to specify the color of the pixel. For this reason, one of the first pieces of information potentially contained in a picture – that may be left out in the design of an imaging sensor for a UAV – is color.

The primary data compression approach for video data is to take advantage of redundancies in the picture to reduce the average number of bits required to describe a pixel. Pictorial data is highly redundant in the sense that neighboring pixels are not independent. For instance, if the picture includes a patch of clear sky all of the pixels in that part of the scene are likely to have the same brightness. If one can find a way to specify a single value of gray scale for all of these pixels without actually repeating the number for each pixel, then the average number of bits/pixel for the complete scene can be reduced.

Even for parts of the scene that contain objects, there tends to be a correlation from pixel to pixel. Except at edges of shadows or high-contrast objects, the gray scale tends to vary smoothly across the scene. Therefore, the difference in gray scale between adjacent pixels tends to be much less than the maximum difference that is allowed by the 6- or 8-bit range in the scale.

15.3.2 Encoding of Gray Scale

The difference in gray scale between adjacent pixels can be exploited by using difference coding, in which each pixel is described by its difference from the preceding pixel, rather than by its absolute value. Since it is very convenient, if not essential, to use the same number of bits for each pixel, difference coding usually requires that all differences be represented by some small, fixed number of bits. For instance, the algorithm might allow only 3 bits to describe the difference. This would allow differences of $0, \pm1, \pm2,$ or ±3. If the raw video is digitized at 6 bits, it can have absolute gray-scale values from 0 to 64. A black-to-white transition, at the edge of a shadow, for instance, could have a difference of 64 and would be severely distorted by a system that could only record a difference of 3. To deal with such transitions, the allowed relative differences of 0 to ±3 are assigned absolute values, such as described in Table 15.1.

The actual values in the "absolute difference" column are selected based on statistical analyses of the types of scenes that are expected to be transmitted. This scheme will clearly result in some distortion of the gray scale in the picture and smoothing of sharp transitions. Therefore, it compresses the data at the cost of some loss of fidelity in the reconstruction on the ground. The

Table 15.1 Encoding of gray scale

Relative Difference	Absolute Difference
0	0 to ±2
±1	±3 to ±8
±2	±9 to ±16
±3	±17 to ±32

compression in the example given is from 6 bits/pixel to 3 bits/pixel, only a factor of 2. It is possible to go as low as 2 bits/pixel with difference-coding schemes.

Further compression is possible with more sophisticated approaches. Many of these approaches are based on concepts similar to Fourier transformation, in which the picture is converted from displacement space to frequency space and the coefficients of the frequency-space representation are transmitted. This tends to reduce the number of bits required because most of the information in a typical picture is at relatively low spatial frequencies and the coefficients for higher frequencies can be discarded or abbreviated.

There is a great deal of potential for clever design in the algorithms for transforming the picture into frequency space and for deciding which coefficients to transmit and which to discard. The picture is normally broken up into sub-elements with dimensions on the order of 16 × 16 pixels prior to being transformed. It is possible to tailor the number of bits used for each sub-element to the content of the sub-element. This allows using a very small number of bits for a sub-element of clear sky or featureless meadow and a larger number of bits for a sub-element that includes detailed objects.

Using a combination of difference and transformation coding, it is possible to transmit recognizable pictures with an average of as few as 0.1 bits/pixel. This would represent a factor of 60 compression from 6 bits/pixel and a factor of 80 from the example worked at the beginning of this section that assumed 8 bits/pixel. At 0.1 bits/pixel, one could transmit a 30 fps video at 640 × 480 resolution with less than 1 Mbps. Unfortunately, the reconstructed picture at 0.1 bits/pixel has reduced resolution, compressed gray scale, and artifacts introduced by the transformation and reconstruction process.

15.3.3 Effects of Bandwidth Compression on Operator Performance

Testing performed to support RPV programs in the Army and other services has explored the effects of bandwidth compression on operator performance. Results of a number of experiments are summarized in Reference [40]. Figures 15.1 and 15.2, redrawn from figures presented in this Reference, show measured performance for various levels of compression using a combination of difference coding and a cosine transformation. The targets in the study were armored vehicles and artillery pieces seen from typical RPV viewing angles and ranges.

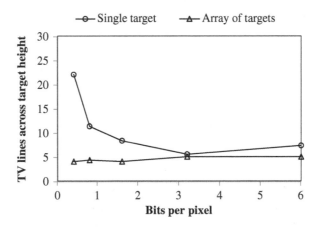

Figure 15.1 Effect of compression on probability of detecting targets

15.3 Video Data

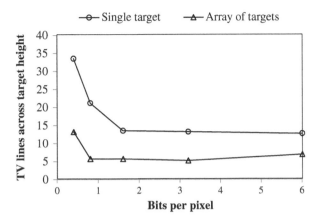

Figure 15.2 Effect of data compression on recognition of targets

The measure of performance was the number of TV/screen lines across the minimum target dimension that was required for the operator to achieve detection and recognition. A larger number of lines correspond to a need to "zoom in" on the scene in order to succeed in performing the function. When dealing with detection, this means that the sensor instantaneous field of view would be reduced in both height and width. This might increase the search time for a given area on the ground at a rate approximately proportional to the square of the number of lines required.

An interesting feature of the experiment was that performance was measured both for single targets and for arrays containing ten targets. An example of an array of targets might be a half-dozen people walking across a field, as compared to one person in the same field. The existence of several targets improved the detection probability for most levels of compression, which is intuitively satisfying as it seems reasonable that if there were four targets it would be more likely that the operator would see at least one of them and then look for more in the vicinity of the one that has been detected.

The results shown in the figures indicate that the level of compression did not affect target detection capability, for arrays of ten targets, down to the lowest number of bits/pixel used in the experiment (0.4). For single targets, however, detection capability began to degrade at 1.5 bits/pixel and was seriously degraded below 1 bit/pixel. For recognition, there was no degradation in performance down to 0.8 bits/pixel for arrays of targets, but significant degradation at 0.4 bits/pixel.

For single targets, the results for recognition were similar to those for detection. These measurements suggest that compression to 0.4 bits/pixel may be acceptable for some applications (e.g., searching for major enemy units well behind the lines or large herds of animals in a range area), assuming that once targets are found, it will be possible to look at them with a narrow field of view that provides enough magnification to allow recognition despite the degraded performance at low bits/pixel. It appears that something between 1.0 and 1.5 bits/pixel should be acceptable for most missions.

It must be noted that the quality of the picture is a function of the particular algorithms used in the transformation and results for one implementation should not be assumed automatically to be universal. Reference [40] reviewed several experiments and concluded that the robustness of operator performance down to 1.0–1.5 bits/pixel was present in all of them. This seems to provide an upper limit on the number of bits required to transmit an acceptable video.

On the other hand, it is not clear that there is any fundamental reason why the number of bits/pixel could not be further reduced by clever application of processing and encoding techniques.

348   15 Data-Rate Reduction

This area offers a potential for further technology development that might make compressions as low as 0.1 bits/pixel acceptable, at least for some applications. It may also be desirable to consider variable compression ratios under operator control, so that picture quality can be traded off against other parameters during various phases of a mission.

15.3.4 Frame Rate

Once the number of bits/pixel has been reduced as far as possible, it becomes necessary to reduce the number of pixels that are transmitted. This requires truncation of the data, rather than compression. For video data, the simplest way to reduce the number of pixels per second is to reduce the frame rate, stated in frames per second or "fps." Thirty frames per second were selected as a video standard based on a need for flicker-free pictures. Nothing on the ground moves very far in 0.033 seconds, so there is little new information in each frame. Flicker in the display can be avoided by storing the frame and refreshing the display at 30 Hz, whatever the rate of transmission of new frames of video.

Most observers will not recognize a reduction in frame rate to 15 fps unless it is called to their attention. At 7.5 fps, there begins to be obvious jerkiness if something in the scene is moving or the vantage point of the sensor is changing. At lower frame rates, an observer can clearly perceive the frames as they come in. However, some functions can be performed just as well at very low frame rates as at 15–30 fps.

Reference [40] reports several experiments that determined that the time required to detect a target within the field of view of the sensor was not affected by reduction in the frame rate down to 0.23 fps. This is consistent with estimates that it takes about 4 s for an operator to completely search a scene displayed on a typical RPV video screen [41]. If searching is performed by holding the sensor on one area for about 4 s and then moving to another area (a so-called "step/stare" search), it would appear that frame rates of 0.25 fps should be acceptable for this particular mission.

15.3.5 Control Loop Mode

Many flight mission activities require a closed-loop control (see Figure 5.7) that involves: (1) the UAV payload/sensor, (2) data link, and (3) operator in the GCS. For example, the operator must be able to: (1) move the sensor to look at various areas of interest (coarse slewing), (2) point at particular points or targets (precision slewing), and (3) lock an auto-tracker on to a target so that the sensor can follow it, as when performing laser designation, or manually track a target. With some UAVs, the operator manually participates in landing the air vehicle while observing a video from a TV camera or FLIR that has been fixed to look down at the end of the runway. In all of these cases, a reduction in frame rate causes delays in the operator seeing the results of his/her commands.

It is important to note that long transmission delays, such as might be expected if the data link uses satellite relays to reach partway around the earth or uses a large network that has significant "packet" delays due to transmission through multiple nodes, have an effect on the operator and system performance that is very similar to a reduced frame rate. In either case, the operator is presented with information that is "old" when he or she first sees it and the operator responses to this information, in the form of commands to be sent via the uplink, is "out of date" by the time that it reaches the actuators on the AV. If a frame rate of 1 Hz causes problems, then a total latency due to delays in transmission (round trip) of the order of 1 s is likely to cause similar problems.

Experience with Aquila and MICNS clearly proved that closed-loop activities are affected by delays caused by frame rate reduction. The effects of the delays can be catastrophic if the control

15.3 Video Data

loops are not designed to accommodate them. Reference [40] reports measurements of performance for precision sensor slewing for three different types of control modes as a function of the frame rate: (1) continuous, (2) bang-bang, and (3) image motion compensation.

"Continuous" control mode is a simple rate input from the operator. The operator pushes a joystick and the sensor moves in the direction indicated at a rate proportional to how far or hard he or she pushes/pulls, continuing to move at that rate until the operator stops pushing/pulling.

"Bang-bang" control uses discrete operator inputs similar to the cursor control keys on the keyboard of a personal computer. The operator can make discrete inputs of up, down, right, or left and the sensor moves one step in the indicated direction for each input. If the operator holds down the control, the system generates a "repeat" function and takes repeated steps in the indicated direction.

The third control mode, "Image Motion Compensation" (IMC), uses information from the air vehicle and payload/sensor gimbals to compute where the sensor is pointing and display this information on the scene presented to the operator without waiting for the new video to be received. When the operator commands the sensor to slew to the right, for instance, at a low frame rate, a cursor moves across his/her screen to the right, showing where the sensor is pointing at any particular instant relative to the video presently displayed. This might go on for several seconds at very low frame rates while the operator places the cursor just where he/she wants the sensor to point. Then, when the next new frame is transmitted, the center of the new picture is wherever the cursor was located in the old frame.

It is clear from the results, shown in Figure 15.3, that continuous and bang-bang control fails catastrophically at frame rates much below 1 fps. Continuous control is seriously degraded even at 1.88 fps. However, IMC continues to perform well at frame rates as low as 0.12 fps. Extensive experience with Aquila/MICNS, which started out with a form of continuous control and later implemented a form of IMC, confirms these results, at least at frame rates at and above 1 or 2 fps.

The data in Figure 15.3 apply to precision slewing and auto-tracker lock-on for stationary targets. If the target is moving, it is necessary for the operator to manually track it, at least for an instant, to lock an auto-tracker on the target rather than the stationary background. To avoid the need to track the target, the operator might try to predict where the target is going and set up the sensor on a point ahead of the target, and then catch it as it passes through the center of the field of view. This approach was tried with Aquila/MICNS and had a fairly low success rate. This experience led to the conclusion that locking an auto-tracker on a moving target requires frame rates similar to manual tracking.

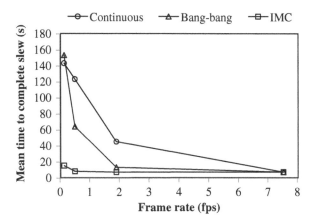

Figure 15.3 Effect of frame rate on time to complete a fine-slewing task

Manual target tracking is the most difficult closed-loop activity likely to be required for a mini-UAV. Reference [40] reports data indicating that manual tracking of a moving target suffers little degradation down to 3.75 fps, but rapidly becomes very difficult and, eventually, impossible as the frame rates goes below that value.

The effects of reduced frame rate on closed-loop control functions are primarily due to the loop delay introduced by the lower frame rates. That is, the operator is responding to old images and data and does not see the results of his/her control inputs until long after those results have occurred. Similar effects would be expected if the link delay were caused by transmission time, as in a satellite-based global communications channel used to control UAVs that are physically half a world away from the operator's location. Unless steps are taken to compensate for these delays, it should be expected that the performance of auto-tracker lock-on or manual tracking of moving targets will be poor.

Some other functions are less sensitive to the type of control loop. Figure 15.4 shows the probability of a successful target search as a function of the frame rate [40]. The same three control modes described above were used for this experiment, which required coarse slewing of the sensor to get the target in the field of view. No major differences between the three control modes were found for this activity. The data show a clear break point at 1.88 fps for coarse slewing. It should be noted that the search task used in this experiment was a manually controlled search of a large area. This tested the ability to control the payload/sensor.

Experience with Aquila indicates that area searches probably should be controlled by software that slews the sensor automatically (using a step/stare technique) and ensures that the search is systematic [41]. That type of search would be characterized by the detection performance shown in Figures 15.1 and 15.2 and should not be seriously degraded down to at least 1 fps.

15.3.6 Forms of Truncation

Two other forms of truncation have been used in UAV data links: reduction of resolution and field-of-view truncation. In the first case, adjacent pixels are averaged to produce a picture with 1/2 or 1/4 as many pixels in either the horizontal or vertical directions (or both). There is some evidence cited in Reference [40] that reducing the resolution by 1/2 in each axis for a factor of 4 data-rate

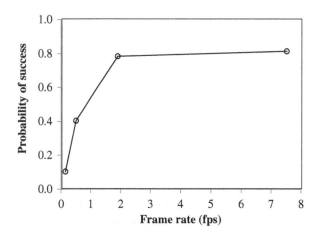

Figure 15.4 Effect of frame rate on probability of success for a manual search

15.3 Video Data 351

reduction is preferable to going from 2 bits/pixel to 0.5 bits/pixel by data compression for the same factor of 4.

However, standard sensor performance models suggest that reducing the resolution by a factor of 2 will typically reduce the maximum ranges for target detection by the same factor of 2 [41]. If this is true, resolution reduction has no net benefit, since the sensor will have to reduce its field of view on the ground by the same ratio as it reduces its resolution in order to perform the same function. The same effect could be achieved by simple truncation of the field of view by a factor of 2 in each axis, which is the other form of truncation sometimes used.

Either resolution reduction or field-of-view truncation can be used when the lowest frame rates will not support the function to be performed. For instance, consider a situation in which a moving target must be tracked, requiring at least 3.75 fps, and the data link cannot support that frame rate at its lowest value of bits/pixel. To achieve a transmittable data rate, the field of view could be reduced by a factor of 2 or 4 by truncation. As an alternative, the sensor could be set to a narrow field of view that has more resolution than is required to track the target, and the excess resolution could be discarded by reducing the resolution of the transmitted picture.

These approaches are the least desirable way to reduce data rate, but there are instances in which their use is appropriate and can improve, rather than degrade, system performance.

15.3.7 Summary

In previous sections, a number of techniques for video data rate reduction and their advantages and disadvantages were presented. In summary, the available data indicate that the compression or truncation data rate reduction techniques – provided in Table 15.2 – is recommended, and may be acceptable for video data.

It should be emphasized that these results are all sensitive to details of specific implementations and also depend on how the operator's task is structured. The factors of 10 or 100 in data rate between 15 fps and 1.5 fps or 0.12 fps, combined with a multiplicative factor of 2.5 or 10 between 1 bit/pixel and 0.4 or 0.1 bit/pixel can have a major effect on data-link cost and AJ capability.

Table 15.2 Recommended data rate reduction techniques for UAV video data

No.	Mission	Recommended/Acceptable Data Rate Reduction Technique
1	Searching for isolated, single targets	Data compression to 1.0–1.5 bits/pixel
2	Searching for arrays of targets (such as convoys of trucks, large groups of people, compounds having several buildings, or tactical units in company strength)	Data compression to 0.4 bits/pixel or lower
3	Automated target search, precision slewing, and auto-tracker lock on for stationary targets	Frame-rate reduction to 0.12–0.25 fps
4	Manual tracking and auto-tracker lock on for moving targets	Frame-rate reductions to 3.75 fps
5	Special cases	Reduction of resolution or field-of-view truncation

There may be significant room for improvement in basic technology (compression algorithms), although there has been a large amount of work in this area to support things such as digital cameras and camcorders, and the commercial market for these functions over the last decade, at least, may have driven the compression algorithms to nearly their practical limits.

There probably is still a potential for test-bed development of approaches and techniques for using lower frame rates for specific UAS functions and for improved IMC functions to aid the operator in compensating for data-link delays.

The whole area of the effects of data-rate reduction on operator performance and system control loop performance is closely linked to training and operator task structures and is ideally explored with operators using ground and airborne test-bed hardware.

15.4 Non-Video Data

It is beyond the scope of this chapter to identify and analyze all of the non-video forms of data that might be transmitted from a UAV to the ground. Some of the payloads and sensors that have been proposed include jammers, EW intercept systems, radars (imaging and non-imaging), meteorological packages, and chemical, biological, and radiological (CBR) sensors.

Some possible data sources have inherently low data rates (compared to a TV camera or FLIR video). Examples of this class include meteorological sensors, CBR sensors, and some kinds of EW payloads, such as simple jammers, that only have to report their own status, rather than collect and report external data.

Some other possible payloads could have very high raw data rates. One example would be a radar interception and direction-finding system. The raw data from such a sensor might involve information about tens of thousands of pulses from dozens of radars each second. In this case, the tradeoff that must be considered is onboard processing to reduce the thousands of data points to a few dozen target identifications and azimuths versus a data link that can transmit the raw data to the ground for processing. As with video data, if the data link must provide significant AJ capability, then the onboard processing may be the best choice.

Another example is an SLAR system that achieves enhanced resolution by coherently combining signal returns from multiple locations as the AV moves, thus synthetically enlarging the receiving aperture. This is a very computationally intense process and almost certainly requires that the raw data be processed on the AV and only the resulting "images" be transmitted to the ground.

The kind of onboard processing suggested for the radar intercept system mentioned above is a form of data compression that is not feasible at present for video data but probably is feasible for at least some types of non-video data. This processing performs correlations of data over time and extracts the significant information from the raw data. It is already being performed in fielded threat warning receivers. The video equivalent would be to automatically recognize targets onboard the UAV and transmit down only an encoded target location instead of the whole picture of a piece of the ground.

Data compression in the same sense as for video data is also feasible for most other kinds of data. A simple example is to use exception reporting – sending data to the ground only when something is happening or something changes. More sophisticated types of compression, analogous to transformation coding of video, can be explored for each type of data based on its particular characteristics.

Truncation is also possible. For non-video data, it might take the form of recording very high data rates for short times and then sending the data down the link over a longer period of time. The

15.5 Location of the Data-Rate Reduction Function

Given that data-rate reduction is required for most payloads and sensors, the question arises of where that function should be performed within the total UAS architecture. Data-link designers tend to believe that it should be done within the data link. For instance, MICNS included the video compression and reconstruction functions, accepted standard TV video (non-interlaced standard), and presented standard 30-Hz refreshment-rate TV video to the ground-station monitors. This simplified the specification of interfaces between the data link and the rest of the system.

On the other hand, the expertise to design compression and reconstruction algorithms that are well matched to the sensor data and minimize the loss of information may be located at the sensor designer rather than the data-link designer. There is no point in designing a sensor that produces data that simply will be truncated by the data link. This leads to useless cost and complexity in the sensor. Therefore, one might argue that data compression and truncation should be performed in the sensor subsystem before the information is passed to the data link for transmission.

This argument is stronger if the data link must deal with a variety of sensors, each of which may require different approaches to compression and truncation. Even a TV and an FLIR are different enough that slightly different algorithms are optimum for video compression transformations. The differences between an imaging sensor and an EW system are much greater. A universal data link would need many different modules (software and/or hardware) to deal with different kinds of data.

A counter argument is that if the compression is handled in the sensor, then there must be a matching reconstruction algorithm in the interface between the ground end of the data link and the operator displays and data recording system. This requires the integration of a module or software from every sensor system into the ground station. Clearly, this could be simplified if standard compression and reconstruction algorithms were available. An example of standard algorithms is one used to compress and reconstruct the JPEG files commonly used in cameras and other imaging systems.

If the compression, truncation, and reconstruction are handled by the sensor subsystem, the data link would be specified as a pipeline that accepts and transmits a digital data stream with certain characteristics. Whatever processing is required to conform to those characteristics would be provided by the sensor and by a reconstruction unit provided by the sensor supplier.

In either case, the UAS integrator must understand the implications of the data-rate restriction, data compression, truncation, and reconstruction required to use the data link, including any control-loop delays introduced by these processes. The system must provide the command capability and software required to adapt the data rate to jamming conditions and to change the mix of compression and truncation as needed for various phases of the mission.

The authors are inclined to believe that the data-rate reduction function should be part of the sensor subsystem rather than the data link, particularly in multipayload systems. However, this decision should be made for each system based on the particular situation, as part of the top-level system engineering effort.

Questions

1) What do fps, Mbps, FLIR, SLAR, EW, CBR, JPEG, and IMC stand for?
2) Define data rate.
3) There are two ways to reduce the data rate. What are they?
4) How does data compression work to reduce the data rate?
5) Is information lost in a data compression technique? Explain.
6) Is information lost in a data truncation technique? Explain.
7) What is the typical video data rate in fps?
8) Can a human operator make use of new information at a rate of 30 fps?
9) Are data rates above about 1 Mbps feasible in a "long-range, moderate-cost, jam-resistant" data link?
10) Provide one example for: (1) essential information and (2) "nice to have" information.
11) In which data rate technique may more information be lost – compression or truncation? Discuss.
12) What is the most common high-data-rate information produced by UAV sensors?
13) What does a numerical value of a picture element (i.e., pixel) correspond to?
14) What does a typical raw video, after digitization, consist of?
15) Consider a data link with the resolution of a picture is 640 pixels horizontally × 480 pixels vertically, at 8 bits/pixel and 30 fps. How much is the raw data rate in Mbps?
16) What is the primary data compression approach for video data? Explain.
17) What is the typical number of bits in each pixel?
18) What is gray scale?
19) What is the difference coding in describing a pixel? What does it usually require?
20) Describe sophisticated approaches based on concepts similar to Fourier transformation in compression of video data.
21) Explain how it is possible to transmit recognizable pictures with an average of as low as 0.1 bits/pixel.
22) What are the disadvantages of a reconstructed picture (after compression) with an average of as low as 0.1 bits/pixel?
23) Consider Figures 15.1 and 15.2. What is the approximate relation between detection of a target – in terms of rate of search time for a given area on the ground – and number of TV lines required across the minimum target dimension?
24) Consider Figures 15.1 and 15.2. What is the relation between "TV lines required across the target height" and "operator success in performing the detection and recognition functions"?
25) Consider Figures 15.1 and 15.2 for single targets. Explain the variations of recognition capability as a function of data compression – in number of bits per pixel.
26) Consider Figures 15.1 and 15.2 for single targets. Explain the variations of detection capability as a function of data compression – in number of bits per pixel.
27) What is the minimum acceptable number of bits/pixel in video data compression for most detection and recognition missions?

Questions 355

28) Is there any fundamental reason why the number of bits/pixel – for video data compression – could not be further reduced below 1? Explain.

29) For video data, what is the simplest way to reduce the number of pixels per second?

30) How many frames per second were selected as a video standard based on a need for flicker-free pictures?

31) How can flicker in the display be avoided?

32) Briefly compare the following video data frame rates for most observers: (a) 30 fps, (b) 15 fps, (c) 7.5 fps, (d) lower fps.

33) On average, how long does it take for an operator to completely search a scene displayed on a typical RPV video screen?

34) Based on experiments, what is the lowest frame rate that the time required to detect a target within the field of view of the sensor is not affected? How about a "step/stare" search?

35) Provide one UAV operator function example for: (a) coarse slewing, (2b) precision slewing.

36) Provide at least three cases where a reduction in the frame rate causes delays in the operator seeing the results of his/her commands.

37) Reference [40] reports measurements of performance for precision sensor slewing for three types of control loops as a function of the frame rate. What are these three types?

38) Describe how an operator in a GCS is controlling a UAV payload/sensor using "continuous mode."

39) Describe how an operator in a GCS is controlling a UAV payload/sensor using "bang-bang mode."

40) Describe how an operator in a GCS is controlling a UAV payload/sensor using "image motion compensation mode."

41) Using Figure 15.3, state which of the following control modes continues to perform well at low frame rates as low as 0.5 fps: (a) continuous, (b) bang-bang, and (c) image motion compensation.

42) What is frame rate (in fps) by which the manual tracking of a moving target suffers little degradation down to this value, but rapidly becomes very difficult and, eventually, impossible as the frame rates go below that value?

43) Based on Figure 15.4, what is the probability of a successful target search – for a manual search – when the frame rate is 1.88 fps?

44) Name three forms of truncation.

45) Explain how "reduction of resolution" is applied for a video data truncation process.

46) Provide one advantage and one disadvantage for "reduction of resolution" in a video data truncation process.

47) Consider a situation in which a moving target must be tracked, requiring at least 3.75 fps, but the data link cannot support that frame rate at its lowest value of bits/pixel. Provide two methods to achieve a transmittable data rate.

48) What data rate reduction technique is recommended for video data, when the UAV mission is to search for isolated, single targets? Please specify.

49) What data rate reduction technique is recommended for video data, when the UAV mission is to search for arrays of targets? Please specify.

50) What data rate reduction technique is recommended for video data, when the UAV mission is an automated target search, precision slewing, and auto-tracker lock on for stationary targets? Please specify.

51) What data rate reduction technique is recommended for video data, when the UAV mission is to manual tracking and auto-tracker lock on for moving targets? Please specify.

52) Provide three examples of payloads (i.e., data sources) that have inherently low data rates (compared to a TV camera or FLIR video).
53) Does a simple jammer report its own status, or collect and report external data?
54) Provide two examples of payloads (i.e., data sources) that could have very high raw data rates (compared to a TV camera or FLIR video).
55) What is the best choice for location of data processing in a data link that must provide significant AJ capability: (a) onboard UAV processing or (b) transmit the raw data to the ground (i.e., GCS) for processing?
56) Explain how "onboard processing suggested for a UAV radar intercept system" is a form of data compression.
57) What is "exception reporting" as a form of data compression?
58) What factors limit the data rate that can be supported by a data link? Provide three items.
59) What are two possible locations for the data-rate reduction function in a UAS?
60) Present pros and cons for locating the data-rate reduction function within the data link.
61) Present pros and cons for locating the data-rate reduction function within the sensor subsystem.
62) What kind of files are commonly used in cameras and other imaging systems?
63) What is the authors' recommendation for locating the data-rate reduction function in a UAS? Is it in the sensor subsystem or in the data link?

16

Data-Link Tradeoffs

16.1 Overview

As discussed in the preceding three chapters (13, 14, and 15), there are many tradeoffs associated with the selection and design of a data link for a UAS.

Most of those tradeoffs involve things that are beyond the boundaries of the data link itself, such as the mission that can be performed, how the mission will be accomplished within the limitations of the total UAS capabilities, requirements for operator training and skills, payload/sensor selection and specifications, ground-station design, and cost, weight, and power requirements in, at least, the AV.

This chapter outlines these tradeoffs, based on the data-link design issues discussed in Chapters 13 through 15.

16.2 Basic Tradeoffs

Operating range, data rate, AJ margin, and cost are strongly interacting factors in data-link design. Data latency due to long-distance transmission or other delays in a distributed communications network can have results similar to those of reduced data rate and must also be considered.

For tradeoffs involving the AJ margin, the effect of range can be considered a step function: one set of design considerations applies to AJ data links that operate within the line-of-sight range from the ground station and a different set of considerations applies for links that must operate beyond that range. Data rate and AJ margin are continuous variables that are inversely related for any given range and cost. Generally, increasing any of the other three parameters will increase the cost of the data link.

Operating range is driven directly by mission requirements and may be the easiest parameter to fix. It is not likely to be available for tradeoff by the system designer. Once it is fixed, it places the data-link design in one of two general regimes: (1) for line-of-sight ranges and (2) for beyond-line-of-sight ranges.

1) For line-of-sight ranges, ground-antenna gain can be substituted for processing gain at reasonable cost (up to 30 or 40 dB) to allow higher data rates for the same AJ margin. This allows a four-way tradeoff of data rate, processing gain, AJ margin, and ground-antenna size and cost (including active antenna processing), with cost as a parameter of the tradeoff.

Introduction to UAV Systems, Fifth Edition. Paul Gerin Fahlstrom, Thomas James Gleason, and Mohammad H. Sadraey.
© 2022 John Wiley & Sons, Inc. Published 2022 by John Wiley & Sons, Inc.
Companion website: www.wiley.com/go/fahlstrom/uavsystems5e

2) For beyond-line-of-sight ranges, ground-antenna gain is not available in the tradeoff unless a large airborne relay vehicle is provided. Using either low frequencies for direct propagation or a small relay vehicle (or both), the tradeoff is limited to three factors: data rate, processing gain, and AJ margin. Even for a moderate AJ margin, it is likely that the available transmission bandwidth will be fully utilized, so that the tradeoff becomes a direct trade of data rate for the AJ margin.

Airborne data and communication relay systems are increasingly considered to be likely features of future battlefields. They might provide the large platforms required for high-gain antennas to provide AJ data-link performance with dedicated UAV data links without the need to field and support those platforms solely for UAV use.

As increasing numbers of military systems become dependent on some sort of networking capability for rapid and universal exchange of data, it becomes increasingly likely that a UAV system will be required to use some distributed network that is not part of the UAV system and largely beyond the control of the UAV system designer. If this is the case, the UAV system proponent must be prepared to ensure that any unique UAV mission requirements are supported by the network. Likely examples of requirements that are unique to UAVs (and, perhaps, unmanned ground vehicles) include closed-loop control of processes that cannot tolerate large data latency in the network.

Operating frequency is involved in the above tradeoffs via its effect on:

- Availability of antenna gain;
- Line-of-sight versus beyond-line-of-sight propagation characteristics;
- Limits on transmission bandwidth and thus on processing gain.

As a general rule, higher AJ margins will require higher frequencies. Higher frequencies will increase hardware costs.

Data rate is the factor in the tradeoff that is most controllable by the system designer and user. Onboard processing and capitalizing on advances in electronics can significantly reduce the volume of data that must be transmitted for a given information content. Appropriate design of control loops and system software can accommodate time delays due to data-rate reduction and data latency. Finally, choices of how to use the UAV system that are made with an awareness of data-link limitations can allow successful operations despite those limitations.

With these factors in mind, it is possible to describe a hierarchy of data-link attributes, ranging from those that are easy (low cost) to achieve to those that are extremely difficult (high cost). Table 16.1 provides a hierarchy of data-link attributes during design tradeoff.

Except for the last category (in Table 16.1), there is no question that all of these attributes can be provided in a dedicated data link that can be used with a UAV. The ranking by difficulty represents escalation in complexity and cost. Technical risk is probably not more than moderate for any of these attributes, but schedule and cost risk could be high for the more difficult combinations of attributes.

For calibration, MICNS falls in the "very difficult" category. There is some ambiguity between the "easy" and "moderately difficult" categories, depending on how many of the attributes listed are combined in a single system and on some basic choices in system design. However, there is no doubt that the attributes listed under "very difficult" and "extremely difficult" belong to data links that are at least expensive, if not risky.

It must be noted, however, that if a secure satellite network is available and the UAS missions can be performed with significant transmission delays in the data link, then the ultimate level listed above as "extremely difficult" probably is not only possible but also relatively simple and inexpensive, given that the secure satellite network is provided at little or no cost to the UAS. This

Table 16.1 Hierarchy of data-link attributes during design tradeoff

No.	Level of Difficulty	Data-Link Attributes
1	Easy	• Resistance to unintentional interference • Protection from ARMs • Remote ground distribution of sensor data (without AJ) • Geometrical AJ (antenna gain only) at line-of-sight ranges • High data-rate downlink without processing gain
2	Moderately difficult	• AJ-capable uplink • Resistance to exploitation and deception • Moderate AJ margin on the downlink at 1–2 Mbps and long range • Low-probability-of-intercept uplink • Navigation data at line-of-sight ranges
3	Very difficult	• High AJ margin on downlink for 10–12 Mbps and line-of-sight ranges • Somewhat lower AJ margin on downlink for 1–2 Mbps and beyond-line-of-sight ranges
4	Extremely difficult	• High AJ margin on downlink for 10–12 Mbps and beyond-line-of-sight ranges

is a very significant impact on UAS tradeoffs from the revolution in worldwide communications over the past thirty years. Of course, the infrastructure for that network may be vulnerable to interference, jamming, or "hacking" of various types, and that must be understood and taken into account by the UAS designer.

A "low-cost, jam-resistant" data link probably should fall in the "moderately difficult" category. If so, it should not be expected to have data rates above 1–2 Mbps unless it is limited to line-of-sight ranges.

The discrete change in characteristics that occurs at the transition from line-of-sight to beyond-line-of-sight ranges suggests that a common data link that attempts to cover both of these range requirements will probably be driven to a significantly more expensive design than either of two different data links designed for the different range conditions. This distinction blurs somewhat for the most capable data links, since they have already been driven to the most expensive configurations in order to meet data-rate and AJ requirements.

16.3 Pitfalls of "Putting Off" Data-Link Issues

A UAV system that is designed to make use of high data rates and low data latencies available with little or no AJ margin in a low-cost "interim" data link may be found to reach a dead end if an attempt is made to upgrade the data link at a later time to provide high AJ capability or to use a network that has significant data latency. The choices may be limited to:

• Major redesigns to the UAV system, including major changes to training and mission profiles;
• A very expensive data link using large airborne relays with tracking, high-gain antennas; or
• AJ margins that are not adequate for the EW environment.

To avoid this dead end, it is necessary to take the attributes of the objective data link into account in the original design. This requires determining what AJ margin eventually will be required and determining what implications this will have on data rate and/or what tolerance of data latency will be required. The system, including the manner in which it will be used, must then be designed in such a way that the burden of supporting acceptable mission performance is reasonably partitioned between the various subsystems to produce an affordable objective system that meets all essential requirements.

If a "high/low" mix of data links is to be kept in the field after the objective data link is available, it is probably necessary to provide an interface system that allows the simple, non-AJ data link to emulate the objective data link for training. If this is not done, the operators may require retraining when they transition between the two data links because of the impact of data-rate reduction on operator task performance. An interface of this type probably could be located entirely within the ground station, but the technical challenge of designing it should not be underestimated.

16.4   Future Technology

From a technology standpoint, the highest leverage with regard to data links appears to be in the areas of:

1) Improved onboard processing to reduce data-rate requirements and
2) Better understanding of operator task performance to allow design of procedures that make the best use of available data rates.

It is critical to understand the applicable limitations and options and to select system designs, mission profiles, and operator procedures that allow mission performance within affordable data-link constraints.

Questions

1) What do AJ, EW, and MICNS stand for?
2) Most of the tradeoffs associated with the selection and design of a data link for a UAS involve things that are beyond the boundaries of the data link itself. List at least five items.
3) Write four strongly interacting factors in data-link design.
4) Data rate and AJ margin are continuous variables for any given range and cost. Are they inversely related or directly related?
5) For tradeoffs involving the AJ margin, the effect of range can be considered a step function. Explain the concept.
6) In the tradeoff during the design of a data link for line-of-sight ranges, provide one solution to increase data rates for the same AJ margin at reasonable cost (up to 30 or 40 dB).
7) In the tradeoff during the design of a data link for beyond-line-of-sight ranges, what is needed to make ground-antenna gain available?
8) Discuss the effects of operating frequency in the data-link design tradeoffs for beyond-line-of-sight ranges.
9) Briefly discuss the relation between AJ margins and: (a) frequencies and (b) hardware costs during a data-link design tradeoff.

Questions 361

10) What factor is most controllable by the system designer and user in the data-link design tradeoff?
11) Provide one technique to significantly reduce the volume of data that must be transmitted for a given information content.
12) What data-link attributes are easy to achieve (low cost) during a design tradeoff?
13) What data-link attributes are moderately difficult to achieve during a design tradeoff?
14) What data-link attributes are very difficult to achieve during a design tradeoff?
15) What data-link attribute is extremely difficult to achieve during a design tradeoff?
16) What is the infrastructure for the network of a UAS – that its mission can be performed with significant transmission delays in the data link (i.e., satellite network) – vulnerable to?
17) What is the expected data rate for a "low-cost, jam-resistant" data link?
18) What are the choices to upgrade – for a UAV system that is designed to make use of high data rates and low data latencies available with little or no AJ margin – to provide a high AJ capability or to use a network that has significant data latency?
19) Briefly discuss pitfalls of "putting off" data-link issues.
20) From a technology standpoint, what are the areas for the highest leverage with regards to data links?

Part VI

Launch and Recovery

Launch and recovery are often described as the most difficult and critical phases of UAV operations, and justifiably so. These two flight phases are inherently of accelerated/decelerated motion, since the aircraft speeds at the beginning of launch and at the end of recovery are zero.

To recover an unmanned vehicle aboard a small, pitching, rolling, heaving ship requires precise terminal navigation, a quick response, and reliable deck-handling equipment. Operating out of a small field during ground operations requires many of the same characteristics as at sea. In the latter case, the platform may be stationary, but is usually surrounded by obstacles and subject to the vagaries of the wind. An analysis of the impact of all of these factors would require a separate volume. Here, we will discuss the basic principles and pertinent parameters of launch and recovery that are necessary to make decisions as to the relative merit of individual systems.

For the larger fixed-wing UAVs that are beginning to appear, launch and recovery often is just takeoff and landing, using a runway or the deck of an aircraft carrier. When a UAV has a conventional takeoff and landing, the launch and recovery systems depend mainly on the landing gear. In this case, the only thing that is different about the UAV is that the pilot is not onboard the aircraft. This requires either a manual, remote-controlled takeoff or landing or that there is some level of automation in the process, with or without human intervention from a control station. Some smaller UAVs operate in a similar manner. Given that there are currently automated landing systems for manned aircraft and that landing is considerably more difficult than taking off, the use of runways or carrier decks for launch and recovery is well within the state of the art.

This Part discusses runway takeoff and landing for smaller AVs but concentrates on the types of launch and recovery that are more unique to unmanned systems. These include a number of concepts for "zero-length" launch and recovery that eliminate the need for an open area or large deck, such as catapult launch and net recovery or mid-air recovery. Also addressed are concepts for recovery of VTOL AVs on ships, including smaller ships that can be expected to be rolling and pitching in even moderate seas.

Currently, the US Air Force uses a concept called "Remote-Split Operations," where the satellite datalink is located in a different location and is connected to the GCS through fiber optic cabling. This allows Predators to be launched and recovered by a small "Launch and Recovery Element" and then handed off to a "Mission Control Element" for the rest of the flight. These two elements form the ground control station of the predator.

Introduction to UAV Systems, Fifth Edition. Paul Gerin Fahlstrom, Thomas James Gleason, and Mohammad H. Sadraey.
© 2022 John Wiley & Sons, Inc. Published 2022 by John Wiley & Sons, Inc.
Companion website: www.wiley.com/go/fahlstrom/uavsystems5e

17

Launch Systems

17.1 Overview

In order for any air vehicle to become airborne, it needs the means to accelerate up to a certain speed to takeoff from the ground or sea. Launch is the process of transitioning a UAV from a non-flying state (e.g., stationary on the ground) to a flying state (e.g., climb or cruise). The purpose of a launch system is to place a UAV into a flight path as rapidly as required. Thus, the primary function of a UAV launch system is to make a UAV become airborne.

In general, there are primarily three techniques of launching UAVs: (1) conventional horizontal takeoff and landing (HTOL), (2) a catapulted or powered launch (including hand-launch), and (3) vertical takeoff and landing (VTOL). The first method requires a long runway and is primarily employed by large and medium-weight UAVs. The powered launch is mainly utilized by small UAVs and requires a launcher. The third technique is applicable when the engine thrust is greater than the UAV weight. All multicopters and helicopter UAVs employ the VTOL technique.

If a runway or road is always available for a particular UAV, then the simplest and least expensive launch mode is to take off using wheeled landing gear. There still might be reasons for using one of the other techniques discussed in this chapter, but they would be based on some system-specific requirements. In this chapter, we primarily address the launch of small-to-medium AVs, particularly using concepts that avoid the need for a runway, road, or large, open area.

Launch without a takeoff run is often referred to as a "zero-length launch." In fact, it is generally necessary to accelerate any fixed-wing AV to some minimum controllable airspeed before releasing it from the launcher, and that cannot be done in zero distance. However, the use of catapults or rocket-boosters can achieve a launch distance that is of the order of from one to a few times the length of the AV.

17.2 Conventional Takeoff

The simplest and most cost-effective technique to launch a fixed-wing UAV is the conventional takeoff on a runway. The vast majority of large fixed-wing UAVs employ conventional takeoff and landing, which requires a length of horizontal flat surface (runway). An aircraft – which usually has its maximum weight at takeoff – usually employs its maximum engine power or thrust during

---

Introduction to UAV Systems, Fifth Edition. Paul Gerin Fahlstrom, Thomas James Gleason, and Mohammad H. Sadraey.
© 2022 John Wiley & Sons, Inc. Published 2022 by John Wiley & Sons, Inc.
Companion website: www.wiley.com/go/fahlstrom/uavsystems5e

366 17 Launch Systems

takeoff. Takeoff is considered complete when the aircraft has reached a safe maneuvering altitude or an en-route climb has been established.

Medium to large UAVs such as General Atomics MQ-1 Predator (see Figure 11.1), General Atomics MQ-9 Reaper (see Figure 1.5), and Northrop Grumman RQ-4 Global Hawk (see Figure 1.4) are conventionally taking off using their landing gears. The high-altitude long endurance UAV, Global Hawk, whose turbofan engine thrust is just over 40 kN, requires a ground run of over 600 m.

Predator is following a conventional launch sequence from a semi-prepared surface. The mission can be controlled through line-of-site data links or through satellite links. The Predator B, which employs a turboprop engine with a variable pitch propeller, has a ground roll of about 300 m. Reference [12] provides more information on conventional takeoff, including its segments, control operations, governing equations, and calculations.

17.3 Basic Considerations

The basic governing equations to be considered for launch and recovery are generally developed in "Dynamics." The governing formulae, which are interrelated, are presented in this section. In a linear accelerated motion (starting from rest), the final velocity (V) is a function of acceleration (a), displacement (S), and an efficiency factor (n) as

$$V^2 = 2aSn \tag{17.1}$$

As the vehicle is moving, the kinetic energy (KE) at each velocity is

$$KE = \frac{1}{2}mV^2 \tag{17.2}$$

Work and energy have the same unit (J in the SI system). The work is defined as the force (F) applied on an object along its displacement. The work done on an object is equivalent to its kinetic energy (KE):

$$KE = FS \tag{17.3}$$

Based on Newton's second law, for a constant mass object, the applied force is equal to its mass (m) times its acceleration (a):

$$F = ma \tag{17.4}$$

In the above four equations, V is velocity, a is acceleration (in launch) or deceleration (in recovery), n is the efficiency factor, m is the total mass to be accelerated, F is the applied force, and S is distance over which the force must be applied (launch distance or stopping distance for recovery, also called "stroke"). The UAV velocity at the end of the launch operation should be about 10%–30% greater than the stall speed of the UAV. If this velocity is less than the stall speed, the UAV will stall and crash.

All real systems will have some variations in the acceleration during the stroke and a friction between the UAV and the launcher. The efficiency factor (n) is an empirical adjustment factor that takes these two issues into account. If the acceleration were to be constant, of course, the value of n would be 1, and Equation (17.1) reduces to the familiar equation:

$$V^2 = 2aS \tag{17.5}$$

17.3 Basic Considerations

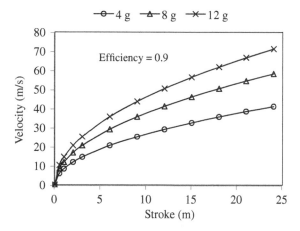

Figure 17.1 Velocity versus stroke

These relationships are shown in Figure 17.1 as a plot of velocity versus stroke for three selected accelerations, expressed in units of g. For the sake of discussion, an $n = 0.9$ is included in Figure 17.1. One can see from Equation (17.1) that, for a given velocity, the loss of efficiency requires a longer stroke to either launch or recover the vehicle at the selected value of acceleration or deceleration.

For ease of calculation, it is assumed that the UAV we are interested in has an all-up weight of 1,000 lb (453.6 kg). For the current discussion, we will merely consider it "the weight" of the vehicle. Actually, the performance of a launcher must consider the "tare weight," not just the AV weight. The tare weight includes the weight of the AV and of all moving parts connected to the shuttle that carries the air vehicle up the launch rail. It is also assumed that the vehicle requires a launch or recovery velocity of an 80-knot (41.12 m/s) True Air Speed, there is no wind, and that the vehicle and its component parts can withstand an 8-g longitudinal acceleration or deceleration.

Referring to Figure 17.1, one can see that the launch (recovery) stroke required for the assumed system with an acceleration of 8 g and an efficiency of 0.9 is about 12 m. Figure 17.2 is a plot of kinetic energy that must be provided to launch a vehicle of a given weight to an 80-knot flight speed. From this plot, we see that to launch the 1,000 lb (453.6 kg) vehicle requires expending

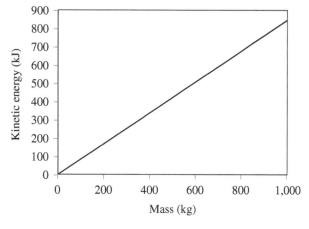

Figure 17.2 Kinetic energy versus mass

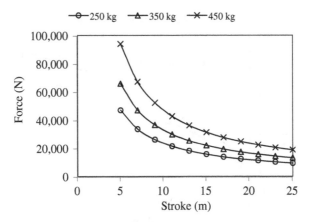

Figure 17.3 Force versus stroke for various vehicle weights

approximately 400 kJ of energy. Conversely, to recover, or stop, the vehicle requires that approximately the same amount of kinetic energy be absorbed. Velocity is the key factor in these calculations, since it is the velocity-squared factor that dominates the energy to be provided or absorbed.

Once the required energy level is determined and knowing the stroke necessary to limit acceleration to the selected value, it is easy to calculate the force that must be applied to an air vehicle over the length of stroke to reach the launch velocity within the g limitation. Figure 17.3 is a plot of force versus stroke for the kinetic energy values for three masses. From this plot, we see that for a launch stroke of approximately 15 m and a 450-kg mass, the force required is about 30,000 N (about 6,750 lb force). It is important to note that this theoretical force must be applied over the entire launch (or recovery) stroke; if not, then the actual stroke must be adjusted accordingly.

While we have accounted for some loss of efficiency in calculating the required stroke in Equation (17.1), we must now look at the force/stroke relationship for the particular power source (or energy absorbing source) to be used.

Remembering that the kinetic energy is the area under the force–stroke curve, Figure 17.4 shows the performance that results from the use of an elastic cord to drive the launcher. Typical of this

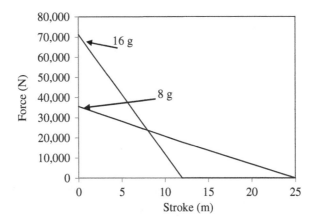

Figure 17.4 Force versus stroke for an elastic cord

17.3 Basic Considerations

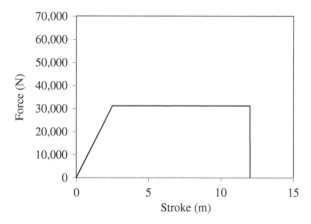

Figure 17.5 Force versus stroke for a pneumatic-hydraulic launcher

type would be a bungee-cord launcher. Force and acceleration are high at the beginning of the stroke and decays as the stroke proceeds. Obviously, the most desirable device would be one that provides a constant force over the necessary stroke distance.

Practically speaking, it is possible to obtain constant (or near constant) force over the stroke. However, to reach the desired force level quickly and efficiently, a rapid rate of change of applied force is necessary and frequently results in force overshooting the desired level. The overshoot, in turn, can lead to excessive "g" forces at the beginning of the stroke or, for recovery, the end of the stroke. To avoid an overshoot problem, the launcher design needs to allow time for a controllable buildup of forces that can be leveled out without significant overshoot. This requires a somewhat longer stroke in order to provide the required level of kinetic energy. Figure 17.5 shows a typical force–stroke plot for a pneumatic-hydraulic launcher with a tailored force that builds up to a desired level and is then constant for the remainder of the stroke.

As previously stated, the foregoing discussion is for basic theoretical considerations. These principles apply regardless of the means of launch or recovery. Of course, there are other practical considerations, which vary depending on the mechanical equipment used. For example, to the casual observer, rocket launch appears to be "zero length," but, in reality, the rocket must impart the required force (as derived from the formulae presented) over the distance calculated, so although the mechanical part of the launch equipment may be "zero length," the UAV must ride the rocket thrust vector over the calculated distance. Similarly, during a net recovery, there are portions of the stopping energy absorbed by the net and purchase line stretch that reduce the amount of energy that needs to be absorbed by a braking system.

Remember that acceleration (a) is defined as the rate of change of speed:

$$a = \frac{V_2 - V_1}{t} \tag{17.6}$$

During a launch operation, the initial speed, V_1, is zero. This equation can be used to determine the duration of launch if the acceleration is assumed to be constant.

Example 17.1

A small UAV with a mass of 320 kg is launched on a rail launcher with a length of 4.6 m. The speed at the end of launcher (i.e., launch velocity) is 25 m/s. Assume the efficiency factor to be 0.97.

a) Determine the UAV acceleration during launch in terms of g.
b) Calculate the launch time.
c) What force is applied during launch?

Solution

UAV acceleration in terms of g:

$$V^2 = 2aSn => a = \frac{V^2}{2Sn} = \frac{25^2}{2 \times 4.6 \times 0.97} = 70 \frac{m}{s^2} \qquad \text{from (17.1)}$$

$$a(g) = \frac{a}{g} = \frac{70}{9.81} = 7.1$$

The launch time:

$$a = \frac{V}{t} => t = \frac{V}{a} = \frac{25}{70} = 0.36\,s \qquad \text{from (17.6)}$$

Applied force:

$$F = ma = 320 \times 70 = 22411.5\,N = 22.4\,kN$$

from (17.4)

17.4 Launch Methods for Fixed-Wing Air Vehicles

17.4.1 Overview

There are many ways in which a fixed-wing UAV can be launched. Some are quite simple in concept, while others are very complex. A number of launch concepts are derived from a full-scale aircraft experience, while others are peculiar to small unmanned air vehicles.

Perhaps the simplest method is the "hand launch," derived from model airplane usage. This method is practical, however, only for comparatively light-weight air vehicles (under about 10 lb) having low wing loading and adequate power. Assume, a typical military male operator may provide a hand force of up to 100 N. Moreover, the length of an extended hand is about 90 cm. Also, simple, but typically requiring a prepared surface, is a normal wheeled takeoff. The small UAV AeroVironment RQ-11 Raven with a mass of 1.9 kg and a wing span of 1.37 m is hand-launched.

Some UAVs, particularly target drones, are air-launched from fixed-wing aircraft. These UAVs typically have relatively high stall speeds and are powered by turbofan/turbojet engines. Such vehicles frequently are also capable of being surface launched using rocket-assisted takeoff (RATO). The RATO launch method will be discussed in greater detail later, but generally requires that the launch force be applied over a significant distance in order to have the vehicle reach flying speed. For this application, the line of action of the UAV propulsion force must be carefully aligned to ensure that no moments are applied to the vehicle, which might create control problems. One way – to accommodate this requirement – is to make sure that the line of action of the UAV propulsion force is aligned with the center of gravity of the fixed-wing aircraft.

17.4 Launch Methods for Fixed-Wing Air Vehicles

Figure 17.6 Truck launch

If one has a smooth surface available, even if too rough for a takeoff on the small wheels of a small UAV, a truck launch is a low-cost practical approach. The larger wheels and suspension of even a small truck can allow driving it at a takeoff speed despite gravel, washboard surfaces, or high grass that would make it impossible for a UAV smaller than a light aircraft to use the surface as a runway. The AV is held in a cradle that places it above the cab of the truck with its nose high to create the angle of attack for maximum lift. Once the airspeed is sufficient, the AV is released and lifts directly upward off its cradle into free flight. Driving a truck at over 88 m/s (60 mph) with a UAV and its supporting structure mounted above the roof might be exciting! Such an arrangement has been used and is illustrated in Figure 17.6.

One novel approach to a UAV launch is a rotary system used with small target drones during World War II and by Flight Refueling Ltd in the United Kingdom for their Falconet target. Figure 11.2 (in Chapter 11) illustrates the concept of a rotary launcher.

In this system, the UAV is cradled on a dolly that is tethered to a post centrally located within a circular track or runway. The engine is started with the UAV on the dolly. The dolly is released and circles the track, picking up speed until the launch velocity is reached. The UAV is then released from the dolly and flies off tangentially to the circle. While this system requires some interesting control inputs at the instant of release, it does work and is relatively easy to operate. The system does, however, require significant real estate and is not mobile.

Another launcher type that has been proposed in the past is the "flywheel catapult." This launcher uses the stored energy in a spinning flywheel to drive a cable system attached to a shuttle holding the UAV. The idea is that the flywheel can be brought up to speed slowly and when "launch" is called for, the flywheel engages a clutch attached to the power train (cables, etc.) and transfers its rotational energy to the UAV. Variations of this launcher type have used mechanical clutches and electromechanical clutches.

While "flywheel" launchers have successfully been built for test and prototyping purposes, most have launched UAVs weighing no more than several hundred pounds and at comparatively low launch speeds. The problem with this concept is the operation of the clutch. Most clutch designs are not robust enough to withstand the rapid onset of energy transfer.

Large UAVs that use runways for conventional takeoffs and landings present some autopilot and control challenges, but otherwise require no special launch and recovery subsystems. The remainder of this discussion concentrates on smaller UAVs using less conventional approaches to launch and recovery.

Many UAV launch systems have a requirement to be mobile, which means being mounted on a suitable truck or trailer. Generally, these systems can be categorized as either "rail" launchers or "zero length" launchers. The material that follows addresses each type separately.

17.4.2 Rail Launchers

A rail launcher is basically one in which the UAV is held captive to a guide rail or rails as it is accelerated to launch velocity. Although a rail launcher could use rocket power, some other propulsion force is usually utilized. The launch angle is often adjusted to be slightly greater than or equal to the UAV maximum climb angle. A typical launch angle is about 15–30 degrees.

Many different designs of rail launchers have been used or proposed for use with UAVs. Bungee-powered launchers have been used for test operations, but this power source is limited to very lightweight vehicles. A typical example of bungee launcher is the one used to launch the Raven RPV in the United Kingdom. For small AVs, the bungee launcher can be configured much like a large slingshot without a rail and may be hand-held.

UAVs that are utilizing rail launchers do not need landing gear for takeoff; thus, they are lighter in weight, which results in a longer range and endurance. Most rail launchers used to launch UAVs in the 500–1,000 lb weight class use some variation of pneumatic or hydraulic-pneumatic powered units. UAVs such as Insitu ScanEagle (see Figure 2.4), Boeing Insitu RQ-21 Blackjack (Figure 17.7), and AAI RQ-7 Shadow are employing rail launchers. It is interesting to note that AAI RQ-7 Shadow is equipped with landing gear, but launched with a rail launcher. Its landing gear is only employed for a landing operation, not for takeoff.

The Boeing Insitu RQ-21 Blackjack, formerly called the Integrator, with a maximum takeoff weight of 135 lb, is a small tactical unmanned air system. It is a prop-driven twin-boomed, monoplane, with a single 8 hp piston engine and a wing span of 4.9 m. This UAV has a maximum speed of 90 knot, a range of 93 km, endurance of 16 hours, and a service ceiling of 19,500 ft. Its first flight was in 2012. The Blackjack is utilized to provide Navy units ashore with a dedicated battlefield

Figure 17.7 Boeing Insitu RQ-21 Blackjack (Source: Lance Cpl. Rhita Daniel / Wikipedia / Public Domain)

17.4 Launch Methods for Fixed-Wing Air Vehicles 373

Intelligence, Surveillance, and Reconnaissance (ISR) capability and its nose-mounted electro-optical/infrared sensor payload provides accurate target locations.

17.4.3 Pneumatic Launchers

Pneumatic launchers are those that rely solely on compressed gas or air to provide the force necessary to accelerate the UAV to launch velocity. These launchers use compressed-air accumulators that are charged by a portable air compressor. When a valve is opened, the pressurized air in the accumulators is released into a cylinder that runs along the launch rail and pushes a piston down that cylinder. The piston is connected to an AV cradle that rides on the launch rail, sometimes by a system of cables and pulleys that can increase the force available at the expense of the stroke or increase the stroke at the expense of a smaller force. The cradle is initially locked in place by a latch. The unlatching process may use a cam to reduce the rate of onset of acceleration. At the end of the power stroke, the cradle is stopped using some type of shock absorbers and the AV flies off the carrier at sufficient airspeed to maintain flight.

Pneumatic launchers are satisfactory for relatively lightweight UAV launches, but operating at low ambient temperatures can be troublesome. Using ambient air at low temperatures, it has been found that pollutants and moisture combined in the compressed air and adversely affect operation. The addition of conditioning equipment to solve the problem presents weight and volume problems.

Another novel pneumatic launcher concept is one using a "zipper" sealing free piston operating in a split cylinder. The cradle or other device, which imparts the driving force to the UAV, is connected to the free-running piston. As the piston moves along the length of the cylinder, the sealing strap is displaced and re-emplaced. The compressed air is held in a tank until "launch" is signaled. At that time, the compressed air is fed into the launch cylinder through a valve that modulates the onset of pressure to reduce initial shock loads, and, in some cases, the valve regulates pressure throughout the stroke in an attempt to achieve constant acceleration. At the end of the power stroke, the piston either impacts a shock absorber or pressure that builds up ahead of the piston brings the piston to a halt.

This type of launcher would have the same drawbacks as exhibited by other pneumatic launchers described above. In one case, an attempt to use a "zipper seal" launcher was made after it had been sitting in rain and drizzle for several days. Although the prescribed pre-pressure was set, the launch velocity achieved was only about two-thirds of that predicted. After several additional attempts, the prescribed velocity was achieved. An investigation determined that moisture was sealing the tape ahead of the piston creating a back-pressure, thus retarding the forward acceleration of the piston. Another possible problem with this type of launcher could be the proper mating of cylinder sections in the event the launcher needed to be folded for transportation.

A third type of pneumatic launcher is one that has been used with the Israeli/AAI Pioneer UAV. In this design, the compressed air, stored as before in a large tank, is discharged into an air motor, which in turn drives a tape spool. This spool, when powered, winds a nylon tape secured to the UAV with a mechanism that releases the end of the tape as the UAV passes over the end of the launch rails. This launcher has no shuttle; rather the UAV is equipped with slippers on the ends of small fins protruding from the fuselage, which ride in slots situated longitudinally along the launch rails. The air storage tank on this launcher contains enough volume to power several launches without refilling or repressurizing. Large tank volume and the effect of increased effective drum diameter as the tape is wrapped on the drum during launch results in a near constant launch acceleration rate, and hence relatively high efficiency.

So far as is known, this launcher was limited to use with UAVs weighing less than 500 lb, with launch velocities of less than 75 knots, and with sustained acceleration rates of 4 g or less. In any event, the launch stroke of units provided to the US Marines has a length of about 70 ft. Based on previous experience with purely pneumatic launchers, the authors would expect that while this launcher appears to operate satisfactorily in a temperate environment, problems could be encountered at low temperatures unless precompression dryers and/or coolers are employed to condition and dry the air. The adaptability of this type of launcher for higher-weight UAVs and higher launch velocities is unknown, but the power requirements for these conditions would involve a significant increase in air-motor size and the volume of air required.

A Boeing Insitu ScanEagle (see Figure 2.4) – with a gross mass of 18 kg and a wingspan of 3.1 m – is launched via a pneumatic launcher. It is catapult launched from a pneumatically operated wedge launcher with a launch velocity of 25 m/s. A number of man-portable pneumatic catapult systems – made of rugged aluminum – were developed by the UAV Factory. For instance, a 4-meter-long catapult – with a pneumatic pressure of up to 16 bars – is able to launch a UAV with a mass of up to 40 kg to a maximum launch speed of 26 m/s.

17.4.4 Hydraulic-Pneumatic Launchers

The power source for a hydraulic-pneumatic (HP) launcher is a closed-loop hydro-pneumatic system, which utilizes both compresses gas (e.g., dry nitrogen) and high-pressure fluid. A hydraulic pump sends high-pressure fluid to store energy by compressing gas in an accumulator. These types of launchers are sometimes referred to as the hydraulic launchers, since the power source is a hydraulic motor (i.e., a pump).

Air vehicles weighing up to at least 555 kg (1,225 lb) have been launched at speeds of up to 44 m/s (85 knots) with this type of launcher. Both full-sized and light-weight variants have been built by All American Engineering (AAE) Company (now Engineered Arresting Systems Corporation (ESCO), a subsidiary of Zodiac Aerospace).

The basic HP launcher concept utilizes compressed gaseous nitrogen as the power source for launch. The nitrogen is contained within gas/oil accumulators. The oil side of the accumulator is piped to a launch cylinder, the piston rod of which is connected to the moving crosshead of a cable reeving system. The cable (in most cases a dual-redundant system) is routed over the end of the launch rail and back to the launch shuttle. The launch shuttle is held in place by a hydraulically-actuated release system. After the UAV is placed upon the launch shuttle, the system is pressurized by pumping hydraulic oil into the oil side of the accumulators, thus pre-tensioning the cable reeving system and applying force to the UAV shuttle.

When the pressure monitoring system indicates that the proper launch pressure has been achieved, the release mechanism is actuated to start the launch sequence. The release mechanism has a programmed actuation cycle that is designed to lessen the rate of onset of acceleration. After release, the shuttle and UAV are accelerated up the launch rail at an essentially constant rate of acceleration.

At the end of the power stroke, the shuttle engages a nylon arresting tape, which is connected to a rotary hydraulic brake, the shuttle is stopped, and the UAV flies off. On some launchers, an optional readout is provided for the launcher end speed. However, variations in end velocity rarely are more than ±1 knot from the predicted value. Unlike purely pneumatic systems, the nitrogen pre-charge is retained, and except for rare leakage, seldom needs replenishment. This allows the use of dry, conditioned air or dry nitrogen in the charge and avoids the problems of using ambient air. The launch energy is provided by the pumps that transfer hydraulic fluid between the accumulators. This type of launcher has a very low visual, aural, and thermal signal.

17.4 Launch Methods for Fixed-Wing Air Vehicles

Figure 17.8 HP 2002 launcher (Reproduced by permission of Engineering Arresting Systems Corporation)

Figure 17.8 is a photograph of an HP-2002 launcher currently produced by ESCO. The HP-2002 is a light HP launcher rated to launch a 68 kg (150 lb) UAV at 35 m/s (68 knots) or a 113 kg (250 lb) UAV at 31 m/s (60 knots). It has a total weight, including a trailer, of 1,360 kg (3,000 lb). Other ESCO HP launchers can be used with AVs up to about 555 kg (1,225 lb).

The hydraulic-pneumatic (HP) launcher concept has been successfully employed in a number of UAV programs. The AAI RQ-7 Shadow 200 is catapult launched from a pneumatic-hydraulic rail launcher.

17.4.5 Zero Length RATO Launch of UAVs

A "zero length" launcher does not use a rail. The AV rises directly from a holding fixture and is in free flight as soon as it starts moving. One of the most common and most successful launch methods is RATO. Rocket assist dates back to the World War II era, when it was used to shorten the takeoff roll required for large military aircraft. In those days, they were called JATOs, for Jet Assisted Take-Off units, a term still occasionally used today.

RATO launch has been routinely used for launching target drones for many years and has been utilized for some of the USAF UAVs such as Pave Tiger and Seek Spinner, for shipboard and ground launch of the US Navy Pioneer and for the US Marine Corps BQM-147 UAVs.

Section 17.5 presents several considerations pertinent to the design of RATO units for UAV applications. The information presented should only be used for preliminary approximations since many factors unique to the particular application and/or AV may significantly influence the RATO unit final design.

17.4.6 Tube Launch

The tube launch system has been recently designed and employed for a number of small UAVs (e.g., AeroVironment Switchblade 300/600 (see Figure 2.7) and Raytheon Coyote). Because the UAV has to fit into a tube, it can no longer have a wide wing and tails, and has a tubular fuselage. These UAVs feature a collapsible wing and tails that fold forward into the body after launch. The

UAV is folded up inside a tube with wings and tails unfolding once it gets airborne. Both sections of the wing are often spring loaded and sweep out upon launching. These types of UAVs are often driven by an electric motor and utilize a twin-bladed propellor in a pusher configuration.

The launch force is provided by a rocket engine, which is separated after launch and recovered by its parachute. To provide a seal chamber, a washer-type ring is placed around the rocket nozzle that fits inside the launch tube. Therefore, the outer diameter of the washer-type ring should be almost equal to the inner diameter of the launch tube, and its inner diameter is almost equal to the outer diameter of the rocket nozzle. This provision is to seal the combustion chamber and to allow the rocket engine to build pressure in the chamber.

17.5 Rocket-Assisted Takeoff

17.5.1 RATO Configuration

Some UAVs are employing rocket engines during launch operations that provide a large force over a short time and a short distance. The rocket engine is just for the launch phase and may be released after takeoff. However, the primary engine for the climb and cruising flight is often a non-rocket engine (e.g., a piston engine).

Rocket-assisted takeoff (RATO) units can be designed to interface with the AV in many different ways, depending on the design of the AV and location of the structural hard points. In some cases, more than one RATO unit is utilized. When a single RATO unit is used, it may be located behind the AV along its longitudinal axis or it may be located below the vehicle fuselage. Where and how the RATO unit is mounted determines its size, its mounting attachment features, and whether its nozzle is axial or canted. In any event, the RATO system is normally designed so that the resultant rocket thrust line passes through the AV center-of-gravity at the time of launch.

As mentioned earlier for "zero-length" RATO launch, the thrust direction must have an upward tilt to support the AV until it is moving fast enough to develop lift.

17.5.2 Ignition Systems

RATO ignition systems can enter the rocket pressure vessel either through the head end or through the nozzle end. Either method is acceptable and can utilize initiators that can be shipped and stored separately and installed in the field just prior to launch.

Several types of initiation have been used. These include a percussion primer actuated by an electrical solenoid for primary initiation and an electric squib built into a remotely actuated rotating safe/arm device. The Pioneer RATO unit used a dual bridge-wire, filter pin-initiator and the Exdrone RATO unit used a percussion-primer-actuated, shock-tube ignition system. Each ignition system was selected to comply with unique system and user requirements. As with munitions, the RATO ignition system will have to meet strict safety requirements to avoid unintentional ignitions.

17.5.3 Expended RATO Separation

The flight performance of most AVs is very weight dependent. It is, therefore, undesirable to carry along expended RATO unit launch hardware for the entire air vehicle flight. Consequently, expended hardware normally is separated from the AV by aerodynamic, mechanical, or ballistic means. Selection of the separation system will depend on how rapidly and in what direction the

17.5 Rocket-Assisted Takeoff

expended hardware must be jettisoned. Care must be exercised, since the falling RATO unit canister becomes an overhead safety hazard for personnel and equipment near the launch site.

17.5.4 Other Launch Equipment

Other launch equipment required for the RATO launch includes a launch stand and usually an AV holdback/release system. The launch stand positions the AV wings level and nose elevated at the desired launch angle. The angle of launch is unique to each specific AV. Normally, it is desirable to minimize the vehicle angle of attack during the RATO unit burn. It is recommended to launch a UAV at an angle that provides a maximum climb rate. The launch stand may provide other features such as deck-tie-down provisions and RATO unit exhaust deflectors, and may also be collapsible or foldable for ease of transport.

The holdback/release mechanism provides a method of restraining the AV against wind gusts and the engine run-up thrust prior to launch; it also provides automatic release of the AV at the time of RATO unit ignition. Systems that have been used include a shear line release for the Pioneer and a ballistic-cutter release for the Exdrone UAV.

17.5.5 Energy (Impulse) Required

The RATO unit designer needs to know the mass of the AV to be accelerated and the desired AV velocity at the RATO unit burnout. These two items determine the energy that must be imparted to the vehicle and will subsequently determine the size of the RATO unit. The required energy, or impulse, is calculated from the impulse-momentum equation as follows:

$$I = m(V_2 - V_1) \qquad (17.7)$$

If the mass (m) of a UAV is entered as kg and the velocity is expressed in m/s, then the calculated impulse will be in units of N·s. For a stationary launcher, V_1 is equal to 0. The above relationship can also be expressed graphically as shown in Figure 17.9. Note that this equation and figure assume that the mass of the RATO unit itself is small compared to the mass of the UAV, since the RATO unit must be accelerated along with the UAV. The RATO unit mass initially includes the mass of the motor grain,

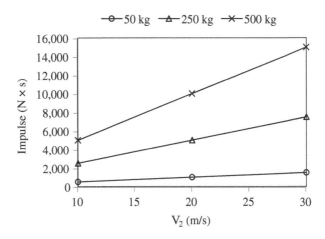

Figure 17.9 Energy requirements for a zero-length launcher

378   *17 Launch Systems*

which burns during the acceleration. As a simple approximation to taking this into account, one might add the mass of the RATO unit to that of the UAV and use the sum as the value of m in the equation.

For example, the Exdrone UAV had a mass of about 40 kg (neglecting the mass of the RATO unit) so for a velocity at RATO burnout of about 15 m/s, it would lie slightly below the line for 50 kg at that value of V_2. This results in a required impulse of about 630 N·s. The Pioneer is significantly heavier, with a mass of about 175 kg with a full set of sensors and for a velocity at burnout of about 40 m/s would require an impulse of about 7,000 N·s.

17.5.6 Propellant Weight Required

The amount of energy, or specific impulse, that a propellant can deliver depends primarily upon the type of propellant used and upon the efficiency of the rocket design. Propellants range from high energy cast composites, such as polybutadiene binders with perchlorate oxidizers, to lower energy slow-burning ammonium nitrates, to extruded single- or double-base formulations. The propellant type will be selected by the designer depending upon the relative importance of such things as environmental conditions, age life requirements, smoke generation, burning rate, specific energy, processability, insensitivity to accidental ignition by artillery fragments and small arms, and cost.

The "specific impulse (I_{sp})" of a propellant is a measure of the amount of impulse that can be produced by burning a unit mass of the propellant:

$$I_{sp} = \frac{I}{W_P} \tag{17.8}$$

The units are impulse divided by weight (note: not mass) of propellant, which comes out to lb·s/lb or N·s/N, which comes out to be just a "second." In general, solid propellants will deliver a specific impulse (I_{sp}) in the range of 180–240 seconds. Thus, the relation between propellant weight (W_P), impulse (I), and specific impulse (I_{sp}) is derived as

$$W_P = \frac{I}{I_{sp}} = \frac{mV_2}{I_{sp}} \tag{17.9}$$

Recall that mass m is the mass of UAV, V_2 is the UAV speed at the end of the launch, and I_{sp} is known for the given propellant.

Example 17.2

The Pioneer UAV with a mass of 175 kg is desired to reach a velocity at burnout of 40 m/s. For the RATO unit, a solid propellant with an impulse of 200 seconds is employed. Find the weight of the propellant that is needed for this launch. Neglect the mass of the RATO unit in the calculation and assume that the total weight of a UAV and RATO unit remains constant during the launch operation.

Solution:

$$W_P = \frac{mV_2}{I_{sp}} = \frac{175 \times 40}{200} = 35\,N = 7.9\,\text{lbf} \qquad\qquad \text{from (17.9)}$$

The RATO unit is providing an impulse of 7,000 m.kg/s (i.e., 175×40).

Rocket design parameters that have an effect on motor efficiency include the operating pressure, the nozzle design, and, to a lesser degree, the plenum volume upstream of the rocket nozzle. Simply dividing the required total impulse by the delivered specific impulse will provide an estimate of the total propellant weight required.

To estimate the overall weight of the RATO unit, one can use the approximation that the RATO unit will weigh roughly twice the propellant weight.

17.5.7 Thrust, Burning Time, and Acceleration

A rocket's energy is delivered as the product of a force or thrust (F) over a duration of impact or a finite time interval (from time t_1 to time t_2):

$$I = F\left(t_1 - t_2\right) \tag{17.10}$$

Acceleration produced on an AV with mass m (or weight w) can be expressed as

$$a = \frac{F}{m} = F\frac{g}{w} \tag{17.11}$$

The maximum acceleration that a vehicle (and onboard subsystems) can withstand is very important and is usually dictated by the structural design of the airframe. Knowing the maximum acceleration and the vehicle weight, the thrust and burn time can be calculated using the above equations.

17.6 Vertical Takeoff

Vertical Takeoff and Landing (VTOL) UAVs, by virtue of their design, need little in the way of launch equipment, especially for ground-based operations. However, logic would dictate that for military operations mobility considerations would require that the VTOL UAV should be operated from a vehicle of some sort. This vehicle would contain devices to secure the UAV during ground transport, and would probably also contain check-out, start-up, and servicing equipment (such as service lifts).

In these UAVs, engines are generating a total upward thrust [42] that should be greater than the UAV weight. After takeoff, the engine's angular velocities or propeller angles of attack are varied to create a forward force needed for a cruising flight. All helicopter UAVs (e.g., Northrop Grumman MQ-8 Fire Scout, MQ-8C Scout, and Yamaha RMAX) and multicopters, including quadcopters (e.g., DJI Phantom 4) are capable of vertical takeoff. In Chapter 20, fundamentals and characteristics of quadcopters are explored.

Questions

1) What do RATO, GCS, HTOL, KE, HP, ISR, and VTOL stand for?
2) What are often described as the most difficult and critical phases of UAV operations?
3) What are the launch and recovery operations for the larger fixed-wing UAVs?
4) Write three prime techniques of launching UAVs.
5) What technique is used in launching a multicopter?
6) What is the simplest and most cost-effective technique to launch a fixed-wing UAV?

7) Define work.
8) Define kinetic energy.
9) What is the similarity between work and energy?
10) What is the primary function of a UAV launch system?
11) What is the typical UAV velocity at the end of a launch operation?
12) What is the UAV tare weight in analyzing performance of a launcher?
13) Briefly describe how to calculate the force that must be applied to an air vehicle over the length of stroke to reach the launch velocity.
14) Briefly describe how to avoid an overshoot problem in the launcher design process.
15) What is the simplest launch method for comparatively lightweight air vehicles (under about 10 lb)?
16) Briefly describe the process for truck launch of a UAV.
17) Briefly describe the features of rail launchers.
18) Interpret how Figure 17.5 shows a typical force-stroke plot for a pneumatic-hydraulic launcher.
19) What are payloads of the Boeing Insitu RQ-21 Blackjack? Where are they located?
20) What is: (a) endurance, (b) maximum speed, and (c) service ceiling of the Boeing Insitu RQ-21 Blackjack?
21) What is: (a) wing span, (b) maximum takeoff weight, and (c) engine power of the Boeing Insitu RQ-21 Blackjack?
22) What is: (a) wing span and (b) maximum takeoff mass of the Boeing Insitu ScanEagle?
23) What are the launch techniques for the following UAVs?
    a) AeroVironment RQ-11 Raven
    b) Northrop Grumman RQ-4 Global Hawk
    c) Boeing Insitu ScanEagle
    d) Boeing Insitu RQ-21 Blackjack
    e) AAI RQ-7 Shadow 200
    f) Northrop Grumman MQ-8 Fire Scout
    g) DJI Phantom 4
    h) AAI Pioneer UAV
    i) AeroVironment Switchblade 300
    j) Yamaha RMAX
24) Briefly describe how pneumatic launchers are operating.
25) Briefly describe how a flywheel catapult launch system is operating.
26) Briefly describe how pneumatic-hydraulic launchers are operating.
27) Briefly describe how the tube launch system is operating.
28) Briefly describe how rocket-assisted takeoff units are operating.
29) Mention at least one drawback of "zipper seal" pneumatic launchers?
30) What is the weight of the HP-2002 launcher?
31) What UAV mass is the HP-2002 launcher rated at?
32) What speed will a 150 lb UAV reach at the end of the launch, if launched by the HP-2002 launcher?
33) Why must the thrust direction have an upward tilt for a "zero-length" RATO launch?
34) Name three separation means for an expended RATO unit launch hardware.
35) What is the function of the holdback/release mechanism in a RATO unit?
36) What is the recommended angle to launch a UAV?
37) Define impulse.
38) Define specific impulse.

Questions 381

39) Consider a Pioneer UAV with a mass of 175 kg. How much impulse is required for launch if the velocity at burnout is 40 m/s?

40) How can a force over-shooting (excessive "g" forces) be avoided at the beginning of the stroke in launching a UAV?

41) Write one example of a high energy cast composite propellant.

42) Write one example of a low energy slow-burning propellant.

43) What is a typical value range for a specific impulse of propellants in "seconds"?

44) A UAV with a mass of 100 kg is desired to reach a velocity at burnout of 25 m/s. For the RATO unit, a solid propellant with an impulse of 220 seconds is employed. Find the weight of the propellant that is needed for this launch. Neglect the mass of the RATO unit in the calculation and assume the total weight of the UAV and RATO unit remains constant during the launch operation.

45) The Pioneer UAV with a mass of 120 kg is desired to reach a velocity at burnout of 20 m/s. For the RATO unit, a solid propellant with an impulse of 240 seconds is employed. Find the weight of the propellant that is needed for this launch. Neglect the mass of the RATO unit in the calculation and assume that the total weight of the UAV and RATO unit remains constant during the launch operation.

46) Name a UAV for each of the following launch powers:
    a) Pneumatic
    b) Hydraulic-pneumatic
    c) Rocket

47) A small UAV with a mass of 280 kg is launched on a rail launcher with a length of 5.3 m. The speed at the end of the launcher (i.e., launch velocity) is 20 m/s. Assume the efficiency factor to be 0.96.
    a) Determine the UAV acceleration during the launch in terms of g.
    b) Calculate the launch time.
    c) What force is applied during the launch?

48) A small UAV with a mass of 150 kg is launched on a rail launcher with a length of 3.7 m. The speed at the end of launch (i.e., launch velocity) is 17 m/s. Assume the efficiency factor to be 0.95.
    a) Determine the UAV acceleration during the launch in terms of g.
    b) Calculate the launch time.
    c) What force is applied during the launch?

49) A 4 lb UAV is hand launched by an operator. What is the maximum launch speed that can be achieved? Ignore air drag and use the typical values for hand force and length of an extended hand.

50) A UAV with a mass of 20 kg on a launcher has reached a velocity of 20 m/s via the RATO launch system.
    a) Calculate the impulse applied to this UAV.
    b) The impulse was applied in 0.3 seconds. Determine the applied force.
    c) Calculate the UAV acceleration during this launch.

18

Recovery Systems

18.1 Overview

At the end of any flight mission, a UAV must fly back to the ground/base and has to land (be recovered). Recovery is defined as transitioning the UAV from a flying state to a nonflying state. Thus, the primary function of a UAV recovery system is to make a flying UAV to become stationary at the ground. UAV recovery can be more challenging than the launch.

The simplest option for recovery is to land the UAV just as one would land a manned aircraft, on a road, runway, smooth field, or carrier deck. For medium-to-large AVs, there are few other options, as the use of nets or parachutes becomes impractical. However, wheeled landing also has been used with many small-to-medium AVs because it is often the least expensive option.

When there is a requirement for "zero-length" recovery, there are a number of options available for small AVs. The most commonly used approaches are identified and discussed in this chapter. Current recovery techniques can be classified into the following items: (1) conventional landings, (2) vertical net systems, (3) mid-air retrieval, (4) parachute recovery, (5) vertical landing, (6) shipboard recovery, (7) break-apart landing, (8) skid and belly landing, and (9) suspended cables. Undoubtedly, there are other schemes unique to specific UAVs and special mission requirements that will not be addressed.

18.2 Conventional Landing

The most obvious fixed-wing UAV recovery option parallels that of full-sized aircraft, that is, runway landing. For all but the smallest AVs, to utilize this option the UAV must be equipped with landing gear (wheels) and its control system must be able to perform the "flare" maneuver typical of fixed-wing aircraft. Experience has shown that directional control during rollout is extremely important, as is the requirement to have some sort of braking system.

One frequently used adaptation to the runway landing technique is to equip the UAV with a tail hook and position arresting gear on the runway. In this way, the need for directional control during rollout and for onboard brakes is minimized. This approach parallels carrier-landing techniques.

There are two types of arresting-gear energy absorbers generally used. One is a friction brake that has a drum, around which is wound cable or tape that is connected in turn to the deck pendant (the cable or line which the UAVs tail hook engages is called a deck pendant even when used on a land

---

Introduction to UAV Systems, Fifth Edition. Paul Gerin Fahlstrom, Thomas James Gleason, and Mohammad H. Sadraey.
© 2022 John Wiley & Sons, Inc. Published 2022 by John Wiley & Sons, Inc.
Companion website: www.wiley.com/go/fahlstrom/uavsystems5e

runway). The second type is a rotary hydraulic brake, a simple water turbine with the rotor attached to a drum, around which is wound a nylon tape. As with the friction brake, the tape is attached in turn to the deck pendant. There is a significant difference between these two braking systems.

With a friction brake, the retarding force is usually preset and fixed and the run-out (the distance it takes to arrest the UAV) varies depending on the UAV weight and landing speed. Rotary-hydraulic brakes, however, are considered "constant run-out" devices, and the UAV will always end up at approximately the same point on the runway even if the weight and landing speed vary. This statement only is true, of course, within limits. A rotary-hydraulic brake system is configured for a "design point" of UAV weight and landing speed and variations of, say, 10%–20% around that design point are readily accepted.

A variation of the classic arresting-gear recovery system is to attach a net to the purchase elements of the energy absorbers in lieu of the deck pendant. The net must be designed to envelop the UAV and distribute the retarding loads evenly on the airframe. Very small AVs may be simply flown into the ground at a shallow angle and allowed to skid to a halt.

Medium to large UAVs such as General Atomics MQ-1 Predator, General Atomics MQ-9 Reaper, and Northrop Grumman RQ-4 Global Hawk are conventionally landing using their landing gears. Reference [12] provides more information on conventional landing, including its segments, control operations, runway, governing equations, and calculations.

18.3 Vertical Net Systems

A logical outgrowth of runway arresting systems, both hook/pendant and net types, is the vertical net concept. In its basic form, the net is suspended between at least two poles with the net extremities attached to purchase lines, which are attached in turn to energy absorbers. The net usually is also suspended by a structure or lines to hold it above the ground and thus suspends the UAV within the net at the end of run-out.

Use of a net generally precludes the use of tractor propellers located at the front of the AV, as they are likely either to damage the net, be damaged by the net, or lead to damage to the engines from forces applied to the propeller and transmitted to the shaft and bearings of the engine. Depending on the configuration of the AV, even a pusher propeller may need to be shrouded to avoid these problems.

When the AAI RQ-2 Pioneer – with a gross mass of 452 kg and wingspan of 5.14 m – is flown and employed by US Navy, it is recovered by a vertical net aboard a ship. The three-pole net recovery system used with Pioneer on the battleship Iowa and with the Lockheed Altair UAV, shown in Figure 18.1, predates most other vertical net systems. In the mid-1970s, the Teledyne Ryan STARS RPV system successfully utilized a three-pole vertical net system. This approach used a "purse-string" arrangement with a reeving system to create a pocket that captures the UAV as run-out proceeds. In some variations, a bungee is introduced to snap up the bottom edge of the net to ensure that the UAV is securely ensnared. Another variation using four poles was used on early Israeli RPV systems.

Key to the success of net systems is the design of the net itself. It should be both strong and flexible to absorb enough energy from the UAV before stopping it to prevent damage. The net should be able to keep the UAV suspended in the net and prevent it from bouncing or falling away from the net after stopping. The net must properly distribute retarding forces on the UAV, and various means have been devised to accomplish this requirement.

Early net systems used simple cargo nets or, in at least one small UAV system, a tennis net. On the Pioneer/Iowa system, early tests using a multiple element net derived from full-scale aircraft work proved to be too heavy, and was easily displaced by the wind over the deck produced by the

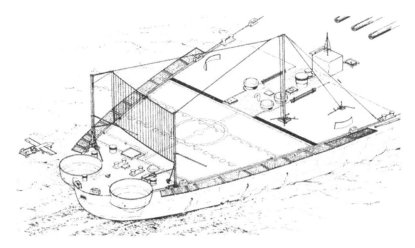

Figure 18.1 Pioneer installation on USS Iowa (Reproduced by permission of Engineering Arresting Systems Corporation)

forward motion of the ship over the water. The final net configuration had a small (15 ft × 25 ft) conventional net at the "sweet point" or aiming point for recovery and triangular members leading to the four corners of the large net. This configuration provided low wind drag and presented a sufficiently large target for manual recovery.

A very important aspect to net design is how accurately the UAV can be flown into the net. AQUILA and Altair used a very accurate automatic final approach guidance system, and variations of only a foot or so from the center of the net were normally achieved. This net was only approximately 15 ft high × 26 ft wide. The net on the Iowa, by comparison, was some 25 ft high × 47 ft wide. These larger dimensions were needed because Pioneer has a larger wingspan and final approach guidance was manual (under radio control with a human operator).

18.4 Parachute Recovery

Parachute terminal recovery systems have a long history of use with target drones and other UAVs. The use of a parachute, of course, requires that the UAV has sufficient weight and volume capacity to accommodate the packed bulk of the parachute. Numerous parachute configurations have been used (e.g., in Mini UAV Orbiter developed by Aeronautics Defense Systems) and they are usually designed to have a relatively low rate of descent in order to decrease the possibility of damage due to the impact with the ground or water. Some UAVs employ inflatable impact bags or crushable structures to attenuate loads on impact. Typical of this approach are the Teledyne Ryan Model 124, which utilized impact bags, and the BAE Systems Phoenix, which inverts to land on a crushable upper structure.

Gliding parachutes have been used in recent years to overcome some of the problems with standard parachutes. The parafoil, or ram air inflated parachute, has been shown to have considerable advantages. Parafoils can be directionally controlled using differential riser control and they trade off forward speed for a low rate of descent. Virtually pinpoint accuracy of landing has been demonstrated using either manual control or by utilizing a homing beacon and an onboard sensor and control system. Parafoils exhibit essentially constant speed characteristics, so if thrust is provided by the UAV engine, the UAV suspended on a parafoil can be caused to climb, maintain level flight, or, with thrust decreased, descend. Directional control under power is generally excellent.

While there are many parachute designs and a wide variation in performance, the cross-parachute design, shown in Figure 18.2, has been particularly successful, having a reasonably low packed bulk and excellent stability during descent. It also has the advantage of providing low forces during deployment.

One disadvantage of the standard parachute configuration is the lack of directional control after deployment. The decent of the UAV after parachute deployment is subject to the vagaries of the wind, and in high surface wind conditions a ground release mechanism is a necessity to avoid damage due to the UAV being dragged along the ground. Parachute deployment at very low altitudes is frequently programmed in order to reduce the drift distance.

Parachute descent into the water requires that the internal systems of the UAV are protected against water immersion or that decontamination facilities are readily available for use after immersion, particularly when descent is made into salt water.

Figure 18.2 Cross parachute (Reproduced by permission of Engineering Arresting Systems Corporation)

For shipboard recovery of parafoil-borne UAVs, some additional ship-based aids are desirable. A promising approach for the parafoil-borne UAV is to use a haul-down system. The UAV drops a line to the ship deck and a winch is used to haul down the UAV. Control of tension by the winch is considered necessary so that the parafoil and haul-down line are not overstressed due to ship motion. The sequence of operation for this approach is pictured in Figures 18.3 and 18.4. Some means of securing the landed UAV is also considered necessary, particularly when shipboard operations at sea states of up to four are considered. (A general discussion of shipboard operational limitations is contained in the following section.)

18.5 VTOL UAVs

A large number of vertical takeoff and landing (VTOL) UAVs have been developed and fielded. These range from a pure helicopter to multicopter to vectored-thrust devices and tilt-wing aircraft. Any of the VTOL UAVs have the distinct advantage of providing for a low relative velocity between the UAV and the ship deck or landing pad. As was explained earlier, this leads to a low relative energy transfer requirement. Vertical landing may be conducted by a remote pilot or by an autopilot with a pre-programmed plan.

All helicopter UAVs (e.g., Northrop Grumman MQ-8 Fire Scout, MQ-8C Scout, and Yamaha RMAX) and multicopters, including quadcopters (e.g., DJI Phantom 4), are capable of vertical landing. In these types of UAVs, the engine thrust – generated via propellers – are upward, so the weight is balanced.

18.5 VTOL UAVs

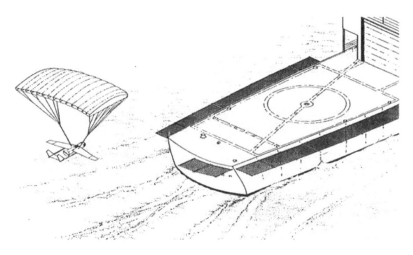

Figure 18.3 Parafoil recovery (Reproduced by permission of Engineering Arresting Systems Corporation)

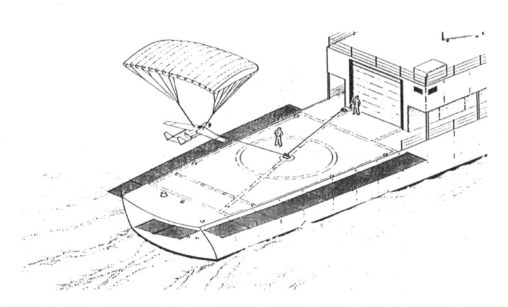

Figure 18.4 Parafoil recovery with winch (Reproduced by permission of Engineering Arresting Systems Corporation)

VTOL UAVs for shipboard operations require a final approach guidance system, an airframe suitable for shipboard operation, and suitable capture equipment to ensure that once landed on the deck, the UAV is secure and will not adversely affect other ship operations.

Figures 18.5 and 18.6 show two possible VTOL recovery techniques in a generic manner. Both involve a launch and recovery platform on the deck of the ship. The platform is on a track that allows it to be moved in and out of a hanger area without risk of sliding around on a heaving deck. There might be multiple platforms with sidings off the main track within the hanger to allow multiple AVs to be stored and retrieved from within the hanger.

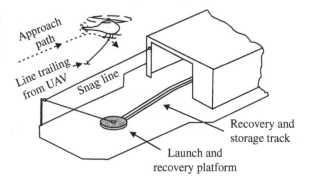

Figure 18.5 VTOL recovery by tether

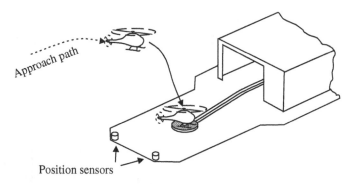

Figure 18.6 VTOL recovery by automatic landing

Figure 18.5 shows a concept where the UAV drops a tether line and hook, which engages a snag line. A clever arrangement of blocks and lines (not shown in detail) allows the snag line to be reeled in by the recovery system to connect the tether line to a winch, similar to the winch recovery associated with parafoil recovery in Figure 18.4. The tether line is then winched in while tension is maintained on the line by the AV. When the AV is in contact with the platform, automatic securing devices lock it down and the AV engine is shut down.

Figure 18.6 shows a tetherless concept in which the AV is automatically landed on the platform using a closed-loop control system based on sensors located on the landing deck. The sensors provide precision information on the location of the AV relative to the recovery platform, which is used to implement a tightly closed loop with the autopilot of the AV and make a precision landing on the platform despite its motion in three dimensions on a heaving deck.

Figure 18.7 provides a more detailed view of a recovery platform, showing a simple concept for securing the UAV after landing. In this concept, hook-shaped clamps are located on tracks on either side of the center of the platform and when the air-vehicle landing skids contact the platform the clamps slide in over the skids and clamp down to secure them to the platform.

18.6 Mid-Air Retrieval

Use of mid-air recovery system (MARS) for UAV recovery provides the opportunity to perform the recovery operation away from the ship and then to fly the UAV as cargo down to the deck, as is done with a normal helicopter operation. It would be possible to equip helicopters currently used

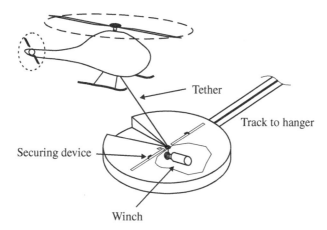

Figure 18.7 Launch and recovery platform

for shipboard operation with a mission kit consisting of an energy-absorbing winch and a pole-operating auxiliary pod that would permit the helicopter to make the mid-air retrieval of a parachute-borne UAV.

If the parachute utilized were of a parafoil type, it would be possible to improve the performance of the recovery operation, since the helicopter pilot would not have to judge the vertical velocity of the parachute-borne UAV when affecting mid-air retrieval if the UAV continued to apply thrust after deployment of the parafoil so that it could continue in slow, powered flight. With power on, the UAV would have a more gradual rate of descent than that of a conventional parachute, giving the helicopter pilot more time to make the recovery. The sequence of operation of the classic MARS recovery is shown in Figures 18.8 to 18.10. For heavy (2,500 lb) UAVs, the main parachute is jettisoned after engagement.

There is a wealth of experience in mid-air retrieval operations within the US Air Force and many thousands of successful recoveries of target drones, cruise missiles, and so on have been made. As an example, the reconnaissance drone program in Vietnam recorded a mission success rate of over 96%. This is one system that does not require a final approach guidance system. MARS requires, however, significant aircrew training, as well as a dedicated, especially configured aircraft. While MARS recoveries generally are only performed with good visibility and during daylight, some experimental night recoveries have been made by illuminating the parachute from below.

The US Army [3] for the first time (in 2020) caught and recovered Area-I Altius UAVs in midair using a quadcopter. For the demonstrations, the Area-I UAVs were launched using a rail atop a truck, as well as from airborne Sikorsky UH-60 and MH-60 helicopters and a General Atomics Aeronautical Systems MQ-1C Gray Eagle UAV. Previously, the Area-I Altius recovery involved belly landing the air vehicle.

18.7 Shipboard Recovery

Safety is the primary concern for UAV shipboard recovery. The type of UAV employed and the means of recovery must not endanger the ship or personnel aboard. This applies to not only the actual recovery action, but also to the recovery equipment installed on the ship, UAV handling, stowage, and all other aspects of the system. Other concerns are reliability and mission effectiveness. In most cases, space aboard the ship is limited; therefore, a high degree of reliability is necessary so that a large supply of spares is not required to keep the system operable.

Figure 18.8 Mid-air retrieval (Reproduced by permission of Engineering Arresting Systems Corporation)

Figure 18.9 Mid-air recovery sequence – snagging (Reproduced by permission of Engineering Arresting Systems Corporation)

18.7 Shipboard Recovery

Figure 18.10 Mid-air recovery sequence – recovery (Reproduced by permission of Engineering Arresting Systems Corporation)

Mission effectiveness means that in addition to safely recovering the UAV, the system must be easily erected and operated by a minimum number of personnel, and must impose the minimum damage to the UAV being recovered. It must not require that the ship significantly deviates from its normal operating conditions in order to affect UAV recovery.

Unlike UAV operations over land, recovery at sea requires that the UAV must perform in spite of the motions of the ship in various sea states and in an environment that is very harsh. The relationship of the sea state to ship motion is complex, and different classes of ship would have different reactions to the various sea state conditions. As an example, the pitch and roll rates of a battleship in a given sea state may be imperceptible compared to those of a frigate.

An example of the conditions under which UAV recovery operations might be conducted is contained in US Military Specification MIL-R-8511A, "Recovery Assist, Securing and Traversing System for LAMPS MK III Helicopter." This specification calls for maintenance of "all required functional characteristics" under stated conditions. The temperature range is $-38\ °C$ to $+65\ °C$ and exposure to relative humidity of 95% where the condensation takes the form of both water and frost. The ship motion conditions are:

> "... when the ship is permanently trimmed down by the bow or stern as much as 5 degrees from the normal horizontal plane, is permanently listed as much as 15 degrees to either side of the vertical, is pitching 10 degrees up or down from its normal horizontal plane, or is rolling up to 45 degrees to either side of the vertical. Full system performance is not required when the ship roll exceeds 30 degrees; however, exposure to ship rolling conditions up to 45 degrees to either side of the vertical shall not cause loss of capability when rolling is reduced to 30 degrees or less."

Complicating the effects of sea-state-induced ship motion on UAV recovery is the fact that most surface vessels create an air wake, or "burble" aft, as a result of airflow past the various superstructures onboard. On some ships, this air burble can significantly affect UAV control as the UAV flies through the burble area on approach to the landing deck. Some data have been collected on this area of concern as an adjunct to ensuring safe helicopter operation from the various ships. UAV designers should take these data into account when considering UAV shipboard recovery and plan to have adequate control when penetrating the burble area, or plan approaches to avoid the area.

Following the Persian Gulf War (1991), the US Navy retired their battleships and lost the shipboard Pioneer UAV capability, which used the Ship Pioneer Arresting System (SPARS) vertical net recovery system. In order to maintain a shipboard UAV capability, the Navy had the SPARS equipment modified for installation on smaller ships designed to support amphibious operations. This system, designated SPARS III, mounted the aft net support poles along the gunnels of the aft flight deck of the ship with the single forward pole mounted slightly off-center on the superstructure. The same basic geometry that was used for the battleship installation was maintained, but the distance between the aft and forward poles was increased. This installation basically took up the entire flight deck, essentially preventing the deck from being available for helicopter operations. Following some initial problems with rigging, the system worked well.

A very significant improvement in SPARS operations has been realized with the introduction of the Common Automatic Recovery System (CARS), which provides extremely accurate final approach guidance. With CARS available, smaller net sizes and overall smaller vertical net systems are feasible. This is important because one of the problems with the SPARS system relates to the area of the erected net and the drag the net imposes on the UAV during recovery. This drag can affect the way the net envelopes the UAV and arrests it without dropping it out of the net.

Another improvement to the SPARS system has been proposed in which four poles are used to suspend the net system, thereby reducing the deck area used for the UAV operation. This modification – coupled with a simplified erection and lowering capability – has the potential for making the ship deck more readily available for helicopter operations between UAV operations. To prevent the remote chance of a UAV undershooting the recovery net, a barrier net can be installed below the primary recovery net.

Vertical-net recovery systems can be adapted to virtually any fixed-wing UAV configuration that does not include tractor propellers. Coupled with CARS or some other automated control system, they provide an effective and reliable recovery system.

Finally, it must be recognized that ship captains are reluctant to have anything resembling a missile aimed at their ship. The UAV recovery method used must have demonstrated a high degree of reliability in its ability to recover the UAV without damage to the ship.

18.8 Break-Apart Landing

Very small UAVs – with a mass less than about a few kg – can be designed for landing without landing gear and any equipment. For this goal, they must be built-in break points so that they break apart on impact. To prevent a catastrophic crash, the UAV needs to glide with a very low angle and with a low speed. Moreover, the structure must be stiff and tough, so that the impact does not influence any onboard system (e.g., autopilot and communications system) and payloads such as a camera. The UAV must be designed and built in a kit form, with the lower surface of the fuselage covered with impact absorbing materials.

To alleviate any impact deceleration, it is recommended that the UAV descends and faces into the wind and is programmed to guide itself and coast all the way toward the ground. After landing, the undamaged break-apart components, such as the wing and fuselage, are attached together for the next deployment and mission. Part of the landing shock is absorbed by the fuselage lower surface, but the remainder will cause the UAV to break apart. The connection parts between the UAV main components should be such that they break by a portion of the impact shock. For instance, if the deceleration is –2g, one g is absorbed by the airframe – mainly the lower skin of the fuselage – and one g will cause the connections to break.

The AeroVironment RQ-11B Raven (Figure 18.11) is designed to snap apart safely while landing and is easily put back together for continued use. It slowly glides in and hits the ground, breaking apart into nine pieces on impact. This UAV is built to break apart in order to absorb the shock of landing without wheels.

Figure 18.11 AeroVironment RQ-11 Raven (Source: Dennis Rogers / U.S. Air Force photo / Public Domain)

18.9 Skid and Belly Landing

In manned aircraft flight operation, belly landing is an emergency landing with the gear in the up position (unextended). However, this undesired operation has been adopted for some fixed-wing UAVs due to its advantages.

In unmanned air vehicle operation, when the UAV is not equipped with landing gear, it may be designed for belly landing. The pilot guides the UAV to land and skid, but it is recommended to land towards the wind and preferably straight into it. This type of landing is not appropriate for medium and large UAVs, but is primarily used by aeromodellers and fixed-wing very small UAVs. Any reasonably flat surface without large obstacles can be used.

In a belly landing, the high friction between the ground and the air vehicle will inevitably slow the vehicle to a complete stop in a short amount of time. Due to the friction damage to the belly, the fuselage should be repaired after a few landings. If the air vehicle is equipped with skid, this component should be repaired, and replaced if necessary, when needed. The lower surface of the fuselage should be covered with impact absorbing materials.

A UAV without landing gear is much lighter compared with when it has the landing gear. Moreover, this UAV will have less challenges in the structural design process, and often incur less cost in manufacturing. For a belly landing, the pilot skill is a significant factor in saving the air vehicle. Moreover, the autopilot can be programmed to conduct an automatic belly landing without remote pilot interference. To alleviate any impact deceleration, the UAV should descend and face into the wind and is programmed to guide itself in a coast all the way toward the ground.

Skid landing has been used successfully, notably by Skyeye, and has the advantage of not requiring a paved surface. Skyeye uses a single skid equipped with a shock absorber and tracks to keep the UAV running straight. The engine is cut on touchdown and friction between the skid and landing surface brings the UAV to a halt. The use of the shock absorber eliminates the need for the flare maneuver; the UAV is merely set up with a low rate of descent and flies onto the landing surface.

Lockheed Martin Stalker with a wing span of 10 ft (3 m) and a maximum takeoff of 17.5 lb is designed for belly landing. This UAV with a mission of "intelligence, surveillance, and target acquisition" is equipped with a prop-driven tractor electric engine. The performance is: (1) maximum speed: 43 knots, (2) endurance: 2 hours, and (3) service ceiling: 15,000 ft.

18.10 Suspended Cables

A suspended cable – also referred to as the arresting line – as a recovery technique is frequently used in fixed-wing small UAVs on ships and aircraft carriers. The system mainly includes a suspended cable and a braking system that are coupled together with a stanchion and a boom. To use this recovery system, the UAV wingtip has to be equipped with a hook in order to be able to latch on to the cable. The cable system can be installed on the deck or on the side of the ship.

Two primary advantages of a suspended cable recovery are: (1) quick recovery and (2) zero-length recovery. In contrast, four disadvantages are: (1) the setup time is long, (2) it suffers from ship motions, (3) rough landing, and (4) the UAV needs a hook.

Small UAV ScanEagle (see Figure 2.4), developed by Insitu and Boeing, had made hundreds of successful shipboard recoveries via a suspended cable. This retrieval system – called SkyHook – utilizes an arresting cable suspended from a 15.2 m boom. The UAV is flown directly to approach the cable and a hook installed in the UAV's wingtip is caught on the cable.

The Integrator shares the same catapult launcher and hooking recovery systems with ScanEagle. Figure 18.12 shows the recovery of a RQ-21A Integrator at sea with a suspended cable. The fielding of the Blackjack (i.e., Integrator) achieved full operational capability in 2019.

Figure 18.12 Recovery of an RQ-21A Integrator at sea with a suspended cable (Source: U.S. Navy / Public Domain)

Questions

1) What do VTOL, MARS, CARS, and SPARS stand for?
2) Define UAV recovery.
3) List typical recovery techniques.
4) What is the primary function of a UAV recovery system?
5) Describe two types of arresting-gear energy absorbers generally used in carrier-landing.
6) Briefly describe the vertical net systems for UAV recovery.
7) What are the requirements for a net in the vertical net systems for UAV recovery?
8) What is the size of a vertical net on the battleship Iowa to recover Pioneer UAV?
9) Write the size of the net on the Pioneer/Iowa system for UAV recovery.
10) Write the size of the net on the AQUILA and Altair for UAV recovery.
11) What are the requirements for a parachute for UAV recovery?
12) Write one disadvantage of the standard parachute configuration.
13) Briefly compare the standard parachute with a cross parachute.
14) Briefly compare the standard parachute with a gliding parachute.
15) What is the requirement for the parachute descent into the water?
16) What are the requirements for the VTOL UAVs for shipboard operations?
17) Briefly describe the technique for the VTOL UAV shipboard recovery (by tether).
18) Briefly describe the technique for the VTOL UAV shipboard recovery (the tetherless concept).
19) What is the sequence of operation of the classic MARS recovery?
20) What is the primary concern for UAV shipboard recovery?
21) Provide at least three concerns for UAV shipboard recovery.
22) What are the ship motion conditions for LAMPS MK III Helicopter recovery according to US Military Specification MIL-R-8511A?
23) Briefly describe the features of SPARS III.
24) Briefly describe the improvements made to SPARS to develop SPARS III.
25) Can the vertical-net recovery system include a fixed-wing UAV configuration that includes tractor propellers?
26) Name at least two UAVs that employ suspended cables for recovery. Provide their images.
27) What are the recovery techniques for the following UAVs?
    a) AeroVironment RQ-11 Raven
    b) Northrop Grumman RQ-4 Global Hawk
    c) Boeing Insitu ScanEagle
    d) AAI RQ-2 Pioneer
    e) Northrop Grumman MQ-8 Fire Scout
    f) Skyeye
    g) DJI Phantom 4
    h) Teledyne Ryan Model 124
    i) Lockheed Martin Stalker

28) What are the challenges of vertical net recovery with regards to the engine propeller?
29) Briefly describe the break-apart landing recovery technique.
30) Briefly describe the UAV requirements for the break-apart landing.
31) Briefly describe how the suspended cable recovery technique operates.
32) Write two primary advantages of a suspended cable recovery.
33) Briefly describe the UAV requirements for a suspended cable recovery.
34) When did the fielding of the Blackjack (i.e., Integrator) achieve full operational capability?

19

Launch and Recovery Tradeoffs

19.1   UAV Launch Method Tradeoffs

In the preceding two chapters, the various methods of UAV launch techniques and equipment were discussed. As a summary, these various methods are listed along with a subjective evaluation of the tradeoffs for each method. For the purposes of this evaluation, only conventional takeoff from a runway or road or other prepared area, pneumatic rail launchers, hydraulic/pneumatic rail launchers, and RATO launchers are considered, since these are the basic types currently in use. Development costs are not considered.

Tables 19.1 through 19.7 present advantages and disadvantages of various launch techniques.

Finally, it might be of interest to compare the operating costs of a typical rail launcher system versus a typical RATO unit launch system. In this tradeoff and, for simplification, personnel costs, transport (truck) costs, and development costs have been deleted, as have any ancillary costs, such as special handling or storage equipment for rockets. Incidental costs, such as for engine fuel, have also been deleted.

Figure 19.1 is a plot representing the costs for each system versus the number of launches. It can readily be seen that for a low number of launches, a RATO system might be attractive, assuming a rocket of the proper size is available without significant Research and Development (R&D) cost. On the other hand, if there is an expectation of a large number of launches, a rail launch would be more attractive.

The UAV developer has a number of launcher options, but must evaluate the advantages and disadvantages of the various launch concepts to determine which is best for a particular AV and set of mission requirements. Above all, the designer should select the launch system early in the design phase, so that the incidence of problems associated with launch can be eliminated, or at least significantly reduced, by producing a total system that integrates launch considerations into all aspects of the design.

Introduction to UAV Systems, Fifth Edition. Paul Gerin Fahlstrom, Thomas James Gleason, and Mohammad H. Sadraey.
© 2022 John Wiley & Sons, Inc. Published 2022 by John Wiley & Sons, Inc.
Companion website: www.wiley.com/go/fahlstrom/uavsystems5e

Table 19.1 Wheeled takeoff using a runway

Advantages	• No hardware needed to implement, and thus no impact of system transportability or cost • Applicable to all sizes of UAVs. • No significant signature.
Disadvantages	• Requires runway, road, or other paved or smooth area • Requires wheeled landing gear that adds to UAV weight and complexity

Table 19.2 Wheeled takeoff using prepared takeoff areas (truck launch)

Advantages	• Allows use of roads or open areas too rough for wheeled takeoff by small UAVs • Avoids need for wheeled landing gear on UAV • No significant signature
Disadvantages	• Applicable only to relatively small UAVs • Requires at least a graded road or flat, open terrain

Table 19.3 Pneumatic rail launchers – split-tube type

Advantages	• UAV held in fixed attitude during launch until flight speed is reached • Relatively low parts count • Low tare weight (total weight of air vehicle, cradle, and all other parts that move up the rail at launch) • Proven concept • Negligible consumables and recurring costs • No significant signature
Disadvantages	• Applicable only to relatively small UAVs • Adds a relatively costly subsystem to the UAS • Degraded performance under adverse weather conditions • Relatively long pressurization time • Need for conditioning system for pressurizing air • Possible sealing problems with folding cylinder • Significant up-front cost • Significant "footprint"

Table 19.4 Pneumatic rail launchers – air motor type

Advantages	• UAV held in fixed attitude during launch until flight speed is reached • Relatively low parts count • Low tare weight • Proven concept • Negligible consumables and recurring costs • No significant signature
Disadvantages	• Applicable only to relatively small UAVs • Adds a relatively costly subsystem to the UAS • Degraded performance under adverse weather conditions • Relatively long pressurization time • Performance at low temperatures unknown • Significant up-front cost

Table 19.5 Hydraulic-pneumatic rail launchers

Advantages	• UAV held in fixed attitude during launch until flight speed is reached • Demonstrated reliability and performance at environmental extremes • Provides repeatable and accurate velocity • Demonstrated with UAVs to at least 1,000 lb and 85 knots • Short pressurization time • Adaptable to wide range of AV weight, speed, and "G" tolerance • Negligible consumables and recurring costs • No significant signature
Disadvantages	• Applicable only to relatively small UAVs • Adds a relatively costly subsystem to the UAS • Significant "footprint" • Significant up-front cost

Table 19.6 RATO launchers

Advantages	• Small "footprint" • No significant environmental limitations • Small up-front cost • No pressurization time required • UAV can be stored for long periods with RATO installed ("ready to go" concept) • Can be used with large UAVs to allow short-field takeoff
Disadvantages	• Applicable only to relatively small UAVs for zero-length launch • Heat, light and aural signature • Rockets require special handling • Alignment with center of gravity of UAV is critical • Safety considerations • Significant recurring cost

Tables 19.7 Vertical takeoff (VTOL AV)

Advantages	• Soft and accurate takeoff. • No supplemental equipment is required.
Disadvantages	• Challenging control system. • VTOL AVs less efficient for cruise.

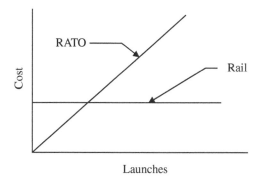

Figure 19.1 Cost tradeoffs for rail versus RATO launch

19.2 Recovery Method Tradeoffs

As with launch, there are a number of options for the recovery of a UAV. A subjective tradeoff between the primary types of recovery is presented in Tables 19.8 through 19.16.

Shipboard recovery has some specialized restrictions, but can use conventional (arrested) landing on a carrier deck, vertical-net recovery, parafoil recovery, or VTOL AVs. A ship-launched AV could also use mid-air retrieval if the ship of some other ship in the formation had manned aircraft.

Ditching in the water is not included in the tradeoffs above, but is a possibility for ship-launched AVs. It is an inexpensive option but has a high probability of damage to the AV and/or its payload.

Table 19.8 Wheel landings using a runway

Advantages	• No supplemental equipment except for arrested landings • Gentle retrieval of sensor equipment
Disadvantages	• Requires either manual or automated landing control • Prepared landing site necessary • Landing sites not easily hidden

Table 19.9 Skid and belly landings (prepared landing sites)

Advantages	• No supplemental equipment necessary • Minimum or no landing site preparation required • Landing sites more easily hidden
Disadvantages	• Hard landing more probable.

Table 19.10 Vertical-net systems

Advantages	• "Zero-length" recovery
Disadvantages	• Adds a relatively costly subsystem to the UAS • Relatively hard landing • Landing site visible to enemy

Table 19.11 Parachute recovery

Advantages	• Easily deployed
Disadvantages	• Adds volume and weight to the AV • Relatively hard landing • Hard to control exact landing site

19.2 Recovery Method Tradeoffs

Table 19.12 Parafoil recovery

Advantages	• Soft landing
	• Accurate control of landing site possible (with control system)
	• Easily deployed
Disadvantages	• Some landing site preparation necessary

Table 19.13 VTOL – vertical landing

Advantages	• Soft and accurate landings
	• No supplemental equipment required
Disadvantages	• Expensive AV
	• VTOL AVs less efficient for cruise

Table 19.14 Mid-air retrieval

Advantages	• Recovers system intact
Disadvantages	• Requires significant manned aircraft assets and significant cost per recovery for the retrieving aircraft sortie
	• Requires additional communications and coordination

Tables 19.15 Break-apart landing

Advantages	• No supplemental equipment required
	• Prepared landing site not necessary
Disadvantages	• Relatively hard landing
	• UAV may incur minor damage
	• UAV has to be reassembled after each landing

Table 19.16 Suspended cables

Advantages	• "Zero-length" recovery
	• No runway is required
Disadvantages	• Supplemental equipment required
	• Requires significant cost per recovery
	• Requires additional coordination
	• Relatively hard landing
	• UAV may incur minor damage

19.3 Overall Conclusions

There are many effective and reliable methods to launch and recover UAVs. The tradeoffs between the possible launch and recovery approaches are summarized above individually, but the total tradeoff for a UAS must include both, since both are required for most UAV systems. They are not independent. For instance, if a wheeled takeoff is selected, then there is no additional cost for using the same wheels to land. On the other hand, if a rail launch is selected, then wheeled landing gear would probably need to be retractable in order to avoid interference with the rail launch, which adds weight and cost to the AV. The launch and recovery tradeoffs must be done at the same time as all the other basic design tradeoffs for a complete UAS.

No single launch or recovery technique will be suitable for all UAV designs. Mission requirements and airframe design will dictate which technique is most suited to a particular program. In particular, fixed-wing AVs that are in the size class of the Predator really have only one option, which is conventional landing and takeoff, possibly assisted by RATO or the large catapults of an aircraft carrier during takeoff and arrested by gear suitable for large aircraft.

Above all, however, UAV developers need to keep in mind that consideration of launch and recovery issues must be recognized at the very beginning of a UAV program or one runs the risk of having to compromise not only the air-vehicle design but also the entire mission.

Questions

1) What do RATO, VTOL, and R&D stand for?
2) Write advantages of wheeled takeoff using a runway.
3) Write disadvantages of wheeled takeoff using a runway.
4) Write advantages of wheeled takeoff using prepared takeoff areas (truck launch).
5) Write disadvantages of wheeled takeoff using prepared takeoff areas (ruck launch).
6) Write advantages of pneumatic rail launchers (split-tube type).
7) Write disadvantages of pneumatic rail launchers (split-tube type).
8) Write advantages of pneumatic rail launchers (air motor type).
9) Write disadvantages of pneumatic rail launchers (air motor type).
10) Write advantages of hydraulic-pneumatic rail launchers.
11) Write disadvantages of hydraulic-pneumatic rail launchers.
12) Write advantages of RATO launchers.
13) Write disadvantages of RATO launchers.
14) Write advantages of vertical takeoff.
15) Write disadvantages of vertical takeoff.
16) Write advantages of wheeled landing using runway.
17) Write disadvantages of wheeled landing using runway.
18) Write advantages of skid and belly landings using prepared landing sites.
19) Write disadvantages of skid and belly landings using prepared landing sites.
20) Write advantages of vertical-net systems.
21) Write disadvantages of vertical-net systems.
22) Write advantages of parachute recovery.
23) Write disadvantages of parachute recovery.
24) Write advantages of parafoil recovery.

Questions

25) Write disadvantages of parafoil recovery.
26) Write advantages of vertical landing.
27) Write disadvantages of vertical landing.
28) Write advantages of mid-air retrieval.
29) Write disadvantages of mid-air retrieval.
30) Write advantages of break-apart landing.
31) Write disadvantages of break-apart landing.
32) Write advantages of suspended cables.
33) Write disadvantages of suspended cables.
34) Discuss cost tradeoffs for rail versus RATO launch.
35) Discuss why launch and recovery systems are not independent for most UASs.
36) Is there a single launch or recovery technique suitable for all UAV designs? Why?

20

Rotary-Wing UAVs and Quadcopters

20.1 Overview

The ability to vertically take off and land and to hover is a valued performance feature for UAVs, which requires a vertical thrust. For UAVs with prop-driven engines, the rotating wing (such as in helicopters) is a conventional method to provide the vertical thrust. One of the main weaknesses of Vertical Take Off and Landing (VTOL) UAVs is the low and limited cruise speed, since at such speeds the rotary wing will stall. Another weakness is a much higher drag for VTOL UAVs, due to a less aerodynamic fuselage and no fixed wing. VTOL UAVs are not able to glide, thus engine failure – either due to any one of the individual motors or the power supply – would lead to an immediate uncontrolled crash.

In the beginning of development of unmanned flight (for about 20 years), fixed-wing UAVs were the only option, since the stability and controllability issues had already been resolved. However, growing interest in VTOL operation and hover allowed the rotary wing configuration to attract attention and research funds. VTOL UAVs – which use a single engine – are mechanically complex and heavy. The first group of rotary wing UAVs were helicopters, which required complex non-fixed-pitch rotor blades to achieve thrust changes and control.

To remove this complication, the next significant improvement for rotary-wing UAVs was achieved, when multiple electric motors were utilized and incorporated. Removing the mechanical transmission system was another benefit, when the complexity and challenges of flight control via varying pitch of rotor blades were removed. The new idea was to have the fixed-pitch rotor blades to achieve vertical thrust changes on each rotor by changing its angular speed.

VTOL UAVs are inherently unstable; thus, an electronic closed-loop control system is required to stabilize the flight. A considerable weakness of electric motors is the limited available energy via heavy batteries, which resulted in short range and endurance and also low speed. However, advances in modern batteries and constantly increasing energy density have allowed these performance indices to improve.

Thus, with regards to the wing attachment to UAVs, there are three alternatives: (1) fixed-wing, (2) rotary wing, and (3) hybrid configuration. In the past few years, the hybrid configuration (combination of fixed-wing and rotary wings) has gained some interest, and a number of UAV manufacturers have begun to develop and design such UAVs.

Introduction to UAV Systems, Fifth Edition. Paul Gerin Fahlstrom, Thomas James Gleason, and Mohammad H. Sadraey.
© 2022 John Wiley & Sons, Inc. Published 2022 by John Wiley & Sons, Inc.
Companion website: www.wiley.com/go/fahlstrom/uavsystems5e

This chapter is dedicated to VTOL UAVs, including rotary-wing UAVs, hybrid UAVs, and particularly quadcopters. Moreover, it presents the configuration, technical features, advantages and disadvantages, hover, and control of the quadcopter configuration.

20.2 Rotary-Wing Configurations

In the past 30 years, multiple rotary-wing configurations for UAVs (see Figure 20.1) have been developed, manufactured, and tested. Most notable configurations are: (1) single rotor, (2) twin co-axial rotors, (3) twin tandem rotors, (4) quad-rotor, and (5) multi-copter. The characteristics of these configurations are presented in the following sections, but quadcopters are treated in much more details in Section 20.4.

20.2.1 Single Rotor

The single rotor is the simplest configuration for a rotary wing UAV. For the manned aircraft, this configuration is the earliest, most popular, and widely used, mainly due to the simplicity in mechanical transmission. However, due to Newton's third law, if no means is provided, the body tends to react and rotate in the opposite direction to the main rotor. In order to prevent this body reaction, a countertorque is generated by a smaller, side-thrusting, tail rotor. This solution poses a number of disadvantages: (1) the tail rotor consumes about 10% of the engine power, (2) the aircraft is asymmetric, (3) complication of rotational control, and (4) the UAV is prone to losing its control during flight.

In these configurations, the rotor-head control system (e.g., Yamaha unmanned helicopter) is utilized, which applies both cyclic and collective pitch changes to the rotor blades. The cyclic stick is used to control forward, backward, right, and left movements, while the collective stick is used to maintain altitude. The throttle is used to control rotor angular velocity to change thrust. The interaction of control sticks makes balanced hovering difficult, since an adjustment in any one control requires an adjustment of the other two, creating a cycle of constant correction.

Single rotor unmanned helicopters are not easy to fly by a remote pilot, since both the main rotor and tail rotor must be controlled simultaneously. Thus, both left and right sticks of the GCS are to be deflected, in order to apply any change in the flight operation. The throttle and collective pitch

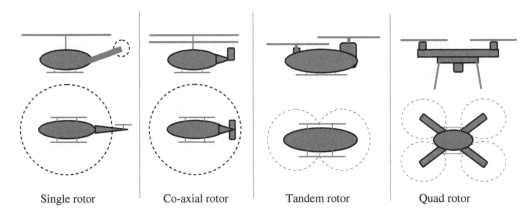

Figure 20.1 Rotary-wing configurations

20.2 Rotary-Wing Configurations **407**

(up and down) are controlled by the same stick motion, while cyclic movements (forward, backward, left, and right) are controlled by another stick. Moreover, the stick that controls the thrust (forward and backward) will be used to control the tail rotor (left and right) as well.

The Yamaha R-MAX (with a maximum takeoff mass of 94 kg, a length of 3.63 m, and an endurance of 1 hour) is an unmanned helicopter developed in the 1990s. The two-cylinder gasoline-powered aircraft has a twin-blade single rotor and is remote-controlled by a line-of-sight user. It was initially designed primarily for agricultural use, but has been utilized as a platform in various guidance and automatic control research projects.

In 2021, Anduril unveiled [3] Ghost 4 UAV as a multimission capable and man-portable unmanned helicopter. The single rotor Ghost 4 with a rotor diameter of 2.274 m has a cruise speed of 52 knot (96.6 km/h). This highly intelligent VTOL sUAV with all-electric powertrain provides more than 100 minutes of flight at a full mission payload with a near-silent acoustic signature. Moreover, Ghost 4 is a modular, customizable platform containing multiple waterproof payload bays.

The Anavia first unmanned helicopter HT-100 with an endurance of up to 240 minutes can carry a payload of 65 kg. The UAV mission can include mapping and surveying, surveillance, logistics, search and rescue, defense, and precision agriculture.

20.2.2 Twin Co-axial Rotors

The twin co-axial rotors (or coaxial double-rotor) configuration was first developed by Kamov, a Russian helicopter manufacturing company. In this configuration, two rotors have opposite directions, which provides zero torque to the body. Dual rotors are mounted one above the other on concentric shafts (one shaft is hollow to allow the second one to be co-centric), with the same axis of rotation. Since all power is consumed for thrust generation, no power is wasted. However, the counter-rotating propellers will have a fair amount of aerodynamic interferences, which causes a lower propeller efficiency.

There is a large degree of coupling between control inputs of two rotors. One basic method to deal with couplings is to regard them as external disturbances. Since no tail rotor is required, it allows for a perfect UAV aerodynamic symmetry, which creates a lower drag. This configuration has the potential to be converted to a hybrid, when a pusher engine (pro-driven or even jet) is added for cruising flight. Both piston engine and electric motor have been employed to power such UAVs.

In this configurations, cyclic and collective pitch changes are applied to the rotor blades to control the UAV during flight. This configuration requires two separate gear boxes and two parallel control systems including two controllers.

Sprite UAV – with a range of about 4 miles and flight time of 12 minutes – uses a compact and enclosed vertical tube airframe. This UAV has a coaxial double-rotor configuration. Its video camera – with a 2-axis stabilized gimbal – is housed in the base (i.e., lower body). Another UAV that has this configuration is the *Epson* micro-flying robot as the world's smallest and lightest UAV with an all-up mass of 12.3 g. This UAV with a propeller diameter of 136 mm has a power consumption of 3.5 watts and an endurance of about 3 minutes.

The *Ingenuity* unmanned helicopter – a product of AeroVironment and NASA (Figure 20.2) – that successfully flew on Mars in 2021 is equipped with twin co-axial rotors. The blades are made of a carbon fiber foam core to provide sufficient lift in the thin Mars atmosphere. The UAV with a rotor diameter of 1.2 m and a mass of 1.8 kg is provided with an electric engine with a power of 350 W. The UAV has a black-and-white navigation camera and a color terrain camera.

 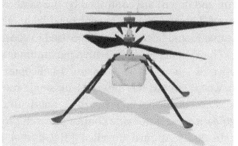

Figure 20.2 NASA and AeroVironment Ingenuity unmanned helicopter (Source: NASA)

20.2.3 Twin Tandem Rotors

In this configuration, twin smaller tandem engines (with two rotors) are employed instead of one large engine (i.e., rotor). Both configurations create the same amount of lift, but, in general, a single engine configuration (with one propeller) is lighter in weight, lower in cost, and provides a higher efficiency than twin engines (and rotors). Moreover, there will be an aerodynamic interference between dual rotors, which requires special attention. Similar to twin co-axial rotors, twin tandem rotors are contra-rotating to cancel the overall torque about the z-axis. Often, one rotor is located in front of another, along the fuselage centerline, to provide UAV symmetry.

The major reason/advantage for a twin tandem rotor is to remove the side-thrusting tail rotor. However, a challenge for this configuration is the complexity of the control system for two rotors to provide UAV trim/control. To partially reduce this complexity, twin engines and rotors are selected to be similar in size and power. Similar to previous configurations, cyclic and collective pitch changes are implemented to the rotor blades as the means of aircraft control.

This configuration has been developed and employed for manned aircraft, but the authors are not aware of any current operational twin tandem rotor UAV. However, the Kaman K-MAX Titan UAV (as a medium-to-heavy unmanned helicopter) with a counter-rotating rotor system [3] to lift up to 6,000 pounds (2,722 kg) had the first flight in 2021.

20.2.4 Multicopters

By multirotor or multicopter UAVs, we mean any number of rotors more than 2; however, it is recommended to have an even number of rotors (e.g., 4, 6, and 8). This is to provide rotational balance without external means (e.g., vertical tail rotor), by using only counter-rotating rotors.

One of the goals in developing multirotors is to remove the complication of cyclic and collective pitch changes to the rotor blades. Moreover, in this configuration, the need for a mechanical transmission system is removed. Thus, the new idea is to have the rotor blades all fixed in pitch. To achieve flight control, engine thrust is independently changed on each rotor by changing its speed of rotation. This configuration is primarily employed by small UAVs, so electric motors are often the best available options.

Three of the most popular multirotor configurations are: (1) quadrotor or quadcopter (four rotors), (2) hexacopter (six rotors, as in Aurelia X6 Standard UAV), and (3) octacopter (eight rotors, as in Aurelia X8 Standard UAV). A multirotor UAV is an appealing platform for amateur model aircraft and aerial photography. Multirotors are generally more expensive, larger, heavier, and harder to construct, but can carry more and heavier payloads. In particular, more details for quadcopters including fundamentals of control, performance, and capabilities are presented in Section 20.4.

20.3 Hybrid UAVs 409

In 2021, German Wingcopter has debuted [3] a new all-electric vertical take-off and landing (eVTOL) fixed-wing drone for multiple destinations per flight. This Wingcopter 198 with eight electric engines, a payload capacity of 13 pounds, and a range of 47 miles is capable of triple-drop deliveries.

20.3 Hybrid UAVs

In general, rotary-wing UAVs have much less performance capabilities (e.g., maximum speed) than fixed-wing UAVs, due to a higher drag. On the other hand, fixed-wing UAVs are unable to vertically take off and land. To overcome both weaknesses/disadvantages in one air vehicle, a number of hybrid configurations have been developed, which combine the capabilities of both rotary- and fixed-wing UAVs: (1) convertible or tilt rotor, (2) tilt-wing, (3) thrust vectoring, and (4) fixed-wing quadcopter combination.

20.3.1 Tilt Rotor

A successful technique to employ a prop-driven engine in both vertical flight (e.g., hover and VTOL) and cruise operation is to allow the engine to rotate its plane of the rotating propeller. At takeoff, the propeller is rotating vertically, while during cruise it is converted to rotate horizontally. For this objective, a prop-driven engine (i.e., rotor) is mounted on to each tip of a fixed-wing (thus, a total of two engines).

The rotors are horizontal in vertical/hover flight, but are converted and tilted forward through 90°, to become propellers for cruising/climbing flight. As the air vehicle gains speed, the rotors are slowly tilted forward, with the propellers eventually becoming perpendicular to the flight direction. Thus, the wing is retained fixed horizontally to the fuselage, but the rotors, with their engines, are tilted relative to the wing.

The tilt mechanism involves the mechanical complexity in transferring the drive from a fixed engine to a tilting rotor system. Moreover, tilting the engines requires the engines to be operational over an angular range of at least 90° (i.e., for both regimes of flight) and creates some complication in the fuel system.

This hybrid configuration is expensive in acquisition, due to an extra mechanism needed for conversion. A major challenge for this configuration is the need for a safe and smooth transition between two regimes of flight, which requires a highly accurate and reliable closed-loop control system. This configuration has been developed and employed for manned aircraft (e.g., Bell Boeing V-22 Osprey and AgustaWestland AW609, formerly the Bell/Agusta BA609), but the authors are not aware of any current operational UAV with a tilt-rotor configuration.

20.3.2 Tilt Wing

The tilt wing is another hybrid configuration to employ prop-driven engines in both vertical flight and cruise operation. It is very similar to the tilt-rotor configuration, except the entire wing is tilted through 90°, to allow the engine to rotate its plane of the rotating propeller. As the air vehicle gains speed, the wings (including engines and rotors) are slowly tilted forward, with the propellers eventually becoming perpendicular to the flight direction.

This configuration requires more power to tilt two components (wing and engine), compared with one component (just engine). Furthermore, the mechanical complexity involved in transferring the drive will be lower. The transition between full hover and full cruise flight is a big challenge. The problem lies in aircraft longitudinal control and maintaining flight altitude. To achieve adequate control, the airflow over the wing should remain attached, while the balance of aerodynamic forces and moments, as well as thrust, is constantly varied. During the tilting process, the wing angle of attack is gradually decreased until it is less than the stall angle. During this period, the wing is stalled; thus, wing lift and aerodynamic pitching are not fully available and the control surfaces are ineffective.

As the aircraft accelerates, the vertical force of the thrust is gradually reduced, until the airspeed is higher than the stall speed. In general, this configuration is not recommended, due to its lower structural safety and functional reliability. Within the tilt wing configuration, an option is to fold the wing during flight. This innovative technique is a challenge for engineers and is currently under development by PetroDynamics [3] to transition between rotorcraft and fixed-wing configurations.

20.3.3 Thrust Vectoring

A thrust vectoring (or ducted fan) hybrid configuration is primarily for a UAV using a jet engine (e.g., turbofan), by adding rotating nozzles through which the engine's fan and core airflows exhaust. This addition will allow the engine's thrust smooth transformation from straight down (for hovering) to directly aft (in cruising flight). The engine nozzles are vectored to allow high speed flow of compressed air to provide the desired movement in pitch, roll, or yaw. During VTOL, the nozzles are angled downwards, directing (i.e., vectoring) the engine thrust downwards. To maintain flight control about all three axes (longitudinal, lateral, and directional), four rotating nozzles are recommended.

A nozzle rotation mechanism should be coordinated such that all nozzles operate in unison – any difference would lead to a rapid loss of aircraft control. In addition to the vectoring nozzles, the UAV requires a method of controlling its attitude during hovering flight, when the aerodynamic surfaces and moments are ineffective. For instance, a group of reaction control nozzles in the fuselage nose, wingtips, and tail may be fitted to the vehicle. These nozzles may be supplied with high pressure air bled from the jet engine.

The rotational control of the vehicle can also be achieved by tiltable vanes in the slipstream. However, this may cause a problem to the attitude control of the body, as the vanes may lack sufficient force or response to ensure a controllable, stable flight.

This configuration features high vehicle cost, heavy structure, complex nozzles rotation mechanism, high engine fuel consumption, and challenging closed-loop control architecture. The vectored thrust configuration has been successfully implemented in the Harrier manned fighter aircraft with a single-engine ducted fan.

This configuration has been developed and employed for manned aircraft (e.g., ground attack aircraft McDonnell Douglas (now Boeing) AV-8B Harrier II and stealth tactical fighter Lockheed Martin F-22 Raptor), but the authors are not aware of any current operational UAV with thrust vectoring configuration. The thrust vectoring hybrid configuration is prone to fatal accident and damage beyond repair. For instance, during service of AV-8B Harrier with the US Marine Corp, the Harrier has had an accident rate three times that of the Corps' F/A-18s. The majority of the accidents had happened during takeoff and landing, which are the most critical phases in flight for this configuration.

20.3.4 Fixed-Wing Quadcopter Combination

There are a variety of innovative hybrid configurations to allow simultaneous VTOL and cruising capabilities. One innovation is to combine a quadcopter with a fixed-wing configuration. The UAV

Figure 20.3 UAV Factory Penguin C Mk2 VTOL UAV (Courtesy of UAV Factory)

has a fixed wing with one or more engines (jet or prop-driven) that is only operational during cruising flight. In addition, the UAV is equipped with four vertical engines (similar to a quadcopter) that is primarily utilized during takeoff, landing, and hovering flight. Thus, the UAV has at least five engines, where four engines are only employed during VTOL operation and are shot down during cruising flight.

To reduce the drag during cruise, the propellers of vertical engines are feathered along the fuselage central line. The blade angle of attack is changed to the point that the chord line of the blade is approximately parallel to the incoming airflow. Example UAVs with hybrid fixed-wing quadcopter configuration are BlueBird Aero ThunderB-VTOL, L3Harris FVR-90, and UAV Factory Penguin C Mk2.

The BlueBird Aero ThunderB-VTOL tactical UAV has successfully implemented [3] this configuration with hundreds of flights. The UAV provides rapid GPS-marked high-definition video, photogrammetric tactical mapping on demand, and other intelligence assets.

The US Navy has selected L3Harris Technologies [3] with its FVR-90 VTOL UAV to participate in a demonstration to identify and evaluate UASs capable of operating in austere deployed environments without additional support systems. This state-of-the-art technology is hoped to give the Navy the needed control to accomplish its missions. The FVR-90 UAV with a wing span of 4.8 m, a length of 2.48 m, and a maximum gross take-off mas of 53 kg can be operated by a pilot and a maintainer. The drone has the capacity to carry a maximum payload of 10 kg including a stabilized multisensor, multispectral imaging system and EO/IR cameras.

The UAV Factory Penguin C UAV (Figure 20.3) has a maximum takeoff mass of 32 kg and a 4.5 kg payload capability. Depending on the payload, it offers up to 14+ hours of flight endurance. This UAV has a hybrid configuration: a quadcopter and a fixed-wing with an inverted V-tail, and a pusher propeller piston engine.

20.4 Quadcopters

20.4.1 Overview

Due to the growing popularity and uniqueness of quadcopters in aviation and commercial applications, this section is dedicated to quadcopters. In the past decade, the popularity of quadcopter UAVs has skyrocketed, mainly because of its unique features. The popularity of the quadcopters is increasing as the sensors and control systems are becoming more advanced, smaller, lighter, and less expensive. Moreover, the simple control via propeller rotation speed, and the possibility to scale down/up make them more attractive. There are mainly five reasons for the popularity of quadcopters: (1) the ability to hover, (2) size and portability, (3) low cost, (4) quick in capturing live events, and (5) safety.

There are a number of similarities and differences between helicopters and quadcopters. A single-rotor helicopter has one main rotor, a gearbox, hub mechanism, and one tail rotor. However, the quadcopters have no gearbox and no moving parts, except for the rotating motors and propellers. The quadcopter is also less prone to vibration and is more flexible when it comes to the placement of the center of gravity. Due to the smaller size of the rotors, they are safer to fly indoors. Table 20.1 provides general characteristics of a quadcopter.

The electric DC motors are the standard propulsion system utilized in quadcopters. The required electric energy is provided by onboard batteries. Quadcopters are capable of capturing vital information from diverse angles; for example, in movies or capturing sport events. They have the potential to change the nature of journalism in the near future. For many different types of missions, the quadcopter configuration with four electric engines is an optimal choice for very small UAVs. The engines – each turns a propeller – are placed in an X-configuration (Figure 20.4), with two rotors spinning clockwise and two counterclockwise.

Table 20.1 General characteristics of a quadcopter

No.	Feature	Specification
1	Number of engines	4
2	Type of engines	Electric
3	Configuration of engines	All upward in an X-shape
4	Rotation of propellers	Two clockwise, two counterclockwise
5	Takeoff and landing	Vertical
6	Hover	Yes
7	Landing gear	Fixed – skid
8	Speed	Slow
9	Drag	High

Top view

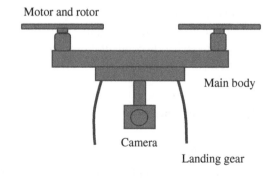

Front-view

Figure 20.4 Quadcopter configuration

20.4 Quadcopters **413**

All motors are generating an upward aerodynamic force – the total force will have two components: (1) vertical and (2) horizontal. The only way to induce a lateral/directional/longitudinal motion is to tilt the entire body. When the entire body is tilted, a horizontal component of the total force will push the vehicle in the horizontal direction, while the vertical component will be equal to the quadcopter weight.

A quadcopter has a high drag, since: (1) four engines have no covers, (2) the main body is not aerodynamically optimal, and (3) the landing gear is fixed. Thus, a quadcopter has a low maximum airspeed compared with other UAV configurations. The vehicle carries the payload (e.g., camera) under the main body; the payload can be faired or un-faired. Compared with single-rotor UAVs, quadcopters have much shorter propellers and blades, which are easier to construct.

20.4.2 Aerodynamics

The lift – to keep the UAV in a desired altitude – in a quadcopter is generated via rotating propellers. A propeller is a means to convert the engine power to an aerodynamic force and thrust. The aerodynamic equations and principles that govern the performance of a wing (see Chapter 3) are generally applied to a propeller.

The technique that each propeller (as a rotating wing) is generating the thrust force is basically very similar to the way that a fixed non-rotating wing is working. As indicated in Chapter 3, the lift (Equation (3.1)) generated by a fixed wing is

$$L = \frac{1}{2}\rho V^2 S C_L \tag{20.1}$$

where ρ is air density, V is airspeed, S is wing planform area, and C_L is the lift coefficient. This equation is revised for a rotary wing (here, the propeller) to generate an aerodynamic force (F_P):

$$F_P = \frac{1}{2}\rho V_{av}^2 S_P C_{L_P} \tag{20.2}$$

where S_P is the propeller planform area and C_{LP} is the propeller lift coefficient. The parameter V_{av} is the average airspeed at the propeller, which may be assumed to be about 70% of the propeller tip speed (V_{tip}). The reason is that the airspeed at the propeller center is zero and is increased as we move toward the prop tip (see Figure 20.5). Each arrow represents a linear airspeed for flow passing under and over the propeller blade. Typical values for the propeller lift coefficient range from 0.2 to 0.4. The limiting factor for the prop diameter is the speed of sound, to prevent any shock wave at the prop tip.

However, during a hover (prop rotation), the local airspeed (V) is proportional to the rotational velocity (ω) of the propeller:

$$V = r\omega \tag{20.3}$$

where r is the radial distance between the local element of the prop to the shaft. Thus, the magnitude of the force (F_P) is proportional to the prop's rotational speed (ω) squared:

$$F_P = k_1 \omega^2 \tag{20.4}$$

where k_1 is a proportionality constant. When ω is in rad/sec, and F_P is in N, a typical value for k_1 for a small quadcopter is 10^{-6} N/(rad/s)^2. This is a simplified mathematical model for the aerodynamic force generated by a rotating propeller.

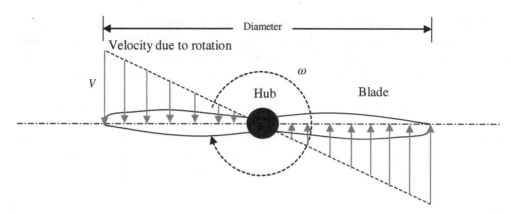

Figure 20.5 Propeller and rotary wing (top view)

The quadcopters are frequently equipped with fixed-pitch propellers, so they are not always operating at their highest efficiency. The propeller angles of attack are either optimized for hover or cruising conditions. Typical propeller efficiency for a commercial fixed-pitch propeller is about 60%–80%. This implies that about 20%–40% of the electric motor power is wasted, mainly due to propeller drag.

In a cruising flight, the local airspeed of a blade is a function of two components: (1) local airspeed due to spinning and (2) UAV cruising speed. The path of motion of each element of the propeller is a circular path ($R\omega$) plus translational path (V_∞), of the propeller. The airflow at the blade leading edge is the vector sum of the UAV free-stream velocity and the propeller circumferential velocity at that station. The propeller speed is given by

$$V_p = \sqrt{V_\infty^2 + V_\theta^2} \tag{20.5}$$

where V_p is the local airspeed at the propeller blade, V_∞ is the aircraft forward airspeed, and V_θ is the local linear velocity of the rotating prop at the blade elements ($R\omega$).

20.4.3 Control

Unlike single-rotor helicopters, quadcopters do not have cyclic and collective pitch changes to the rotor blades. The motion control is implemented by directly altering the rotational speeds of the individual motors via control commands. Each rotor is individually driven by an electric motor mounted at the rotor head. This in turn generates the rotational torque differences to control the flight path. Engine propellers are responsible for both lift and thrust; hence the vertical component of the engine force is lift and the horizontal component is the thrust, or forward force.

All engines have similar power, size, and propellers. During vertical takeoff, landing, and hovering flight, they are all spinning at the same angular velocities. The net yawing rotation is zero, since two propellers are rotating clockwise and two counterclockwise. This means that there is no need for a tail rotor as used in conventional helicopters. For a cruising flight, the UAV should be tilted to generate a horizontal force (i.e., thrust).

For any non-hovering flight, the rotational velocities of rotors will be different. This is to create a desired torque for an intended maneuvering flight operation. Unlike fixed-wing aircraft, the quadcopters do not have any conventional control surfaces (e.g., elevator, rudder, and ailerons). Thus, for any rotational motion, an external torque is created instead of aerodynamic moments (e.g., rolling, pitching, and yawing moments).

20.4 Quadcopters

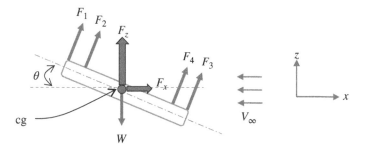

Figure 20.6 Lift and thrust forces (side view)

For the aircraft to move forward, the rotational speed of the two rear rotors is increased to pitch the aircraft nose-down and direct the resulting thrust forward. At the same time, the total thrust must be increased to prevent loss of height and the rotor speeds must again be harmonized.

This format guarantees that, for both counterclockwise (say, positive) rotation and clockwise (say, negative) rotation, the motor torque direction is correctly incorporated. The subscripts 1, 2, 3, or 4 for force F in Figure 20.6 refers to the motor number. To ensure that the overall forces (all thrust forces in the level configuration) are upward, in summing all force thrusts, another negative sign is added to two negative values. To include the direction of rotation in the thrust force calculation, the following overall force (F) equation is derived:

$$F = F_1 + F_2 + F_3 + F_4 = k_1 \left[|\omega_1|\omega_1 - |\omega_2|\omega_2 + |\omega_3|\omega_3 - |\omega_4|\omega_4 \right] \tag{20.6}$$

Equation 20.6 is valid for level flight only (i.e., no pitch angle (θ) and no bank angle). When the UAV has a pitch angle (e.g., during a cruising flight), the z and x components of this overall force are

$$L = F_z = F \cdot \cos(\Theta) \tag{20.7}$$

$$T = F_x = F \cdot \sin(\Theta) \tag{20.8}$$

where L stands for overall lift and T for the overall thrust.

Rotation of any rotor will induce an opposite torque on the quadcopter body. If the prop is rotating counterclockwise, then the induced torque is clockwise, acting on the quadcopter main body. To cancel out such torques, two propellers should rotate clockwise, while the other two are rotating clockwise.

A pitch, bank, or yaw angle is generated by increasing the angular velocity of one rotor (or two adjacent rotors) and decreasing the angular velocity of the diametrically opposite rotor. The control algorithms to achieve this are complicated and should take into account the wakes and aerodynamic interference patterns between the propellers.

Various control algorithms and laws have been implemented for quadcopter control, but one of the successful, effective, and simplest control laws is PID (Proportional–Integral–Derivative). The commercial autopilot Micropilot MP21283X employs the PID control law in its feedback control system, while PID gains can be defined and varied by users.

Onboard artificial intelligence (AI) algorithms have been used by Anduril Ghost 4 – a packable rotary-wing UAV – to identify and track people, missiles, and battlefield equipment, and to survive any environment. The battery-powered UAV also has more than 100-minute flight endurance, and a cruise speed of about 52 kt (96.6 km/h).

In 2021, the US Air Force flew an artificial intelligence system (Skyborg autonomy core system) onboard a subsonic autonomous UAV (Kratos UTAP-22 Mako) for the first time. The service intends to field a family of UAVs capable of taking on missions too risky for human pilots and operating in a team with fighter aircraft like the F-35 and F-15EX (i.e., manned–unmanned teaming).

An important control feature in UAVs is an onboard detect-and-avoid (DAA) system that helps to advance the safety for UAVs while simultaneously expanding the flight envelope. Particularly, this feature is a requirement for Beyond Visual Line of Sight (BVLOS) flights in uncontrolled airspace. Sagetech Avionics is working with embedded computing specialist Hover [3] to provide an onboard detect-and-avoid system for UAVs. This particularly vital technology to avoid collisions is intended for installation on UAVs flying above 400 feet, beyond the visual line of site, in the National Airspace System (NAS).

A Detect and Avoid (DAA) system and Command and Control (C2) solutions for Beyond the Visual Line of Sight (BVLOS) operations are currently integrated in many open-source autopilots including uAvionix George. Skytrac has developed Integrated Mission System 350 (IMS-350) to provide UAVs with Command and Control capabilities for enhanced mapping, photogrammetry, LiDAR scanning, and logistical operations.

The Iris Automation's onboard detect-and-avoid system (Casia) has received the Special Flight Operations Certificate for Beyond Visual Line of Sight flights in uncontrolled airspace to conduct commercial missions over power lines in Canada. The system provides commercial UAVs with automated collision avoidance maneuvers.

Due to their configuration, quadcopters are not inherently stable; thus, they require a closed-loop control system that provides artificial stability. Hence, this configuration is naturally less gust-sensitive than the fixed-wing UAVs, and its control response is expected to be faster. Moreover, the first-person view (FPV) UAVs are notorious for crashing; thus, adding a parallel command augmentation system is very crucial in saving such UAVs in hard maneuvers.

Questions

1) What do eVTOL, AI, PID, DAA, BVLOS, C2, and FPV stand for?
2) What was the configuration of the first group of rotary wing UAVs?
3) Are VTOL UAVs inherently stable?
4) Write one considerable weakness of electric motors for UAVs.
5) Write three alternatives to UAVs with regards to the wing attachment.
6) Write notable rotary-wing configurations for UAVs.
7) Briefly describe features of a single rotor configuration for rotary-wing UAVs.
8) Briefly describe features of a twin co-axial rotors configuration for rotary-wing UAVs.
9) Briefly describe features of a twin tandem rotor configuration for rotary-wing UAVs.
10) Briefly describe features of a quad rotor configuration for rotary-wing UAVs.
11) Briefly describe features of a multicopter configuration for rotary-wing UAVs.
12) Write four disadvantages of a single rotor configuration for rotary-wing UAVs.
13) Explain how change in pitch and roll is applied to rotary-wing UAVs with a single rotor configuration.
14) What is the primary function of the tail rotor in single rotor configuration UAVs?
15) Provide the names of three UAVs that have a single rotor configuration.
16) What is the endurance of (a) Anduril Ghost 4 UAV and (b) Anavia HT-100 unmanned helicopter?

Questions 417

17) Write (a) max takeoff mass of Yamaha R-MAX, (b) payload mass of Anavia HT-100 unmanned helicopter, and (c) cruise speed of Anduril Ghost 4 UAV.
18) What air vehicle is the only unmanned rotary wing UAV to fly in another planet?
19) Identify the configuration of the following rotary-wing UAVs: (a) NASA and AeroVironment Ingenuity, (b) Yamaha R-MAX, (c) Kaman K-MAX Titan, and (d) Wingcopter 198.
20) Write (a) engine power of NASA and AeroVironment Ingenuity, (b) endurance of Sprite UAV, and (c) payload capacity of Wingcopter 198.
21) Name (a) a rotary-wing UAV with four rotors, (b) a rotary-wing UAV with six rotors, and (c) a rotary-wing UAV with eight rotors.
22) Write at least four hybrid configurations for UAVs.
23) Briefly describe features of a tilt-rotor hybrid configuration for UAVs.
24) Briefly describe features of a tilt-wing hybrid configuration for UAVs.
25) Briefly describe features of a thrust vectoring hybrid configuration for UAVs.
26) Briefly describe features of a fixed-wing quadcopter combination hybrid configuration for UAVs.
27) Provide the names of two aircraft with a tilt-rotor hybrid configuration.
28) Provide the names of two aircraft with a thrust vectoring hybrid configuration.
29) Provide the name of three UAVs with a fixed-wing quadcopter combination configuration.
30) For UAV Factory Penguin C Mk2 UAV, write (a) the maximum takeoff mass, (b) the payload capability, and (c) endurance.
31) Write the main reasons for the popularity of quadcopters.
32) Briefly describe the similarities and differences between single-rotor helicopters and quadcopters.
33) Provide general characteristics of a quadcopter.
34) Provide reasons why a quadcopter has a high drag.
35) Briefly describe how lift is generated in a rotary wing.
36) What is a typical value for the propeller efficiency?
37) Briefly describe how control is applied (i.e., pitch, bank, or yaw angle is generated) in quadcopters.
38) Briefly describe how a quadcopter is moving forward.
39) What control algorithm has been used by Anduril Ghost 4?
40) Name a UAV that is equipped with an onboard detect-and-avoid system.
41) Name an open-source autopilot that features a detect-and-avoid system.
42) Name an open-source autopilot that employs PID control law.
43) Are quadcopters inherently stable? How is flight stability provided?

References

1 Wagner W and Sloan W, *Fireflies and Other UAVs*, Aerofax Inc., Tulsa, 1992

2 Mazzarra A, *Supporting Arms in the Storm, Naval Proceedings*, Vol. 117, United States Naval Institute, Annapolis, November 1991, p. 43

3 Daily Launch, American Institute of Aeronautics and Astronautics, Reston, VA, 2020, 2021

4 Kinzig B, *Global Hawk Systems Engineering Case Study*, Air Force Institute of Technology, 2010

5 Federal Aviation Regulations, Part 107, *Operation and Certification of Small Unmanned Aircraft Systems*, Federal Aviation Administration, Department of Transportation, Washington DC, 2016

6 Kemp I (editor), *Unmanned Vehicles, The Concise Global Industry Guide*, Issue 19, The Shephard Press, Slough, Berkshire, UK, 2011

7 Tennekes H, *The Simple Science of Flight from Insects to Jumbo Jets*, The MIT Press, Cambridge, MA, 1996

8 Anderson J D, *Fundamentals of Aerodynamics*, Fifth edition, McGraw-Hill, 2010

9 Abbott I H and Von Donehoff A F, *Theory of Wing Sections*, Dover, 1959

10 Houghton E L and Carpenter P W, *Aerodynamics for Engineering Students*, Fifth edition, Elsevier, 2003

11 Simons M, *Model Aircraft Aerodynamics*, Argus Books, Hemel Hempstead, England, 1994

12 Sadraey M, *Aircraft Performance Analysis: An Engineering Approach*, CRC Press, 2017

13 Alexander D, *Natures Flyers: Birds, Insects, and the Biomechanics of Flight*, Johns Hopkins University Press, Baltimore, Maryland, 2002

14 Anderson J, *Aircraft Performance and Design*, McGraw-Hill Book Company, New York, 1999

15 Hale F, *Introduction to Aircraft Performance Selection and Design*, John Wiley & Sons, New York, 1984

16 Roskam J, *Airplane Flight Dynamics and Automatic Flight Controls*, DARCO, 2007

17 Stevens B L, Lewis F L, and Johnson E N, *Aircraft Control and Simulation*, Third edition, John Wiley, 2016

18 Farokhi S, *Aircraft Propulsion*, Second edition, John Wiley, 2014

19 Hughes A and Drury B, *Electric Motors and Drives: Fundamentals, Types and Applications*, Fifth edition, Newnes, 2019

20 Mattingly J, *Elements of Propulsion: Gas Turbines and Rockets*, AIAA, 2006

21 Megson T, *Aircraft Structures for Engineering Students*, Fifth edition, Elsevier, 2012

22 Budynas R G and Nisbett J K, *Shigley's Mechanical Engineering Design*, Eleventh edition, McGraw-Hill, 2019

Introduction to UAV Systems, Fifth Edition. Paul Gerin Fahlstrom, Thomas James Gleason, and Mohammad H. Sadraey.
© 2022 John Wiley & Sons, Inc. Published 2022 by John Wiley & Sons, Inc.
Companion website: www.wiley.com/go/fahlstrom/uavsystems5e

23 Peery D, *Aircraft Structures*, McGraw-Hill Book Company, New York, 1949

24 Federal Aviation Regulations, Part 107, *Operation and Certification of Small Unmanned Aircraft Systems*, Federal Aviation Administration, Department of Transportation, Washington DC, 2016

25 Salyendy G, Handbook of Human Factors and Ergonomics, Third edition, John Wiley, 2006

26 Dorf R C and Bishop R H, *Modern Control Systems*, Thirteenth edition, Pearson, 2017

27 Bhatta B, *Global Navigation Satellite Systems; New Technologies and Appliceations*, CRC Press, 2021

28 Kendoul F, Survey of advances in guidance, navigation, and control of unmanned rotorcraft systems, *Journal of Field Robotics*, Vol. 29, no. 2, pp. 315–378, 2012

29 Steedman W and Baker C, Target Size and Visual Recognition, *Human Factors*, Vol. 2, August 1960, pp. 120–127

30 Naftel J C, *NASA Global Hawk: Project Overview and Future Plans*, NASA Dryden Flight Research Center, 2011

31 Simon C, Rapid Acquisition of Radar Targets from Moving and Static Displays, *Human Factors*, Vol. 7, June 1965, pp. 185–205

32 US Department of Transportation, *Federal Aviation Administration; Advisory Circular 36–1H: Noise Levels for US Certified and Foreign Aircraft*, May 25, 2012

33 Richards M A, Scheer J A, and Hilm W A, *Principles of Modern Radar: Basic Principles*, SciTech Publishing, 2010

34 *Operations for Urban Air Mobility Concept of Operation*, Federal Aviation Administration, June 2020

35 Daily Launch, American Institute of Aeronautics and Astronautics, Reston, Virginia, June 22, 2021

36 Hershberger M and Farnochi A, *Application of Operator Video Bandwidth Compression Research to RPV System Design*, Display Systems Laboratory, Radar Systems Group, Hughes Aircraft Company, El Segundo, CA, Report AD 137601, August 1981

37 Frenzel L, *Principles of Electronic Communication Systems*, Fourth edition, McGraw Hill, 2015.

38 Skolnick M, *Introduction to Radar Systems*, Second edition, McGraw Hill, New York, 1980

39 Friedman D, *et al. Comparison of Canadian and German Weather*, System Planning Corporation, Report 566, Arlington, Virginia, March 1980

40 Hershberger M and Farnochi A, *Application of Operator Video Bandwidth Compression/Reduction Research to RPV System Design*, Display Systems Laboratory, Radar Systems Group, Hughes Aircraft Company, El Segundo, CA, Report AD 137601, August 1981

41 Bates H, *Recommended Aquila Target Search Techniques*, Advanced Sensors Directorate, Research, Development and Engineering Center, US Army Missile Command, Report RD-AS-87-20, US Army, Huntsville, 1988

42 Kohlman D, *Introduction to V/STOL Airplanes*, Iowa State University Press, Ames, Iowa, 1981

Index

Adverse yaw, 105
Aerodynamics
  boundary layer, 60
  center, 50, 56
  coefficients, 50, 52–3
  control, 101
  drag, 51–2
  efficiency, 57, 66–7
  forces, 49
  induced drag, 58
  lift, 52
Airfoil, 49–56, 58, 60, 62, 63, 66, 92, 110, 126, 128, 138, 148, 267, 268
Amazon Prime Air UAV, 287
ANAFI Ai, 307, 311
Angle of attack, 50, 52–5, 59, 62, 65, 76, 77, 90, 95, 96, 100–102, 104, 114, 129, 144, 191, 192, 197, 238, 371, 377, 410, 411
Antenna
  airborne, 295–6
  array/directional, 308
  effective radiated power, 318, 337
  gain, 316, 318
  location, 296, 298, 302
  types of, 307
Anti-jam
  antenna gain, 318
  definition, 300
  frequency hopping, 323
  margin, 315, 327
  processing gain, 322
  spread spectrum, 322
  techniques and approaches, 315

Anti-radiation munitions, 298–9
Aquila
  data link, 12, 294, 302
  history and significance, 3, 5, 12
  launch, 14
  recovery, 14, 385
  search, 15, 235
Artificial intelligence, 162, 195, 206–208, 273, 415
Aspect ratio, 20, 57, 67, 75, 77, 78, 110, 220, 226, 233
Autonomy, xviii, 159, 162, 187, 189, 206–208, 270–73, 287, 288, 415
Autopilot, xviii, 8, 35, 64, 89, 99, 100, 102, 105, 144, 159, 162, 183, 186, 188–91, 193–200, 204–208, 217, 221, 249, 295, 301, 304, 371, 386, 388, 392, 393, 415, 416
Autotracker, 255, 348

Batteries, 122–3
Boresight, 242
Boundary layer, 60
Bridges and gateways, 173
Buckling, 136, 138–40, 153, 154

Canard, 91, 97
Centrifugal force, 93, 104, 119
Chemical, biological, and radiological (CBR) detection
  biological detection, 204
  chemical detection, 204, 281
  radiological detection, 204, 282
Command link, 295. *See also* Uplink

Introduction to UAV Systems, Fifth Edition. Paul Gerin Fahlstrom, Thomas James Gleason, and Mohammad H. Sadraey.
© 2022 John Wiley & Sons, Inc. Published 2022 by John Wiley & Sons, Inc.
Companion website: www.wiley.com/go/fahlstrom/uavsystems5e

Construction techniques, 135, 148–9
Control, aerodynamic
  actuator, 195
  autopilot, 190
  control axes, 193
  controller, 194
  control process, 193
  directional control, 104
  lateral control, 105
  longitudinal control, 102
  pitch control (*see* longitudinal control)
  sensors, 193, 196–7
  yaw control (*see* directional control)
Control-loop delay. *see* Data link
Control, mission
  autonomy, 206
  levels of control, 184
  mission control, 204
  payload control, 183, 201
  piloting, 183–4
Coordinated turn, 104

Data link
  antenna
    gain, 316
    gain, applications of, 318
    types, 307
  attributes, 296
  commonality, 305
  data rate reduction
    compression, effects of, 346
    compression of data, 343–4
    frame rate, effects of, 348
    non-video data, 352
    truncation of data, 343
    video data, 344
  data rate restrictions, 302
  deception, 299
  delays, in control loop, 302–303, 348
  digital, 301
  down link, 322
  frequency hopping, 323–4
  functions, 295
  interchangeability, 305
  interfaces, 302
  interoperability, 305
  low probability of intercept, 298
  margin, 322
  noise, 337
  processing gain, 322
  propagation, 334
  security, 298–9
  signal budget, 336
  signal-to-noise, 337
  spread spectrum, 322
  tradeoffs, 357
  transmitter power, 316
  uplink, 295
Deep penetrators, 5
Development Setback, 18
Dihedral angle, 99, 100, 269
DJI Phantom, 35, 94, 121, 379, 386
Doppler effect, 278
Drag Polar, 49, 57, 66
Drag, total, 63
Dutch roll, 101
Dynamic stability, 100

Effective radiated power, 318, 337
Efficiency
  aerodynamic, 49, 66–7
  propeller, 75, 78, 80, 82, 128–9, 407, 414
  propulsion, 109
Electromagnetic interference (EMI), 175, 252,
    256–7, 298
Electromagnetic spectrum, 265
Electronic warfare, 10, 280
Endurance. *see* Performance
Engine. *see* Propulsion
Environmental sensors, 282
Ergonomics, 179–80
Ex-Drone, 6
Expendable air vehicles, 11, 42, 248, 251

Fail-safe structure, 149
Fatigue, 145, 155, 179, 282
Federal Aviation Administration, 29, 144,
    179, 287
Federal Aviation Regulations, 73, 81
Fireflys, 5
Flight envelope, 89–90, 95, 101, 143, 257, 416
Flight operations. *see* Control, mission
FLIR video, 343. *See also* Imaging sensors
Fly-by-mouse, 189

Index

Four-cycle engine. *see* Propulsion
Frequency hopping, 323–4
Fuel cells, 126–7

Gas turbine. *see* Propulsion
Global Hawk, 7, 16–18, 32, 41–2, 53, 67, 73, 77–9,
 104, 120, 137, 138, 145, 152, 169, 185, 208,
 217, 237, 238, 269–70, 280, 283, 309, 311,
 319, 322, 327, 366, 384
Global positioning system, 163, 198–9
GPS. *see* Global positioning system
Gray Scale, 344–6
Ground control station, 161
 functional description, 161
 interfaces, 169
 subsystems, 161
Ground data terminal, 296
Gyro
 attitude, 193, 197
 rate, 192–3, 197–8
Gyroscope, 196–8, 238–40, 286

Hydraulic/pneumatic launchers, 374

Image motion compensation, 349
Imaging sensors, 203, 216–7
 detection, 213, 217–31
 identification (*see* detection)
 infrared imaging, 213, 217–31
 recognition (*see* detection)
 TV (*see* visible and near-IR)
 visible and near-IR, 213, 217–31
Ingenuity unmanned helicopter, 407, 408
Inner and outer loops, 191

Jam resistant. *see* Anti-jam

LAN. *see* Local area network
Launch, 365
 catapult stroke, 366–7
 energy required, 367–8
 environmental effects, 373, 378
 fixed wing, 370
 hydraulic/pneumatic, 374
 methods, 365
 other launch equipments, 377
 pneumatic, 373

rocket assisted, 375
 tradeoffs, 397
 VTOL, 379
Levels of communication, 172
Lift distribution, 58–9, 105, 140–41, 153
Load factor, 90, 105, 139, 143
Loads, 35
 dynamic, 139, 143, 155
 maneuver, 143
 stress, 141
Local area network, 172
Low probability of intercept, 298
Ludwig Prandtl, 58

Materials, 145
 aluminum alloy, 145, 147
 balsa wood, 145, 147
 composite, 145–8
 reinforcing materials, 147–8
Meteorological sensors, 202, 282
MICNS. *see* Aquila, data link
Micro-Electro-Mechanical System, 199
Micro-UAV, 34–5
Mid-air retrieval, 388
Minimum resolvable contrast (MRC), 221
Minimum resolvable temperature (MRT), 221
Mission control. *see* Control, mission
Missions, 29
Motor. *see* Propulsion
MPCS. *see* Ground control station
MQ-8B Fire Scout, 39
MQ-9B SkyGuardian, 19
MQ-1C Gray Eagle, 19–20, 389
MQ-8C Scout, 379, 386
MQ-1 Predator, 19–20, 30, 32, 41, 53, 67, 78, 116,
 167, 206, 252–3, 280, 366, 384
MQ-9 Reaper, 7, 19–20, 30, 73, 77, 79, 104–105,
 120, 185, 208, 237, 252, 280, 366, 384

NACA, 52–4, 66, 92
Nacelle, 63, 138
NASA, 7, 17–18, 52–3, 121, 232, 283, 286, 288,
 309, 311, 407, 408
National Airspace System, 22, 179, 416
Navigation, 199
Net systems, vertical, 384
Neutral point, 96–7, 103

Newton's laws, 102, 110, 239, 366, 406
Non-video data, 352
Nuclear radiation sensors, 202, 282

OSI standard, 175

Parabolic reflectors. *see* Antenna, types of
Parachute recovery, 385
Parafoil, 39
Pathfinder, 121, 286
Performance
  climbing flight, 80
  cruise, 71, 73, 76
  endurance, 78
  gliding flight, 82
  range, 73
PID, 194–5, 415
Pilot-rated operator, 179
Pioneer, 6–7, 40
Pitch control, 102
Pitot tube, 63, 64, 192–3, 197
Pneumatic launchers, 373
Pointer, 6
Precipitation losses, 336
Predator, 7, 250
Pressure distribution, 49, 54–6, 63, 139–40
Processing gain. *see* Data link
Propeller efficiency, 75, 78, 80, 82, 128–9,
    407, 414
Propulsion, 109
  batteries, 122
  electric power, 121
  four-cycle engine, 116
  fuel cells, 126
  gas turbine, 119
  powered lift, 111
  rotary engine, 118
  thrust generation, 110
  turbine engine, 119–20
  two-cycle engine, 116
Pseudo-Satellites, 7–8, 277, 283–6

Radar, 277
  bands, 266
  synthetic aperture, 10, 17, 20, 41, 208, 217,
    280
Radar Cross-Section, 267

Radar horizon, 321
Rail launchers, 372
Range. *see* Performance
RATO ignition systems, 376
RATO launchers, 375–9
RC plane, 8, 27, 110, 114, 145, 159, 166,
    194–5, 310
Reaper, 7, 19–22, 30, 41, 73, 77, 79, 104–105, 120,
    185, 208, 237, 252–3, 280, 309, 322,
    366, 384
Reconnaissance, 30
Reconnaissance/surveillance payloads, 215
Recovery, 383
  mid-air retrieval (MARS), 388
  parachute recovery, 385
  parafoils, 39, 386
  recovery systems, 383
  shipboard, 389
Remote ground terminal (RGT), 13
Reynolds number, 50, 52–4, 61–3
Rotary engine, 118
RQ-21A Integrator, 394
RQ-21 Blackjack, 220, 372
RQ-4 Global Hawk. *See* Global Hawk
RQ-5 Hunter, 102, 104
RQ-2 Pioneer, 5, 33, 38, 167, 384
RQ-1 Predator, 7, 19
RQ-11 Raven, 67, 73, 77, 79, 82, 121, 168, 238,
    370, 393
RQ-170 Sentinel, 41
RQ-7 Shadow, 104, 167, 372, 375

Safety factor, 149–50, 154–5
Sandwich construction, 146
ScanEagle, 31, 37, 64, 169, 185, 206, 208, 213,
    218, 238, 283, 310–11, 372, 374, 394
Search process, 231
Semimonocoque, 136
Sensor
  autopilot, 196
  biological, 204
  chemical, 202, 281
  meteorological, 202, 282
  radiological, 202, 282
  reconnaissance/surveillance, 215
Side-looking airborne radar, 280
Signal intelligence (SIGINT), 10, 280

Index

Signature, 258
  acoustical, 259
  emitted signals, 269
  infrared, 264
  laser radar, 258–9
  radar, 265
  visual, 263
Solar cells, 124
Sperry-Curtis Aerial Torpedo, 4
Stability, 89
  augmentation systems, 188, 191
  derivative, 96, 98, 100–101
  dynamic, 65, 94–5, 100–101
  lateral, 95, 99–100
  longitudinal, 95
  modes of operation, 192
  static, 94
Stabilization, design, 238
Stabilization, sensors, 191
Stabilized platforms, 238
Static stability, 94
Stealth, 31, 41, 138, 252, 258–60, 264, 269, 410
Structures, 135
  construction techniques, 148
  materials, 145
  sandwich construction, 146
Switchblade UAV, 43, 307

Target detection, recognition, and identification, 217–31. *See also* Imaging sensors
Target location, 163, 199
Thermal design, 241
Trim, 89–97, 99, 101, 103, 238
Turbine engine, 39, 109, 114–15, 119–20, 122, 260, 262, 264
Turbofan, 16–7, 41, 67, 73, 77, 79, 109–110, 120, 127, 237, 260, 366, 370, 410
Turbojet, 41, 109–10, 113, 120, 127, 370
Turboprop, 20, 21, 73, 77, 79, 82, 109–110, 120, 127, 237, 260–62, 366

Turboshaft, 109
Two-cycle engine. *see* Propulsion

UAV Manufacturers, 21, 405
Unintentional interference, 297–8
  antenna gain, and, 318
  frequency hopping and, 325
  processing gain, and, 322
Uplink, 295
  anti-ARM, 299
  anti-jam, 300, 334
  delays in, 304
  low probability of intercept, 298
  resistance to deception, 299
  security, 298–9

Vertical takeoff and landing, 379, 386
Viking, 232

Weapons
  autonomy, 270
  compared to UAVs, 248
  Hellfire, 250
  history, 249
  laser designation, 241, 250, 255, 270
  mission requirements
    electrical, 255
    payload, 253
    safe separation, 257
    signature reduction, 258
    structural, 253
  missions, 252
  Predator, 7, 250
Weight and balance, 90

X-45, 30
X-47, 30
XQ-58 Valkyrie, 30–32, 269

Yamaha RMAX, 379, 386